THE REALITY OF THE UNOBSERVABLE

BOSTON STUDIES IN THE PHILOSOPHY OF SCIENCE

VOLUME 215

THE REALITY OF THE UNOBSERVABLE

Observability, Unobservability and Their Impact on the Issue of Scientific Realism

Edited by

EVANDRO AGAZZI

Department of Philosophy,
University of Genoa, Italy

and

MASSIMO PAURI

Department of Physics (Theoretical Division),
University of Parma, Italy

KLUWER ACADEMIC PUBLISHERS

DORDRECHT / BOSTON / LONDON

A C.I.P. Catalogue record for this book is available from the Library of Congress.

ISBN 0-7923-6311-6

Published by Kluwer Academic Publishers,
P.O. Box 17, 3300 AA Dordrecht, The Netherlands.

Sold and distributed in North, Central and South America
by Kluwer Academic Publishers,
101 Philip Drive, Norwell, MA 02061, U.S.A.

In all other countries, sold and distributed
by Kluwer Academic Publishers,
P.O. Box 322, 3300 AH Dordrecht, The Netherlands.

Printed on acid-free paper

Printed in the Netherlands.

TABLE OF CONTENTS

EVANDRO AGAZZI AND MASSIMO PAURI

INTRODUCTION

Observability and Scientific Realism

It is commonly thought that the birth of modern natural science was made possible by an intellectual shift from a mainly abstract and speculative conception of the world to a carefully elaborated image based on observations. There is some grain of truth in this claim, but this grain depends very much on what one takes observation to be. In the philosophy of science of our century, observation has been practically equated with sense perception. This is understandable if we think of the attitude of radical empiricism that inspired Ernst Mach and the philosophers of the Vienna Circle, who powerfully influenced our century's philosophy of science. However, this was not the attitude of the founders of modern science: Galileo, for example, expressed in a famous passage of the *Assayer* the conviction that perceptual features of the world are merely subjective, and are produced in the 'animal' by the motion and impacts of unobservable particles that are endowed uniquely with mathematically expressible properties, and which are therefore the *real* features of the world. Moreover, on other occasions, when defending the Copernican theory, he explicitly remarked that in admitting that the Sun is static and the Earth turns on its own axis, 'reason must do violence to the sense', and that it is thanks to this violence that one can know the *true* constitution of the universe.

From these remarks, it follows that *ancient* natural science was mainly based on *observation*, understood as a careful report of sense data, while *modern* science was based, to use again a famous expression of Galileo, on 'sensible experiences and mathematical demonstrations'. The 'sensate esperienze' of Galileo are not so much observations, but rather 'experimental investigations', such that one is entitled to translate this expression as 'sensible experiences' (rather than 'sense experiences'), something that is by no means alien to the Italian meaning of 'sensate' even at the time of Galileo, and has become very common subsequently. In this light, one can see that perceptual experience must be guided, interpreted, integrated in a theoretical framework, in order to be scientifically significant. All this amounts to saying that modern science is different from ancient science in that it is experimental, and this is even terminologically clear, for example, in Newton, who speaks of 'experimental philosophy' when he wants to denote the new natural science.

What these people wanted to exclude from science was the admission, by means of pure speculation, of alleged 'intimate essences' or 'hidden qualities' of things: but these expressions (of Galileo and Newton respectively) did not have the meaning of 'unobservable' features or entities. The correct understanding of these expressions would show that they are related to other epistemological and ontological issues (such as the distinction between essence and accidents, and the power of intellectual intuition, for example), but we shall not dwell on this issue here.

E. Agazzi and M. Pauri (eds.), The Reality of the Unobservable, 1-30.

In addition, if we consider the rapid and impressing growth of modern science in the domain of mechanics, and then in other domains, we must admit that only a limited amount of new observations stimulated this growth, while its advancements chiefly consisted in the introduction and development of many powerful *idealizations* of an intellectual kind: progress of science was not the result of looking hard, but of thinking hard.

The foregoing reflections by no means aim at minimizing the role of observation in science, but rather at making clear what this role precisely is, and to this end one must stress that observation certainly *includes* sense perception, but is not *reduced* to it. This becomes particularly clear if one considers that the observations on which the advancements of all empirical sciences have been based are of an *instrumental* kind. Such a fact is often interpreted according to two correct and obvious characteristics: (a) instruments allow for a much greater *accuracy* than simple unaided sense perception; (b) instrumental observation is standardized, and therefore practically free of those *subjective* features that inevitably accompany any sense perception. These are certainly significant advantages, but they can only be correctly admitted when instrumental observation and unaided sense perception are about the *same things*, and this occurs only seldom in science. In most cases instrumental observations that amounted to real breakthroughs in scientific knowledge were those that permitted the 'observation' of something that was not 'perceivable' by pure sensation. Of course, one must 'see' something in or through the instrument (this is why perception is *included* in observation) but the not unreasonable question is whether this something 'really' exists, or is rather a simple image produced by the instrument.

Certain scholars (such as Cremonini) who refused to acknowledge Galileo's astronomical discoveries obtained through the telescope, precisely relied on such a legitimate doubt. The classical methodological maxim *non fit scientia per visum solum* (science is not based on sight alone) was based on the awareness of common sense illusions, or of artificially created optical illusions (like those of disfiguring mirrors), and required that for a sense perception to be accepted as true, it had to be in keeping with an already existing, accepted and sound *theoretical* framework. In the case of the opponents of Galileo's observational astronomical discoveries, such an accepted theoretical framework was a metaphysical cosmology in which the number of celestial bodies, their trajectories, their intrinsic motion, etc. were allegedly determined in a cogent way, and exactly corresponded to the evidence provided by the unaided sense perception. Since the new sense perceptions provided by the instrument were in contrast with such a framework, they had to be rejected as illusions, like several well-known optical illusions. The most obvious way of rejecting such objections is that of convincingly demonstrating the *reliability* of the instrument, but it would be very naive to imagine that such a demonstration be provided by a simple comparison of perceptions in certain favorable cases. What is needed is a *theoretical* justification of this reliability. In our example this could only partially be provided by a correct optical theory of the telescope: sooner or later, the general theoretical framework that made the new observations unacceptable would have to be replaced. This meant that a new cosmology had to be constructed (and this was what historically actually occurred).

Such a complex but propitious solution, however, may not always be available. Historians of science had noted that in several ancient books on microscopic anatomy there appeared carefully drawn figures of structures that we do not actually observe today. The mystery was explained when certain scholars decided to *repeat* the observations by using the same methods of preparation of the tissues and the same microscopes that were used by the ancient scientists: the mysterious figures appeared again, and it was not difficult to discover that this was due to the fact that such old microscopes were not acromatic, and therefore produced 'aberrations' that could not be detected as being such at that time. In this case, it was a technological advancement that allowed for an improvement of the reliability of the instrument (and, at the same time, permitted to explain-away the initial observations which were 'untenable'). Something similar occurred at the beginning of our century when histological methods were introduced in the study of the brain and started providing 'observational evidence' in favor of the neuronal theory. Such evidence was possible thanks to very special procedures (having mainly to do with the particular kind of 'impregnation' adopted for the tissues), but for a while there remained questions as to whether the observed structures were simply 'brought to light' by the method in question, or rather 'produced' by it.

The discussions and examples provided lead us to formulate a few fundamental questions. Some contemporary philosophers of science equate observation and perception, and maintain that we cannot, for example, observe electrons because we cannot perceive them. We can only perceive the traces they allegedly produce in a cloud chamber or on a photoplate. Therefore, these scholars conclude: *admitting* the existence of electrons is simply a consequence of *accepting* a given physical theory, and of *giving* to this theory a particular ontological purport. It is clear that this position is not characterized by something that could appear as a mere nominalistic 'simplification' (that is, the equation of observation and perception), but by attributing, to perception alone, the privilege of having an ontological purport, since it is (allegedly) free of theoretical commitments. What perception *ostends* exists, while what a theory *claims* might not exist. This position can be challenged for several reasons. First, because it apparently accepts a moderate or minimal level of realism, to the extent that features attested by direct *observation* are considered to be real. In the second place, however, observation is reduced to perception, thus annihilating almost all the advantages science has obtained from instrumental observation, since such an observation is devoid of ontological purport. In the third place, the *reason* why perception alone is ontologically committed is because it does not imply *theory*. Each of these points raises serious questions.

The minimal level of realism accepted in this position relies upon a simple (or simple-minded) analysis of the notion of existence: we unproblematically say that chairs, tables, mountains, etc. exist, *because* we observe (perceive) them. We do not want to deny that science can speak of existing things, but demand that the *meaning* of existence remain the same. Therefore, we can attribute existence, at least in principle, to those entities spoken of by science which might possibly be brought within the range of accessibility of our observation (perception), like the features of a very distant planet that we can see only with a telescope, but that would actually be perceived if we could look at them directly from a spaceship one day. A *theory* cannot produce the same result, since it may at best give a sufficient logical explanation of certain observational (perceptual) states of affairs, but never a *necessary* (that is cogent and univocal) one.

Some questionable aspects of this perspective can be reduced to the following. There is the reasonable intention of preserving *common sense realism*, but this could possibly degenerate into some form of the Berkeleyan idealism: *esse est percipi* (real things are reduced to bundles of perceptions). However, serious scrutiny of the actual ways in which existence is predicated, even in common language, would demonstrate a much broader spectrum of application. A second aspect concerns the role attributed to theory. Such a role is not really specified, but allusion is simply made to the well-known fact that any theory is always underdetermined as well as, in a certain sense, overdetermined with respect to a given set of empirical data. Therefore, the most we can require from a theory is that it be 'empirically adequate' (that is, logically compatible with the empirical data), but not that it be *true*. From the point of view of the predicate of existence this amounts to saying that, in order to apply a theory (for cognitive or practical purposes) we 'do not need to believe that the entities it introduces are real'. This alleged difference between perception and theory is questionable, for we do not really 'need to believe' that the contents of perception are real either. The essential point though, is that 'we do not have any reason to believe that they are *not* real', but this condition could even apply to theories. A third aspect is strictly related to the second, and concerns the disqualification of instrumental observation, because the entities that are said to be 'observed' through the instruments are only something that has 'produced' certain perceived features 'according to a certain theory'. And, if a theory is no warrant for the existence of its affirmed entities, it follows that the instrument cannot ensure the existence of what is observed through it. If we adopt, however, the more sensible attitude mentioned above, we should say that the objects we observe through an instrument must be held to exist, unless we have *reasons* for believing (or at least doubting) that the theory on which the instrument is based is wrong.

Still a much more radical question must be raised: is equating observation and perception in itself correct? In fact, even within the domain of features that can be grasped by the unaided use of the senses, pure and simple perceptions are *mental states* which are not, by themselves, *judgments or propositions*, and for this reason do not represent any *knowledge* in a proper sense. This fact is very well represented by the basic non-existence of a private and phenomenal sense-data language: a putative language that would allow a primary introspective experience, of one's sensations in an immediate verbal form. As a consequence, the most spontaneous move is that of almost automatically translating the content of perception in a proposition, but then the error becomes possible, because in a proposition we engage ourselves in stating 'how things really are', we take position with regard to reality, stating that it conforms or does not conform with the perceptual content. So in the classical example of the stick appearing broken in water, we can translate the visual perception by saying: 'the stick appears broken', but we do not commit ourselves ontologically unless we say, for example: 'the stick *is* broken'. In this case we formulate a false statement, but simply because the *concept* of 'broken' consists of more characteristics than those contained in the relative visual *perception*. And if we are obliged to recognize (through additional observations) that such characteristics do not obtain we conclude, instead, for the proposition: 'the stick is *not* broken', but at the same time we are able to say *why* (due to the particular medium of our initial visual perception) it could and must appear. What this means is that in stating a proposition, we implicitly presuppose a set of normal or standard circumstances that must surround our judgment in order for it to apply to reality. This has nothing to do with perception, but is a body of 'available knowledge' that includes some rudimentary *theory* as well. Similarly, the solar disk appears larger at sunrise and sunset than at noon, nevertheless not even people entirely relying on commonsense say that the sun is larger at sunrise and sunset than at noon. Why? Because all tacitly admit the (theoretical) assumption that the mass of the sun remains constant, as well as other (theoretical) reasons that *account* for the optically perceived different size: some of them could be wrong (such as, for example, those attributing this variation to the different distances of the sun from the Earth during the day); others could be right (such as those attributing this variation to the differentiation in thickness of the atmosphere through which the sun's beams must pass in order to reach the observer at different times of the day).

Now *observation* is precisely this: it is a complex procedure in which perceptions and concurring theoretical prerequisites are combined to produce a judgment or proposition. Therefore, even if we concede that perception is theory-free, perception cannot be identified with observation, because observation (even in those cases in which it seems to be almost coincident with perception), is necessarily theory-laden; and this for reasons that are far more deeply rooted than those elaborated on the grounds of linguistic analysis. In this case, however, the following consequence is inevitable: if one maintains that ontological reference can be granted to perceptions only, it follows that any observation is devoid of ontological import, and this independently of its being immediate or instrumental.

A more radical question deserves to be asked: why has perception been endowed with the privilege of being a faithful witness to reality? The answer given by classical or traditional epistemology was this: inasmuch as perception of something is simply the *presence* of this something within the mind, it is therefore *identical* with this something in a specific sense, that is, not *ontologically* identical (for the mind does not become, say, a tree when it perceives a tree), but *intentionally* identical (since the mind's perceptual capacity which, in itself, is pure potentiality, becomes concretely identical with the actual image of the tree). According to this epistemology the situation is quite clear: the mind 'contains' the world, not in any spatial sense, but in the *intentional* sense: the world is *present* within the mind in the form of representations. Representations are the world itself inasmuch as it is partially *known*. Therefore *that which is known* is the world, while the representations are *that through which it is known*. At the beginning of 'modern' philosophy a different (and completely ungrounded) way of conceiving knowledge appears. According to this conception of knowledge, our *representations* are what we know, but we do not know whether they 'coincide' with the world or not: such concurrence must be *proven*. This position, which has been appropriately called 'epistemological dualism', has produced the most famous problem of modern epistemology, that of the 'origin' of representations (called 'ideas' in the philosophical development of the 16th-18th centuries). That ideas should come 'from' somewhere is actually a mere gratuitous presupposition, but it was not recognized as such at the time, and philosophers had roughly two ways of answering this ill-raised question. Some of them maintained that ideas are innate (*innatism*) and placed in the mind directly by God, who secures their coincidence with the world. Others maintained that ideas are produced in us by the 'external world', and were thus confident of their coincidence with this world. Both currents admitted a certain capacity of the mind to combine and manipulate the 'internal' world of ideas.

The second of these two modes of thought was that which is usually called 'empiricist', and its most significant characteristic is not so much that of having attributed to 'experience' the preeminent role, but rather that of having used the conceptual framework of modern natural science for explaining the content of experience. Indeed the cornerstone of experience (and of knowledge in general) was indicated through sense perceptions, presented as the *effect* of *causal* actions performed on our sense organs by external objects according to the patterns of causal mechanical actions which were about to be displayed by modern science.

Two interesting aspects can be pointed out. Here, we find the conceptual ancestors of today's fashionable 'causal theory of perception'. At the same time, we already find the main reasons for its appeal and its shortcomings. Its main appeal is that, according to this theory, perceptions are, so to speak, produced by things and not generated by us, and therefore they can truly warrant the existence of things. However such a warrant is very weak indeed. In fact, such a production is largely imagined, and can be spelled out only in part, and this solely with the joint aid of several complex *scientific theories*. This is a first serious blur to the alleged uncontaminated force of perception, since its reliability is *justified* on grounds of theories, and is therefore no more sound than the soundness of these very theories. But let us ignore this incoherence and try to see what perceptions could tell us about the world. They seem at least to be able to grant that the 'external' world does *exist*. Unfortunately, this is not true: the existence of the external world is *presupposed* by any causal theory of perception, and is by no means *shown* by it (precisely as it happens with natural science, that presupposes the existence of nature and tries to know it). Let us also ignore this point, and ask whether this causal theory can at least make us confident that perceptions give us the 'real features' of the world: again, this is unbelievable, since there is no reason that an effect should be identical or even isomorphic with its causes (if I press a piano key with my finger and this action causally produces a sound, it would be ridiculous to say that this sound 'resembles' my finger). To conclude, a causal theory of perception is the strongest support for the thesis that we are inexorably prevented from knowing how the world is.

It is not by chance that the most brilliant representatives of modern philosophy reached precisely this conclusion. Hume was far from denying the common sense beliefs concerning the existence and the features of the external world, or the causal link between events, but honestly recognized that we cannot *justify* them according to the criteria of his empiricist philosophy, and left them in the status of beliefs (this is the exact sense of his 'skepticism'). Kant explicitly declared that the world of *things in themselves*, is unknowable and, though maintaining that sense perceptions are the starting point of our knowledge, considered them as passive modifications of our senses and, as such, *pure appearances*, that could not even be called *knowledge*, since knowledge demands the intervention of the *active* functions of our intellect.

All this does not totally disqualify the causal theory of perception, but simply indicates its real merits, as well as what it cannot provide. Its merits consist in the clarification of several *physical procedures* that those physical beings, which are able to perceive, concretely use in their formation of certain representations of the world. However, this does not clarify the *representational* character of these procedures, something that, as we have already seen, implies the intervention of a capacity that is within the knowing subject and not in the physical world. Since any physical object whatsoever is immersed in the totality of interactions which in principle links it to the whole universe, which specific physical object is the observer observing at a particular moment ? Which of the various stages of the causal process does the observer actually observe, and how should he or she distinguish between them ? It is clear that a mere causal account conflates perception with physical interactions and measurements. What is missing in any putative causal explanation is, first of all, the acquisition of non-inferential beliefs, and then, the function of conceptual categorization, in other words an autonomous subjective and epistemic content. Thanks to this capacity, the subject is able to express judgments that concern the state of affairs of the world, and not the results of physical interactions. If these judgments are legitimately called observations, one cannot see any reasonable objection to also calling observations other much more complex causal interactions that (through certain physical instruments) can be presented as pointing to certain *specific* features, which can be assigned to the world on account of certain reliable theories. In other words, the causal theory of perception, in its validity, provides support for the legitimacy of equating observation with instrumental observation.

If we consider the development of the natural sciences until the end of the 19th century, we see that their rapid progress was based on that combination of observations and theories that had been inaugurated by Galileo and Newton and was summarized under the denomination of experimental method. The general attitude of scientists at the time was *realistic*, in the sense that they intended to investigate how reality actually is, and were confident to be able to fulfill this task, though readily admitting that the domain of the unknown was still vast, and awaiting scientific exploration. It is noteworthy not only that the rich display of 'observations' on which scientific knowledge relied, was largely constituted by instrumental observations, but also that those features of nature (thought to be genuinely 'real') were, mostly, not even 'observable'. In fact, the most significant achievements of science were estimated to consist in the discovery of *laws*, considered as general ontological structures of nature that cannot be 'observed', but are 'discovered' through *intellectual* research.

The role of observation was not underestimated, but was confined to the important task of pointing out *phenomena*. The importance of this role lay in the understanding of phenomenon in a scientific sense, not just as a perception, but as the result of a careful, repeated, intellectually conscious and theory-guided process of observation. In addition, phenomena, in the scientific sense, are by no means synonymous with 'pure appearances': they rather mean 'manifest' features of reality (to use Newton's vocabulary), or 'observable' features of reality, as is stated above. Even in the positivistic age, when a mistrust of metaphysics and its possible infiltration in science led to the rejection of the expression 'law', and its replacement with the seemingly more neutral expression 'regularity among phenomena', phenomena were considered in a realistic sense.

The peacefulness of this picture became ruffled though at the end of the 19th century, due to several developments in the domain of *theories*. Radically different theories were successfully proposed to account for the same set of observations (for example, corpuscular and wave theory of light, 'continental' and Maxwellian electrodynamics, different formulations of classical mechanics, etc.). Even the most complete theories proved unable to account for newly observed phenomena (in both the sense of understanding as well as explaining them): several logical difficulties surfaced concerning the method of using and combining standard physical concepts, and so on. Many such difficulties may be seen as the consequence of a radically new situation: Newtonian mechanics was in keeping with common sense because, from many points of view, it could be seen as an 'idealization' of common sense itself, and the subsequent developments of this mechanics (including its extensions into the modeling of new classes of non-mechanical phenomena) continued to remain within this idealization. In particular, the micro-world was basically understood as a miniaturization of the macro-world, keeping the fundamental features of this last unaltered. The newly discovered phenomena destroyed this illusion. One possible (correct) interpretation of this fact is the following: reality has different aspects, or *levels*, and it is not reasonable to pretend that they share the same characteristics, apart from the pure order of magnitude. Therefore, having discovered certain inadequacies is not at all scandalous. A second possible interpretation is a consequence of the 'epistemological dualism' we already described: we cannot know reality in itself, but only phenomena (understood as appearances), therefore theories - which are meant to describe reality as it is - are obliged to show their cognitive inadequacy sooner or later. The remedy is not to look for better theories (in a cognitive sense), but rather to delimit cognition to phenomena understood as *perceptions*, leaving theories to other more modest pragmatic tasks. This was Mach's proposal, and is at the root of many present anti-realistic positions in the philosophy of science, as well as that of the equation of observation and perception we amply discussed, in which observation (thus understood) is credited with some realistic purport and, in any case, with the privilege of being the only basis of cognition. Carnap's program of the *Aufbau*, in which the whole apparatus of our cognition should be 'constructed' by purely logical devices starting from perceptual predicates, was the most serious effort to give a realization to this view, and has inspired much of the contemporary empiricist philosophy of science.

The central importance of the notion of observability for realism has always been emphasized, and sometimes overemphasized. This notion is generally considered to represent a danger for realist positions, and realists have often tried to reduce its impact by weakening the force of the distinction between observable and unobservable or even challenging its very *possibility*. In this way, those who believed in the reality of observable entities had to extend their belief to unobservable entities as well, since both are essentially inseparable. This goal has been pursued, for example, by over-emphasizing the theory-ladenness of observations, or by developing some 'continuity argument'. The latter argument relies on the existence of a continuous spectrum leading from unaided observations of everyday life to more and more instrumentally aided observations of scientific practice, so that any clear boundary or discontinuous change within this spectrum turns out to be arbitrary. While the first argument more or less amounts to asserting that nothing is purely observable, the second maintains that, somehow, everything is observable. In connection with this, one should take into consideration the specific form that the issues of scientific realism and observation assume in quantum theory. For example, it's worth emphasizing that the crucial novelty brought about by quantum theory, as far as these results are concerned, has nothing to do with the empirical discovery of more and more entities that are 'unobservable', in the naive sense of the term. For, apart from the particular and limited technical issue of quark unobservability, a specific 'quantum unobservability' already arises from the basic tension between the symbolic structure of the theory and the causal spatio-temporal language in which measurements and empirical findings have to be necessarily described. Theoretically, this tension is already established by the *atomization of processes*, discovered by Planck. Such a limitation in the observability of quantum stuff is radically different from the kind of unobservability arising in any other physical theory.

SECTION SUMMARIES

This lengthy introduction has attempted to show just how complex and intricate the problem of observability is, and in what ways *it may* be connected with the problem of scientific realism.

The present book does not pretend to illustrate this problematic in its entirety. It simply aims to single out certain important factors and aspects of the rich theme evoked above, in an effort to unite its scientific and philosophical sides. The papers contained in this book analyze certain aspects of this vast domain, and are organized according to a specific conceptual order. Since many historical considerations have already been sketched out in this Introduction, only one paper will be devoted to the history of scientific realism in its more technical sense. It will follow a part devoted to the presentation of certain general philosophical preliminaries (mainly of an epistemological nature), after which the issue of scientific realism will be analyzed under different aspects. A third part will be specifically devoted to the topic of observation from the philosophical point of view. Since the present debate about scientific realism has been mainly fed by the philosophical discussions on quantum theory, a fourth part will deal with this discussion. In the last part some concrete ways in which problems of observation and realism are addressed in quantum theory (rather than in reflections about this theory) will be presented.

A summary of the different texts will provide the reader with a preliminary overview of the structure and content of this volume.

In his *The Origin of Scientific Realism, Boltzmann, Planck, Einstein*, Erhard Scheibe claims that scientific realism was invented by physicists rather than philosophers. Scheibe defines scientific realism as a philosophical thesis concerning, essentially, the progress of scientific theories. As such, realism has little to do with the issue of the veridicity of our perceptions, as in classical philosophical arguments from Plato to Descartes or G. E. Moore. In order to illustrate the essential contributions offered by physicists to the question of scientific realism thus understood, he broaches the relevant philosophical views of Boltzmann, Planck and Einstein. Scheibe remarks how in each of these natural philosophers' intellectual developments, the influential antirealism of Ernst Mach has played a pivotal role, if only as a constant - imagined or real - source of critical objections to a naively conceived faith in the existence of hypothetical or 'theoretical' entities. In defending the kinetic theory of gases, Boltzmann had, for instance, taken a rather liberal view of scientific theories, one that somewhat mediated between a naive belief of atoms as explanatory devices in the kinetic theory of gases and a pure phenomenalism *à la* Kirchhoff. Like Hertz, Boltzmann believed that scientific theories are pictures (Bild) or 'signs of reality', that are intentionally construed to go beyond experience via idealizations or hypotheses. Although Boltzmann seems to oscillate between what we would call realism and instrumentalism, he equated the issue of the existence of atoms with the question of whether the atomistic theory of gases was empirically successful, and foreshadowed the account of realism focused on by Scheibe. In his 1908 Leiden Lecture, Max Planck - who had begun his career as a follower of Mach - related the idea of scientific realism to scientific progress more explicitly than Boltzmann, and tried to link both notions to the important concept of 'the unity of physics'. Planck's main theses are that (i) not only does the history of physics illustrate this particular development toward unity, but also that (ii) such a development can only be explained by the hypothesis that physics is concerned with a real world, progressively regarded as existing independently of 'human-historical notions'. This deanthropomorphization of our primary experience may be considered a loss, compensated by a corresponding gain in the progressive unity of our extremely disparate phenomenal world, as it is represented in every branch of physics. According to Planck, 'physics aims at a complete detachment of the physical world from the individuality of the creative mind'. Moreover, Planck attacks Mach's principle of economy by pointing out that it did not play any role in the conscious effort of a Copernicus or of a Faraday to fight against the intellectual traditions they inherited. According to Scheibe, the ensuing, harsh controversy with Mach proves, beyond any doubt, how important the question of scientific realism was for Planck himself. Paradoxically, however, it was rather the quantum theory (Planck's major scientific legacy) which called forth a view of scientific theories in which Mach's phenomenalism could be proven correct. For exactly this reason, Einstein disliked quantum theory until the end of his days. After having started as a Machian (like Planck), Einstein in his later years formulated what Scheibe calls 'costructive realism', a version of scientific realism in which the speculative and constructive contibution of the human mind was particularly stressed. In essence, Einstein distinguished between a reality hypothesis and a reality criterion. As to the latter, he defended a coherentist position, typical of idealist philosophers: here 'coherentism' means that the ability of an event to integrate itself, in an already existing theoretical system, is the most important criterion of its reality. Moreover, Einstein, at one point, even identified the thesis that 'the totality of our experiences admits of a logical, conceptual system connecting them' with 'the reality hypothesis'. According to Scheibe, the postulate of logical unity is one of the most repeated requirements of a theory to be found in Einstein's writing. Such a requirement motivated Einstein in his late search for a unified field theory, but did not lead him to deny a realist hypothesis, or to accept some sort of Quinean thesis about the under-determination of theories by data. In the words of Einstein: 'nobody who has gone really deep into the matter will deny that in practice the world of phenomena uniquely determines the theoretical system, in spite of the fact that there is no logical bridge between phenomena and their theoretical principles'. To conclude, according to Scheibe, what Boltzmann, Planck and Einstein have in common is the idea that if scientific realism is wrong, the success of science would be, as Putnam put it, 'a miracle'.

The paper, *Observability and Referentiality*, by Evandro Agazzi tries to explore the reasons for which the requirement of observability has been endowed with such a decisive importance in the philosophy of science of our century. He finds an answer to this question in the simple fact that, in science (and in any cognitive undertaking in general), humans are first guided by a spontaneous 'realist' attitude', consisting of the belief that there exists a world of things ontologically distinct and independent of the knowing subject, and that this world must be 'accessible' in some way. The expression commonly used to denote this reality is 'external world', and it is misleading since it conveys a spatial image which is insufficient to help us understand the real nature of this phenomenon. A much more appropriate term for it would be 'known reality', since it more clearly conveys that reality is, in a certain sense, independent and in another sense dependent on the features of the knowing subject. We can generally say that our sense organs constitute our immediate 'access' to known reality, and this entails certain consequences: First of all, we never know any neutral and so to speak colorless piece of reality, but always a 'connoted reality', that is, a reality endowed with certain *qualia* or characteristics. Symmetrically we never know any such *qualia* by itself, in a separate state, but always as the particular feature *of* an individual entity *in which* it exists. Yet it is perfectly clear that one could not see a red rose without being in suitable lighting conditions and endowed with 'normal' sight. This means that any *quale* exists at the same time *in the* subject and depends on its features. Therefore *qualia* are neither simply objective, nor simply subjective, they necessarily are both things at the same time. We call the way in which a certain ontological feature is present to the subject *representation*, and the world of representations has its own particular ontological status as well, since representations can combine, let it be remembered, in ways that no longer depend on their being the means through which things are present to the subject's mind. This fact explains well known phenomena such as hallucinations, dreams, and sense illusions: in such cases the image or representation seems to be evidence of a state of affairs that does not actually occur, but this can happen precisely because representations have a special kind of existence (an *intentional* existence) of their own, that is related, but not strictly dependent on the ontological existence of things. The relation that the representations have to non-subjective reality is that of *reference*: representations 'point to' non mental reality, and *observation* is one of the tools we use in order to gain warrant of referentiality. In fact, many representations come to us on the occasion of a certain 'presence' of non mental entities, but they can then be combined, thus giving rise to complex constructions that belong to the realm of *intentionality* (and therefore to the realm of *meaning*): as such, they denote a *possible world*, but if we want to know the *real world*, we must force our representations to come into contact with referents, and this is precisely what observation is expected to do. Observation can perform this task because it is by no means identical with perception, and this not only because it entails much more 'activity' on the part of the subject (concentration, intentional focussing on certain details, etc.), but especially because observation, in its proper sense, leads to the formulation of a *judgment* and not to the retention of a pure representation. Even if perceptions are not as passive as radical empiricists maintain, observations are the result of a complex interaction between the state of affairs of the world, the physical procedures of stimulation of our sense organs, and especially the powerful intervention of our 'cognitive apparatus' which does not simply mean (as in the case of perception) a correctly functioning sensory apparatus. Rather, it indicates the need for a substantial amount of 'background knowledge' that enables us to *interpret* the perceptual evidence (according to reliable *theories*) as being the *causal* output of a certain process, such that we must say that we are *perceiving* some, let us say, visual features, but *observing* certain micro-particles, since the intended *referents* of our observation are these imperceptible entities, that are nevertheless 'accessible' through an *instrumental* operation, and not just through an *a priori* speculation.

In his *A New Approach to Human Cognition and its Significance for the Philosophy of Science*, Paul M. Churchland advances a neurocomputational model of perceptual recognition and of cognitive activity in general. Such a model is meant to provide new evidence for a specific view of scientific progress, scientific realism and truth, previously introduced in the work of Tom Kuhn and by evolutionary epistemology. In particular, he holds that there is no such thing as the unique and best description of the world, and that the historical sequence of scientific theories does not converge toward Truth or Perfection, despite the fact that it provides us with better and better results. In order to argue in favor of such conclusions, the first part of his paper introduces us to *connectionist artificial intelligence*. He does this essentially by describing a three-layered neural network designed for the recognition of human faces. The number three in this case is not arbitrarily chosen: due to a very important and general result, any input-output transformation can be approximated by a three-layered network. In the case of our visual system, the three 'layers' above correspond to the input patterns of the retinal stimulation; to the intermediate elaboration of the lateral geniculate nucleus (LGN); and to the final output of the visual cortex in the back of our brains. Any pattern of retinal stimulation can be modeled by specifying a unique list of numbers, called input vectors, indicating how much each cell is stimulated. Each 'retinal cell' of the net is connected to a large number of target, intermediary neurons of the LGN, which, in their turn, are linked to a large quantity of output cells in the final layer. Via their connections (*synapses*), whose remarkable plasticity can be measured by modifiable weights, such net-like devices transform the first pattern into two other patterns, one for each successive layer of neurons. The algorithm to find the right configuration of weights for a given stimulating pattern at the sensory periphery is called 'back-propagation', and works by progressively reducing the difference between the initial output of an untrained net - whose weights are set at random - and the correct answer. After many encounters with the entire set of training images, and thanks to the operation of a serial computer which nudges the weight of each synapse, the net settles into a configuration of weights which accurately yields the correct output vectors for any input image: faces (and their gender) are reliably recognized from non-faces. The crucial ability displayed by such face-recognizing networks, and used by Churchland to illustrate his main thesis about human creativity, is 'vector completion', basically a capacity to generalize to new cases. Surprisingly, even if the networks are presented with different pictures of the same faces, or with the same pictures in which 20% of the information of the original input image is lost, they are still able to perform well. The explanation of this ability points to the presence of hierarchically divided 'regions' in the activating space of the second and third layers. According to Churchland, such partitions correspond to rudimentary concepts, categories or prototypes. By hypothesizing that instances of vector-completion be at work in inductive and ampliative inferences in all living creatures, he claims that the main revolutionary transitions in the history of physics - from Descartes, to Newton and to Einstein - are essentially constituted by some form of recognition of familiar patterns or prototypes in an otherwise puzzling situation.

A significant contribution to the illustration of a concretely existing, and widely adopted, means for humans' acquisition of knowledge (both in science and everyday life), and which does not remain within the limits of the methodology admitted by empiricist epistemology, is presented in the paper of Jean-Pierre Desclés, *Abduction and Non-Observability - Some Examples from Language Science and the Cognitive Science*. Indeed, a suitable way of rendering the permission for 'speculations' more precise, as is advocated in Shapere's paper, is offered by the methods of *abduction*, whose features were presented by C.S. Peirce at the beginning of our century. The specific characteristic of abduction is that it makes it possible for man to state hypotheses that appeal to some directly unobservable entities or facts, starting from certain admitted knowledge (which may consist in observational evidence, but is not limited to that, and can therefore be equated with the 'background information' spoken of by Shapere or other similar notions presented in other papers). Admittedly, Peirce's abduction is more or less explicitly presented as a procedure for attaining 'plausible hypotheses' and increasing their degree of plausibility, which often consists in comparing them with results from adequate testing and, in such a way, conforms with the requirements of the empiricist methodology. However, as is clarified in this paper, in several interesting cases abduction allows for the admission of entities or facts that remain definitely and essentially unobservable, and yet constitute actual conquests in science. The examples provided by the author do not belong to physics, but to the domains of language and psychology. However, this does not diminish their significance, since analogies with what happens in physics are not difficult to be found, and, moreover, this circumstance rather indicates the *generality* of this procedure, which is an advantage, and not a weakness. This, in particular, explains why similar considerations were also developed by other authors, such as in G. Polya's theory of 'heuristic reasoning'.

The paper of Roberto Torretti, *'Scientific Realism' and Scientific Practice* indicates that the core of a realism, which can be denoted as specifically 'scientific', is the claim that science is able to uncover profound entities or aspects of the world that are more 'real' than the 'things' (or *pragmata* to use the Greek term) of our everyday life: an attitude which had already emerged in ancient thinking and was repeatedly adopted by the most distinguished representatives of modern science and modern philosophy. According to this view, the aim of science is that of adequately describing such a profound and immutable structure of reality. The author criticizes this conception, and maintains that science is completely embedded in our *pragmatic* life and is, therefore, 'the continuation of common sense by other means, as required by the overall *pragmatic* situation'. In order to sustain his thesis, the author criticizes, first, certain common arguments used against realism, and then comes to the defense of his thesis through the discussion of a few historical examples. Particularly relevant to his argument are examples in which it is clear that conceptually disparate theories are jointly brought together to bear on the understanding of specific phenomena and the solution of particular problems. One is Hawking's celebrated letter to *Nature* on black hole explosion (1974), in which principles of non-relativistic Quantum Mechanics are used to explain and calculate the 'evaporation' of a black hole, in spite of the fact that black holes are introduced on the basis of General Relativity and that these two physical theories are, at present, incompatible. The second example discusses the problem of the explanation of the 'advancement' of Mercury's perihelion. Using the tools of Newtonian theory applied to Mercury's gravitational interaction with the other planets, it is possible to account for more than 99% of this anomaly, while the remaining 43" of arc per century was explained by Einstein in his new theory of gravity contained in the field equations of General Relativity. However, according to the author, a careful technical analysis of the way in which this theory allows for the explanation of Mercury's anomaly, shows that, *strictu sensu*, the adoption of General Relativity accounts only for the small precession of 43" per century, but not for the greater precession of 530" per century, which had still to be accounted for by means of the Newtonian theory. On a more general scale, the author remarks that in experimental physics the interpretation of the observed results intended to test a new theory depend on old theories involved in the experimental equipment employed. And these theories are often incompatible with the new one. The realist can certainly believe that the entire experiment can be satisfactorily reinterpreted in terms of the new theory, but this is simply to 'live on credit' and is perhaps less rational than accepting (as the pragmatist does) that 'physics, like every other major human enterprise, is a patchy, makeshift affair'.

Developed at the end of World War II by Stanislaw Ulam and John von Neumann as a means of calculating high-dimensional definite integrals, the *Monte Carlo method* was based on a simple idea: provide a criterion to decide whether a point in a multi-dimensional space sits inside the volume we are calculating; feed into it a long enough list of *random* samples; tally the ratio of *hits* to *tries*. Such an algorithm warranted a faster convergence than any previously known method, and gained widespread appreciation. To obtain uniformly distributed random samples, one would initially employ the physically generated *true randoms*, such as counts of alpha decays and cosmic ray events. Soon the impracticality of this led to the diffusion of *pseudo-random* numbers, spewed out by number-theoretical algorithms, and thought to be *for all practical purposes* equivalent to the true randoms. In 1963, however, physicist Joseph Lach noticed that the popular *multiplicative congruential generator* showed subtle yet disquieting correlations in its output that may potentially mar the hundreds of scientific calculations which employed it. True, the algorithm had passed many *tests of true randomness* with flying colors, but it failed at the slightly more imaginative one thought up by Lach. What is more, S. K. Zaremba pointed out that no set of tests will ever be able to single out processes by their stochasticity. Zaremba introduced, instead, the concept of the *quasi-random generator*, satisfying no test except that of providing the most efficient distribution of numbers to tackle the problem at hand. In his *Random Philosophy*, Peter Galison argues that the adoption of quasi-random numbers actualizes three important perspective shifts. First, the *desacralization* of the random: the virtues of the Monte Carlo method had been ascribed to the very concept of randomness, that hence acquired a special metaphysical status, embodying perfect equability and naturalness; from the point of view of *quasi-ists*, on the contrary, true randoms represent only one distribution among many, and not necessarily the most efficient one. The second shift, is known as the *end of simulation as representation*: the Monte Carlo method was often used to model processes that were themselves thought to be random; this was seen to enhance the value of the simulation, on the grounds that, in a way, it was as natural as the physical process; now, instead, *doctored random processes* could perform better than *true ones*, bringing this mystique to an end. The last shift was towards the *typicality* of problems: for instance, to integrate a one-dimensional function the best approach may be neither taking systematic, regular samples, nor confiding in complete, blind randomness, but something in between that may achieve the maximally efficient extraction of information from the integral, skipping bare, uninteresting plains while giving enough attention to jagged heights.

It is generally acknowledged that symbols, notations and formal ways of representing abstract entities play an important role in scientific praxis. Their possible gnoseological relevance, however, has been often undervalued. In *Formal Representation and the Subjective Side of Scientific Realism*, Mario Casartelli examines reasons for this neglect and explores the peculiarities of scientific notations. He points out, for instance, their hierarchical structure (from graphic signs to high-level representations), or their capability of creating a frame within which scientific objects can be captured and understood. This happens even independently of goals, primitive intentions, and psychological attitudes of their inventors. Particular attention is devoted to the fertility of notations. Fertility implies indeed the fallacy of a reductionist scheme that sees representations as the result of an evolutionary process of adaptations to known problems. In this context, the main point is that not only notations often originate or create problems by themselves which would be otherwise inconceivable, but that this creation shares the objectivity of the 'natural' world. This fact suggests the existence of an *intermediate* level of reality between subjectivity and observability, which shares in some way the features once attributed to the world of *universalia* by symbolic realism. The principle of symbolic realism, 'ars imitatur naturam in sua operatione', fits indeed most of those epistemological peculiarities of representations which are coherently disregarded by the historian or the psychologist of scientific discovery. Therefore, after discussing some possible misunderstandings of the status of representations, the author finally remarks that a correct analysis of the problem requires restoring some classical and forgotten issues about the ontology of symbolism.

In his *Convention and Observability - Poincaré once again*, Gerhard Heinzmann focuses on the relation between observations and conventions in Poincaré's philosophy of geometry. He illustrates how Poincaré's conventions are neither merely analytical nor purely synthetic. Rather, by being 'analytical but non-logical propositions', they can be constitutive of our experience, while at the same time are guided by it. In this sense, Poincaré's position escapes Quine's criticism of the 'two dogmas of empiricism', and through the influence of Hermann Helmholtz, amounts to some form of constructivism. While Heinzmann stresses that Poincaré often conceives conventionalism in a linguistic sense - namely as a thesis about the inter-translatability of different geometrical and physical languages - his main concern is to show how it is possible to achieve a mediation between 'rigor' or 'artifice' and 'objectivity'. The role played by these two requirements is investigated by illustrating which conventional elements enter the pre-theoretical aspects of the genesis of a geometrical theory. According to Heinzmann, one needs to distinguish between two kinds of conventions in Poincaré's thought, the first of which (classifications) is necessary for the very constitution of the objects of observations. The conventional element of classifications is given by the linguistic usage of a society, which objectivizes the observational content of a protocol statement. The second type of convention in the philosophy of geometry is in part provided by the concept of a mathematical group, which is the articulation of a scheme of rules bringing together the conventions of a lower order. The concept of group is related to Euclidean motions, and presupposes the idealized notion of a rigid body and its displacements in space. By exploring the role of the concept of group in the constitution of both the notion of space and that of an idealized body, Heinzmann tries to justify Poincaré's famous claim that Euclidean geometry cannot be refuted by experience.

The paper by Ramón Queralto, *Scientific Realism, Objectivity, and 'Technological Realism'* stresses that contemporary science is bound to the application of sophisticated technologies to such an extent, that technology has become a genuine 'condition of possibility' of science. Indeed scientific objectivity is based on technology both in the sense that technological operations secure inter-subjective testing of scientific theories, and in the sense that the 'objects' of contemporary science are determined through technological procedures. This amounts to saying that 'technological compatibility' is a condition for scientific statements, and this justifies the notion of a 'technological realism' as a particular specification of scientific realism.

In his paper *Testability and Empiricism*, Dudley Shapere discusses the capital question of the 'unobservability in principle' that seems to affect many fields of the most advanced branches of contemporary physical science (several examples in present cosmology are mentioned). Since testability has been recognized as the distinguishing mark of science by practically all of the doctrines in philosophy of science of our century, this fact seems to justify the antiscientific attitudes adopted by several scholars today. Especially because such an inability for verification is usually presented as a matter of *principle*, related to the famous thesis that all scientific observations are 'theory-laden', since they are inexorably affected by certain theoretical 'presuppositions'. Shapere points out that this tenet relies upon a confusion: it is true that observation must remain the source of scientific *evidence*, however it is not true that such evidence coincides with the content of *sense-perception*, something which has been taken for granted as being true within the empiricist tradition. Therefore, it is true that scientific observation depends on certain 'presuppositions'.These last, however, are by no means theoretical frameworks that deprive observations of any objective value, since they are actually a system of 'background information' (mostly of reliable scientific information) that justifies a departure from sense-perception, and allows for an 'internalization' of observation as something that receives a specialized (but rationally grounded) meaning in science. This broadening of the concept of scientific observation does not render 'observable' everything we admit as such in science: in particular, many statements of contemporary science (such as those mentioned at the beginning) remain related to unobservable entities. However, on the basis of the background information we have, certain carefully selected and rationally argued *speculations* are admissible (within objectively determinable limits). A much more drastic departure from empiricism is studied in the last part of this paper: Is it admissible that the need for further observational tests will one day become superfluous for an entire area of theoretical investigations? The author maintains that, at least in principle, such a stage could be reached, in the *ideal* case that the background information would have become self-consistent, detailed, constraining of further possibilities, and endowed with a great power of unification, such that it could generate and answer questions about its subject-matter and directions of research. The author provides a series of technical arguments to show that this ideal is in principle very reasonable, and even rather close to be fulfilled.

In *Observing the Unobservable*, Jan Faye invokes the distinction between *epistemic*, *semantic* and *ontic realism*. An epistemic realist is someone who believes in the existence of truth-securing procedures in a certain area of inquiry, while an epistemic anti-realist denies exactly that. For instance, an epistemic anti-realist typically claims that there are no reliable methods that can take us from our sense impressions to the belief in the external world or in unobservable entities. However, epistemic anti-realists like van Fraassen can at the same time defend both a semantic and an ontic type of realism. This means that they can believe that sentences have meaning even if their truth conditions transcend our means of verification (semantic realism) and that there exists something that is independent of our minds (ontic realism), even though this something cannot be known. Semantic anti-realism consists in the thesis that the meaning of a sentence cannot be analyzed in terms of truth conditions that go beyond our means of verification. The link between semantics and ontology is supplied by one's notion of truth, while the relationship between ontology and epistemology is constituted by different approaches to the notion of warranted beliefs. The question here is whether true beliefs must be confined to warranted beliefs, or whether true beliefs may transcend them. Given these distinctions, Faye urges that it is possible to have beliefs about the so-called unobservable entities to the same extent that we have warranted beliefs about observable entities. In order to reach this conclusion, he first presents and then criticizes the causal theory of perception. According to such a theory, the (*secondary*) properties we ascribe to things on the basis of the immediately perceptual acquaintance with them are not their real properties, but the result of information processing (causal chains) in the brain. Since the external world is not as it looks, the causal theorist of perception thinks that it must be attributed to the structure of a hypothetical or postulated entity, not unlike that of a quark or an electron. Consequently, there is no difference in principle between observing an airplane and an electron. For the causal theorist, instruments should be considered as an extension of our senses, because physically there is no difference in principle between the operation of biological instruments (our senses) and technological instruments. In the end, it is science that tells us what counts for observable and what doesn't, and science is the only way to find out which properties of objects exists independently of the mind. To the causal theory, Faye objects that the notions of perception or observation require, unlike that of measurement, a human being, since they require the acquisition of beliefs as a consequence of an act of observing or perceiving. He claims that in any perception, the key role is played by an epistemic (not simply causal) relationship between a perceiving subject and a perceptually experienced object. Moreover, if we did not see something as it is, as the causal theorist has it, we could not assert anything about what we see and our observations could not be used to verify or falsify scientific theories. In contrast to the causal theory of observation, Faye proposes what he calls the *semantic theory of observation*, which stresses that the common epistemic element both in perception and scientific observation is the acquisition of linguistic beliefs. There is, in other words, a close connection between our competence as speakers about a given subject matter and our ability to perceive something as something specific. Consequently, for Faye, there is no significant epistemic distinction between observable and unobservable entities, because observations in both cases presuppose an acquisition of beliefs, essentially mediated by our linguistic competence concerning names of natural or artificial kinds. Thus, a term functions as an observational term, whether it stands for something perceptible or not, if its application can be associated with certain canonical experiments and observational situations. It turns out that any object O is an observable entity if, and only if, O can yield a belief that the circumstances in which it is correct to apply the name of O are realized. Thus, given these circumstances, the belief counts as warranted. In sum, observation may or may not involve instruments, but the question becomes that of whether or not such warranted beliefs are true. The epistemic anti-realist may want to identify the notion of true beliefs with that of warranted beliefs, but the semantic realist or anti-realist must agree that we observe non-perceptible entities.

In *What Does it Mean to Observe Physical Reality?* Giovanni Boniolo resists Quine's suggestions by inviting us to clearly separate the notions of reality, observability and objectivity, while at the same time keeping their validity both at the macroscopic, commonsense level and at the microscopic, physical level. The three main philosophical assumptions of his paper are (1) a Kantian primacy of 'theorizations' or conceptual schemata upon data, (2) a clearly stated distinction between epistemology and ontology, namely between a theory about the conditions for knowing x and a theory about the nature of x, and (3) the idea that both ontological 'idealism' and ontological realism affirm that something exists, so that they are uniquely opposed as to whether this something is mental or material. The overarching aim is to find a mid-way between what Agazzi called 'epistemological dualism' - namely a dualism between what can be known but does not warrant truth and the unknowable, which must be postulated to avoid radical constructivism - and epistemological monism (realistic or idealistic), according to which knowledge is resolved into what is known. According to Boniolo, the perceiving subject does not construct reality but selects, makes meaningful and constitutes certain aspects of reality which he 'subjectivises' by means of his own physiopsychological state. Reality in itself and in its totality cannot be grasped, and is the necessary and indeterminate background of perception and, therefore, of knowledge. The notion of objective entity may be arrived at by a process of emendation from the non-invariant, purely subjective features of perception: what is shared by all perceiving subjects is an intersubjectively ascertainable shape, dimension and position. Any invariance in perceived entities must be perceived in principle by all subjects: the observability of an entity must be at everybody's disposal. But the notion of 'object' presupposes theorizations and is defined in terms of them: an object is a knot of theoretical properties. In this way, Moby

Dick and Pegasus turn out to be objective entities even though they are not perceivable, and the set of intentional objects (objective entities subsisting in a semantic domain) is larger than the set of real, existing objects, just as the latter set, the ontologically obscure background of perception, is larger than the set of observable entities, which only captures aspects of reality.

The paper by Alberto Cordero, ***Realism, and the Case of Rival Theories without Observable Differences*** handles, through a case study, a classical topic of philosophy of science, that is, the role of observation in discriminating between rival theories, and the consequences that the impossibility of applying such a discrimination has on the issue of scientific realism. The case study involves three different approaches to the foundation of quantum physics, that accept the wave function as a physical field: Bohm's quantum theory of motion, the physicalist version of the 'many decohering worlds', and a family of theories that incorporate wave function collapse. They really are different since they predict different phenomena; according to the author, however, it happens (contingently) that such different phenomena are actually not *observable*. Moreover, it also happens that these theories are equivalent even from other points of view that are adopted for comparing theories according to the standard methodology of science. This situation clearly poses serious difficulties for a realistic interpretation of quantum physics. This paper studies the possibility of defending a certain form of realism in spite of these difficulties. The anti-realist interpretation of such a situation is often based on its presentation as a typical example of *incommensurability*. The author criticizes this attitude, which was originally grounded on the overestimation of linguistic holism, and on the claim that such theories cannot have common *data*. On the contrary, there is a lot of data that is commonly accepted, and indeed interpreted the same way, by the advocates of these theories even down to a certain explanatory *level*. Above this level, the models of the three theories converge over a significant stable intersection. Concretely speaking, this common level can be seen as the 'standard version of quantum theory' along with those little improvements that originated, precisely because of the developments of these rival refinements. All this increases the plausibility of a realistic interpretation of quantum physics, and this is correct, provided one does not adhere to certain misguided conceptions of realism itself. One of them is the so-called 'metaphysical realism', that amounts to considering 'reality' as totally independent of any knowledge there may be of it. Its excess has produced, in reaction, several attitudes of 'super-anti-realism'. According to the author, a viable defense of realism consists in focusing the analysis on the credibility of our assertions, in order to see whether this is comparable with the credibility that leads us to admit the real existence of the entities of our everyday experience. The criteria our scientific theories should fulfill in order to satisfy the claims of objective independence (with respect to the individual wishes and dispositions to assert) and descriptive power (that is, not a purely instrumental value), are internal consistency, coherence with the rest of current knowledge, successful critical testing, and coherence with a robust history of past successes. This realism, of course, is relativized in relationship to our current scientific knowledge and ongoing scientific practice, and can do justice to our most rational claims about the observable as well as the unobservable. In addition, this realism corresponds with the fact that we gain insight into the structure of nature even as we fail to come to a conclusion about the 'ultimate nature' of physical reality.

With respect to the problem of the relationship between theory and observation and the issue of the existence of the so-called theoretical entities, Mauro Dorato's contribution,

Measurability, Computability and the Existence of Theoretical Entities, is situated in line with the shift in emphasis from observability and reference to experiment and causal intervention, preached in the eighties by the now dispersed Stanford group (I. Hacking, N. Cartwright, P. Galison). He argues that the natural ontological attitude of realism about entities, typically endorsed by experimental physicists, is not the mere by-product of changing historical fashions, but can be traced back to some 'sufficient conditions', that are nothing but the explication of standards that physicists often adopt only implicitly in their practice. Besides Hacking's criterion of manipulability of an entity as a sufficient condition for its existence, Dorato stresses another, independently sufficient condition of existence for a non-observable entity *x*, namely measurability of some of its magnitudes, to be discussed together with its close cognate, computability. Given the current importance of computers both in very complex theoretical calculations and in the analysis of experimental data, he claims that some new synthesis between what Galison called 'the theoretical and the experimental cultures of physics' could be attained by paying due attention to the notion of computability. Such a notion could in fact supplant the old notions of observability and empirical adequacy, in the context of an operationalistic philosophy of science in which the operations 'with pen and pencils' would correspond to those performed by recursive functions. With this aim in mind, it becomes important to develop the analogy between computing a number via an algorithm and measuring a magnitude of a physical system in the laboratory. Of course, measuring only one property of an entity *x* may not be sufficient to attribute existence to *x*, and the history of science shows how physicists have sometimes misidentified the natural kind to which the entity *x* belongs. However, if more than one property of *x* has been measured, perhaps by using independent techniques, experimentalists are usually confident that *x* exists, especially when *x* may be subsequently used to inquire into the properties of other, more unfamiliar domains. By trying to clarify the relation between the highly indirect causal intervention realized by measurements of distant, non-manipulable objects (like stars), and the peculiar kind of 'virtual' manipulability, instantiated by computer simulations in current laboratory practice, Dorato defends a realistic position about entities, separating the latter kind of realism from the riskier and more controversial realism about theories. However, a necessary presupposition of the idea that measuring some properties of *x* is sufficient to regard *x* as existent is that quantities or magnitudes can be considered to be mind-independent attributes of events and things, a question that has received too little attention in literature.

Peculiar features of observability, unobservability and their impact on the issue of scientific realism surface in cosmology. Considering the universe as a scientific object *per sé* already implies a strong philosophical commitment. Applying general relativity - as a theory of gravitational-inertial and metric field - to physical spacetime as a whole entails, in addition, a substantial conceptual leap and a number of meta-empirical assumptions, which are hidden nowadays within the nearly axiomatic way of presenting relativistic cosmology. Also, there are more specific and technical issues of observability (better 'unobservability') related to black-hole and space-time singularities, time warps and the like. Jesús Mosterín, in his *Observation, Construction and Speculation in Cosmology*, believes that the standard big-bang cosmological model (or family of models) is a theory without working alternatives and is observationally well-grounded by three fundamental facts: the expansion of the universe, the abundance of chemical elements, and cosmic background radiation. On this basis he claims the existence of a strong demarcation line between the standard achievements of the established theory and the speculative cosmological extensions which, since the seventies, have been introduced by elementary particle physicists into the old and respectable cosmological community (mainly a 'reserve of astronomers and general relativists'). It's difficult to resist Mosterín's arguments against the recent speculative extensions of the standard big-bang model like: (i) the explanation of the decoupling of the strong and the electroweak forces at 10^{-35} sec. after the big-bang, through some (putative) Grand Unification Theory; (ii) the 'chaotic inflationary universe' model; (iii) quantum gravity regime; (iv) speculations on many universes causally and informationally isolated from ours (which are 'ineffable, beyond science, and beyond any type of checking, by definition'). These speculative extensions originated essentially from three unsolved 'problems', namely the horizon problem, the flatness problem, and the monopole problem. Yet, as Mosterín stresses, there is no direct observational evidence to confirm the inflation theory and, what is more, the present data seem to contradict it. In any case, it builds on untested theories like the Grand Unification Theories (whose predictions are either theoretically untestable, or testable and seemingly proven wrong). The advocates of the inflationary scenario claim that this latter simultaneously solves all three problems noted above. That may be true, but this is so only because the putative physical mechanism at work has the effect of *rendering the initial conditions irrelevant*. This is the crucial point: all these extensions have been essentially motivated by various aspects of the initial conditions of the universe which - in the standard model - are handpicked from the observations and cannot be deduced from the theoretical nucleus of the model itself. Now, it is true that, even before the seventies, some cosmologists were concerned both with the initial parameters of the standard big-bang model in their being just contingent features of the world, and, even more, that many such initial parameters were 'improbable' (whatever this may mean in such a context). Other people, however, do not see any real problem with having some 'just such', not further explainable, initial conditions. It is clear that this quarrel about whether or not issues that have to do with the 'initial conditions' of a unique 'object' like the universe as-a-whole are *problems* or *non-problems*, is in itself non-scientific and has little to do with the real issue of observability or unobservability. Rather it has to do with philosophical conceptions about the status of the scientific description of the world. Although 'philosophers have stopped speculating' (about cosmology), as Mosterín notes, physicists have now supplanted them, even in this field of knowledge.

The paper by Mark Wilson, *Where did the Notion that Forces are Unobservable Come From?* contains a historical analysis that discredits a statement often advanced by the defenders of scientific anti-realism. They say that anti-realism is by no means a consequence of quantum physics, but rather the expression of a critically mature way of considering science, that was already attained by several theoretical physicists at the end of the 19th century, and whose most significant example is the elimination of the notion of force from the foundations of mechanics, due to its being unobservable. Mechanics could be adequately reformulated by means of strict logical analysis, using only observable quantities (which are functions of time and position alone) as basic concepts of mechanics (mass, force, momentum, and kinetic energy were, in such a way, explained away as pure relations). The author does not at all deny that such anti-realist positions were actually expressed at that time, but he tries to demonstrate that, far from being the manifestation of a mature epistemological reflection, they were mainly an *excuse* for evading certain mathematical difficulties that naturally arise in the treatment of continua, that is, certain peculiar puzzles that attend classical treatments of fluids, elastic solids, aether and other forms of continuously flexible bodies. In such a setting the notion of force was subjected to various unexpected tensions, and this inspired the efforts to get rid of it. Therefore the thesis that classical mechanics happily tolerates divergent notions of force because of its recognized non-observationality is 'a completely fictional account' of the facts.

In her *Quantum Mechanics without the Observables*, Nancy Cartwright attacks the Born interpretation of quantum mechanics, on the grounds that it misrepresents the relations between quantum and classical physics. Based on generally accepted understanding, backed by the Born interpretation, the quantities of classical physics play no causal role in nature; they merely serve as the 'observables' that are supposed to supply us with epistemological access to the quantum state. By now, however, our methods for finding out information about the observables are neither more nor less reliable than those for assigning quantum states. Not only does talking about observables generally presuppose a sharp distinction between theory and observation (a separation that was questioned long ago), it also demonstrates an underlying axiomatic view of scientific theories, thus understood as symbolic structures that are in need of an interpretation, something that is supposed to give meaning and reference to uninterpreted symbols. According to Cartwright, this sort of 'Hilbertization' of theories is just another instance of an illegitimate, theory-dominated view of science, a view that neglects experimental techniques and causal interventions. A good theory needs no interpretation, unless we have specific cases and particular instances in mind, and this moral applies also to quantum theory. More precisely, we should not even speak of the quantum theory as if there were such a unitary, monolithic thing, but rather of the quantum theory of the laser, of the quantum theory of molecular bonding, of superconductivity, etc. etc. Using examples from superconductivity and lasers, Cartwright argues that the takeover of classical physics by quantum physics does not succeed: 'classical properties that get relegated to the role of the observables are in no way reduced to facts about the quantum state nor do they supervene on them'. But resisting the quantum take-over does not make one an anti-realist about quantum mechanics. Whether quantum mechanics can be judged true or not depends on the evidence that things have quantum states, and that the quantum states behave as they are supposed to in the relevant quantum theory. It does not commit one to the view that quantum terms pick out all there is to talk about, and quantum claims are all the truth there is.

In *Observation, Contextuality and Realism*, Bernard d'Espagnat contrasts two different attitudes towards the notion of 'reality' and the notion of 'observation': the 'Aristotelians' and the 'Mechanists'. While the former remain closer to sense data by allowing for a great number of different, merely qualitative descriptions of the world around us, the latter organize their conceptual resources in a hierarchical and mathematically precise way, but they sacrifice the link with direct experience. Both worldviews, however, share a commitment to commonsensical notions and concepts, that has been called into question by the general theory of relativity and by quantum theory. Although 19th century physicists, and Fourier in particular, have tried to restrict the aim of science to giving a phenomenological description of the behavior of bodies, without a commitment to ontological questions, such an approach is not available today. After all, to interpret Fourier's equation of diffusion, one need not specify or define the nature of heat, but with quarks or curved spacetime, such a link with common sense is lost. Consequently, d'Espagnat pleads the importance of a distinction between *strong* and *weak* objectivity: by relying on the possibility of referring to the properties of bodies, the former form of objectivity is possible only in classical physics. The latter form concerns not the physical world as it is in itself, but only what we can observe of it. Contextuality - which for d'Espagnat generalizes the form of non-separability tested by Bell-type experiments - leaves us with the following trilemma: (i) *radical idealism*, according to which the notion of knowledge is prior to the notion of existence, (ii) some form of *strong inseparable realism*, which can be regarded as a radically speculative proposal, and (iii) the solution of *weak structural realism*, which is the one recommended by the author. This third approach is Kantian in the sense that physical phenomena can be regarded as neither mind-independent nor as created by us: 'they vaguely reflect something Real which surpasses our cognitive possibilities and that is conceptually prior even to space and time'.

There are many ways in which it is said that quantum theory is linked to the issues of observability and the mind-body relation, and mainly through the so-called 'macro-objectification' problem. Some positions are often justified by the philosophical necessity of 'closing the circle', that is the need to understand the knowing subject as an entity in nature and to assess claims to knowledge in the light of this understanding. This closure is sometimes intended, in a strong physicalistic perspective, as the requirement that quantum theory should even account for *our experience as observers*. These positions more or less saddle a specific, historically contingent, physical theory (namely quantum mechanics) with the task of 'solving', as it were, the mind-body problem itself and almost all other problems of knowledge. Such positions implicitly attribute to this physical theory, as it were, a meta-empirical status. By allowing for an indefinite macro-realm, a subclass of these positions (the so-called 'many-worlds' or 'many minds' interpretations) tries to secure only definite appearances, thereby denying that the macro-world of our experience is, in fact, what it appears to be in observation. Further positions on the matter (like that of Wigner) hold that *consciousness* itself is responsible for state-vector reduction, thus entangling the mind-body problem within the *present form* of quantum theory. Finally, there are the attempts to solve the quantum macro-objectification issue by means of suitable technical *modifications* of the theory. These attempts explicitly resist an involvement of the mind-body issue with the physical theory, according to a traditional self-restrained view of scientific practice. In his *Leibniz, Kant and the Quantum - A Provocative Proposal about Observation, Space-time, and the Mind-Body Issue*, Massimo Pauri proposes that, far from being an issue that has to do with interpretations of quantum formalism or simply with technical modifications of the mathematical structure of the original theory, the macro-objectification problem has all the marks of a profound technical *and* philosophical issue having to do with the possible *forms* of human knowledge of physical reality. The author argues that the radical denial of atomism, implied by the discovery of the *atomization of action* (i.e. of 'processes' instead of 'things') must entail far reaching consequences for the *part-whole relation* and for the concepts of space and time in their relationship to the mind-body problem. Consequences, which, so far, have neither been properly drawn up nor envisaged. In this perspective, he stresses the surviving relevance of the debate about spatial wholes and parts that ideally took place between Leibniz and Kant, and advances a *non-trascendental reinterpretation* of the role of intuition of the pure homogeneous extensions of space and time. The conjunction of the macro-objectification conundrum and the unavoidable conceptual tension existing between elementary quantum processes and spatiotemporal description is specifically exploited with the intent of defending the following provocative thesis. There must be something in the reality of space which is beyond extension and perceivable relations. Corresponding to this, there should exist in the natural constitution of humans (and, to a certain degree, all living organisms as well) something that causes our perception of things in the continuity of extension. This something should be the same something that brings about the 'composition' of elementary processes into spatially extended, and inter-subjective macro-reality. From this point of view, quantum decoherence would not be an effect of the intervention of consciousness - as some positions hold - but rather a specific and necessary pre-condition for our observational experience of spacetime. The author's thesis is tantamount to outlining a new scientific, philosophical and cognitive problem, which is to explain, by means of a putative reformed quantum theory, the 'material' pre-conditions of the pure subjective intuition of the homogeneous continuum in its 'non-propositional formatting' nature. There would be a profound and important sense in which we would *not* be entitled to say literally that tables and chairs are quantum mechanical *compounds* of their elementary constituents *in* an extensional space-time background. To conclude, the author suggests that space (understood as the homogeneously extended object of our sensible intuition) is, as it were, *epistemically prior* but not *ontologically prior* (as for Kant), but that there is something in the reality of space (and time) which is indeed ontologically prior with respect to the intuition of homogeneous extension. If extension is not ontologically basic but only epistemically prior, the ontic structure cannot be intelligibly decomposed into parts in extension, i.e., understood in the *anschaulich* modality which is unique to appearances or phenomena, as perceived and 'observed'. At the same time, the epistemic priority of homogeneous extension would justify the peculiar resilience of scientific imaginery towards the attempts to give up the mathematical continuum in our physical theories or even vindicate an appeal to a primary classical reference in the sense advocated by Bohr. From a theoretical viewpoint, bypassing the epistemological priority of extension would require a *systematic* recourse to a *non-local formalism*, by far more extensive, as well as intensive, than used so far. Pauri's proposal is clearly based on the assumption that any perception of a conscious observer is triggered by an unambiguous macroscopic situation. However, the opposite situation is a-priori conceivable and, from a foreseeable technological point of view, may even be empirically verifiable. That is, *perceptual* processes might be triggered by genuine superpositions of (spatiotemporally distinguishable) states which, when considered by themselves, would give rise to different (spatiotemporal) perceptions. Pauri notes that, in this case, his proposal would be empirically refuted.

In *Efficient and Final Cause as CPT Reciprocals*, Olivier Costa de Beauregard argues that the three most common arguments against the existence of efficient causes - the unreality of the future, the lack of backward causation, and the absurdity of *psychokinesis* (i.e., of the belief that ideas can move matter) - are somewhat superseded by the emergence of the theory of relativity, the time-symmetry of physical laws, and the quantum non-separability. In order to both defend this astonishing conclusion and to give a sound metaphysical interpretation of contemporary physics, Costa de Beauregard proposes relying heavily on the transition amplitude formalism. According to the author, such a formalism has various advantages, among which that of 'being time-symmetric and time-extended between preparation and measurement', and that of giving an adequate formal account of quantum non-separability: the resulting interpretation of quantum mechanics is, in fact, Lorentz and CPT invariant. By analyzing reciprocal symmetrical couples of concepts like 'coding' and 'decoding'; 'gain in knowledge' and 'psychokinesis'; 'entropy' and 'information'; he claims that 'efficient' and 'final cause' are reciprocal under a transformation that takes the mirror image of a system (P), reverses its charges (C) and inverts the temporal order of the processes (T). The three arguments above, against efficient causes, are then tackled as follows. For the first argument, Costa de Beauregard maintains that the reality of the future is an inescapable consequence of the relativity of simultaneity, or, alternatively, the assumption that the speed of light is not infinite. In his second argument, he urges that backward causation can be assumed if advanced electromagnetic radiation and entropy decrease in the vast majority of closed systems are only *de facto* absent or extremely improbable in our universe. According to contemporary physics in fact, the time-reversal of every process is not prohibited by laws, with the exception of *kaon* decay in weak interactions, as a consequence of CPT invariance. As to *psychokinesis*, Costa de Beauregard thinks that it should be admitted, as a consequence of the *negentropy-information* equivalence and of the CPT invariance of the transition amplitude. In this sense, *gain-in-knowledge*, which is accepted by common sense, is reciprocal to *psychokinesis*, and the latter is nothing but the operation of final causes. The question of establishing the consequences of contemporary physics on the fundamental metaphysical problem of 'what there is' is an active and difficult area of philosophical research. Considering relativity, quantum field theory and high-energy physics, Quine for instance has claimed that physics forces us to abandon the commonsensical notion of the material physical object, to be replaced with the four real numbers indicating its coordinates.

Charles Paul Enz, in his *Observability and Realism in Modern Experiments with Correlated Quantum Systems*, argues in favor of what he calls an 'active realistic' view of quantum theory. Enz maintains that philosophy of physics has, so far, been limited essentially to the analysis of free particles and fields, and of the traditional issues of particle-wave duality, including, of course, all of its ramifications with non-locality, complementarity, uncertainty, and the Copenhagen interpretation. In so doing, according to Enz, the epistemological analysis has neglected for too long the fact that the general set of *correlated quantum systems* is much wider than the traditional EPR class. While EPR correlations arise for composite systems obtained by *separating* a quantum system in parts, there are other correlations, which arise from the very *interactions* between parts. Enz stresses that, apart from modern quantum optics, condensed-matter physics is the most promising territory for understanding the intimate *real* quantum world in spite of its mainly nonrelativistic nature. As a matter of fact, recent experimental advances in the so-called *nano-technology* provide powerful tools for analyzing the *spatial appearances* of strongly correlated quantum systems, in particular for probing metallic surfaces on an atomic scale and for locating single atoms or electrons. Enz exploits a variety of modern experiments in order to show that: 1) The idyllic picture of integer charge and individual electrons ceases to hold when correlations between them become important (this correlation in fact became the main theoretical problem of the high-temperature superconductors discovered in 1986); 2) The conventional notions of free particle and of particle-wave duality have very limited validity, due to both correlations and particle-number uncertainty; 3) Correlated electrons may behave in a way in which neither of the naive particle or wave pictures apply, so that the only trustworthy guiding principle remains the *realistic many-particle wavefunction*; 4) Although the old wave function ψ is in general a complex-valued function in a multi-dimensional space and, hence, has a quite abstract and even a symbolic meaning, ψ is itself directly observable, both in its amplitude and its phase, when taken in its *realistic correlated many-body form*, a fact that from the point of view of quantum reality, is one of the most spectacular experimental results. Of course, due to the uncertainty relations $\Delta N \cdot \Delta\phi \geq \hbar$, measurement of the local electron density $\rho = |\psi|^2$ and of the wave function phase ϕ require complementary experimental settings.

In *Quantum Mechanics and the Ultimate Observer*, Bernulf Kanitscheider discusses the interpretive problems of quantum mechanics by reviewing the main solutions to its paradoxes that have been proposed. According to the famous 'cat paradox', for instance, a cat that has been closed in a room in which a poisonous gas can be released by a random atomic decay is in the superposition of a live and dead state until we observe it. Schrödinger surmised that in our commonsense macroscopic world, we would never encounter situations in which such incompatible properties are inextricably entangled. Likewise, Bohr thought that macroscopic bodies could not display quantum behavior and should always be described by classical physics. In view of recent work on superconductivity or macroscopic tunnel effects - that seems to imply that it is now possible to create individual macroscopic quantum objects which can be in a state of superposition - Kanitscheider claims that the validity of quantum mechanics should not be dogmatically restricted to microscopic systems. It then becomes necessary to explain why we don't have contradicting visual impressions about macroscopic objects. With respect to this problem, Kanitscheider discusses both Putnam's internal realism - in which certain properties are really possessed by physical systems like cats only relatively to a framework set by the observer - and Wigner's 'idealistic' solution. As is well known, Wigner argued that a material measurement apparatus and an intelligent organism react quite differently to a superposition, given that only conscious observers have the peculiar capacity of selecting one definite outcome of a measurement out of a range of possibilities. Kanitscheider objects that such an anti-naturalistic position is unreasonable because it attributes to conscious states a special status, contrary to contemporary findings in neurophysiology. Likewise, Barrow's and Tipler's invocation of the Ultimate Observer - located in the final singularity of a closed universe, or at time-like infinity in an open universe and bringing into definiteness or even existence the entire Universe - relies on an implausible *deus ex machina* from outside spacetime. Since singularities cannot be located in spacetime without running into counterintuitive results and mathematical difficulties, the presence of observers at singularities seems to hide some sort of supernatural or eschatological interpretation, or at least a 'tremendous confusion between the mental, the physical and the metaphysical'. Kanitscheider concludes by endorsing either some non-linear modification of the Schrödinger equation (Penrose indicated gravitation as a possible cause for wave collapses realistically intended), or, alternatively, a non-collapse interpretation that, like Gell-Mann's and Hartle's, applies to the whole universe and, consequently, does not consider 'measurements' or 'observers' as fundamental concepts as in Bohr's interpretation.

The statistical predictions of quantum mechanics are usually thought of as referring only to sufficiently large ensembles of equally prepared systems, considered in a post-measurement state. In his *Individualistic and Statistical Interpretation of Quantum Mechanics*, Peter Mittelstaedt argues that the problem of applying the statistical (Born) interpretation of the quantum formalism to individual systems might dissolve into a pseudo-problem, since, as he claims, quantum theory can instead be regarded as being primarily concerned with individual systems, which are described by 'yes-no propositions'. The question of the meaning of quantum probabilities is perhaps the central problem of interpreting quantum mechanics, and it is obviously related to the more philosophical issue of establishing whether such predictions are just an expression of the observer's information about the system, or may be referred to a mind-independent reality. For instance, in order to attach an objective, non-epistemic meaning to $p(\varphi,a)$ - the probability p of finding a value a for the observable A of a single system S, even when S is prepared in a state φ which is not an eigenstate of A - Heisenberg rehabilitated the concept of *potentia*. With this term, he meant to distinguish an *intermediate level* of reality, which is located somewhere between the purely subjective domain of the observer's knowledge and the full definiteness of mind-independent reality. Analogously, Popper developed the concept of *propensity*, an alleged tendency, that single systems would objectively possess to display certain values in measurement interactions. By using two formal results about quantum probabilities, Mittelstaed argues that such metaphysical baggage is hypothetical and unjustified, since, from the above pure state φ, one can deduce all the objective properties of an individual system, including the statistical ones, which correspond to properties of conveniently defined ensembles of equivalent systems. Moreover, when $0< p(\varphi,a)<1$, there is no measurable property of a single system which could be interpreted as a probability, and it is objectively undecided whether the single system S really possesses the property in question. Despite this limitation, Mittelstaedt concludes that some restricted form of realism can be vindicated and quantum mechanics may be considered as making statements about objective properties only, either for individual systems or for compound systems of N equally prepared systems: specifically, the $p(\varphi,a)$'s refer either to properties of an individual system (when $p(\varphi,a)$ is an element of $\{0,1\}$), or to relative frequencies of these properties in ensembles of identically prepared systems.

In her *How to Observe Quarks*, Brigitte Falkenburg discusses the issue of observation in physics with specific emphasis on the observability of quarks. Of course, the key problem in this context is that quarks - unlike electrons, protons or neutrons - have never been detected as 'free' particles, that is, as roughly localized and isolated fractional charges and *quanta* of energy. Nevertheless, the quark model has been regarded as well-established in particle physics for more than twenty years. In order to discuss this puzzling issue, Falkenburg analyzes the current experimental evidence for quarks in the context of the generalized concept of observation proposed by Dudley Shapere in 1982. By invoking a distinction between direct and indirect observation within the framework of naturalized epistemology, Shapere had claimed that the use of the term 'observation' in empirical science should be disentangled from the traditional 'epistemological' concept based on perception. In this way, he had also succeeded in accounting for the theory dependence of scientific experiments. Falkenburg suggests 'translating' Shapere's notion of 'direct observation' in terms of those measurement procedures typical of the theoretical models used to analyze the scattering experiments of particle physics. As a result, she claims that a necessary *reliability condition* for the observability of a corpuscular physical entity is the possibility of singling out an *individual* system carrying the physical magnitudes attributed to the entity. According to Falkenburg, such a necessary reliability condition was not satisfied by the early quark experiments - namely those preceding the discovery of the J/ψ resonance (1974) - and physicists themselves recognized the experimental evidence for quarks as being only 'indirect', in agreement with Shapere's characterization of this notion. Such experiments violated the necessary reliability condition in two ways: first, the relation between a single, measured, quark momentum distribution and one particular collection of permanent properties (like mass, spin, charge and 'flavor') was established only in an ambiguous way. Secondly, the early quark models didn't support a causal story relating a definite part of a measured 'structure function' to an individual quark, since a causal story could only be told about the total quantity or the total effect of several quarks inside a nucleon. It is in this sense that we should draw the conclusion that, previously, quarks had been observed only 'indirectly'. On the other hand, Falkenburg argues that the new experiments that have been performed since 1974 do provide experimental evidence for discrete or individual quarks: at present there is at least one kind of evidence which may count as 'direct observation' in Shapere's sense of the term. Specifically, Falkenburg argues that in high energy e^+e^- annihilation experiments giving rise to the formation of the so-called 'jets', both parts of her reliability condition for a direct observation are fulfilled so that the inference from hadron jets to quarks may be regarded as based on the reliable background knowledge which is admittedly involved in a direct observation of quarks. Since, however, there are experiments in which individual quarks cannot be singled out from experimental evidence, Falkenburg only defends the claim that quarks 'exist' under the specific conditions that are realized in certain kinds of experiment in high energy physics. Finally, she claims that once the structural features of what counts as an observation are clarified, a sharp distinction between observation and measurement is possible: while observation requires an *individual* entity to be observed, measurement requires only *physical magnitudes* to be measured. If observations can confirm the existence of an entity of a certain kind within a certain *space-time region*, measurement interactions can confirm only the validity (or empirical adequacy) of a model expressed in terms of some magnitudes. For Falkenburg, the possibility of making such a distinction runs counter van Fraassen's claims about the vagueness of the concept of observation, despite the explicitly mentioned limitations that quantum theory sets upon a spatiotemporal description of processes, let alone of 'entities'.

The problematic interface between classical and quantum physics, entailing such problems as the onset of decoherence as well as all of the puzzling aspects of the quantum theory of measurement, is tackled by Giovanni Maria Prosperi who discusses a recently proposed approach to the macro-objectification issue in quantum theory in his *Common Experience and Quantum Theory - Observables and Beables*. This paper begins with a comparison of the descriptions of a physical phenomenon as described by classical and by quantum theories. It is well-known in classical physics, that the properties of an object are supposed to be specified in terms of a set of variables having a well-defined value at any time and changing continuously according to strictly deterministic laws. On the contrary, quantum theory consists of a set of abstract mathematical rules, which make explicit reference to observation and allow (assuming certain initial information about the system) the evaluation of the probability that a subsequent experiment furnishes a certain result. The author exploits the double slits experiment to illustrate the essential peculiarities of the theory and, in particular, Bohr's point of view. According to the latter, quantum theory necessarily refers to a double level of understanding of the world: while 'quantum objects' are described by the symbolic language of the theory, experimental set up and results of the experiments have to be intrinsically described in the language of classical physics, considered to be the language of our common experience. Prosperi discusses von Neumann's analysis of the measurement process, according to which quantum theory is also applied to the apparatus. In particular, he deals with the consistency problem which arises from the 'shifty' character of the theory, that is, from the possibility of arbitrarily shifting the borderline between the object (to which quantum theory applies) and the apparatus (which is described classically) and, ultimately, from the role of the human observer. Having pointed out some pitfalls of von Neumann's approach, the author defends the thesis that the consistency problem can be solved in a satisfactory way in the context of the modern, more general, formulation of quantum theory, mainly due to Ludwig, Davies and Holevo. Within this modern formulation, *yes/no* observables are no longer necessarily associated with 'projection operators', but more generally with the so-called 'effect operators' (positive operators smaller than 1), which enable us to consider *imperfect measurements* as well. Correspondingly, the state collapse is no longer represented by a simple projection, but by a more general operation acting on the statistical operator of the system. Then, pure and mixed states are put on the same footing and the collapse can be simply interpreted as the transition from a joint probability to a conditional probability. Still, there exists the need to break, at some level, the von Neumann's chain of 'apparatuses observing other apparatuses'. In this connection, the author discusses the merits and the limits of the classic attempt to solve the problem by considering the apparatus as a many particle system and the consequent approximate suppression of the interference terms between different macroscopically distinguishable states. Finally, Prosperi illustrates the recents attempts to solve this problem made by him and coworkers, attempts which are based on the so-called *theory of continuous observations*. In that formalism, based on the modern formulation of quantum theory mentioned above, it is possible to deal with the case of a quantity which is continuously (even if imperfectly) monitored. Then, by an appropriate change of language, a classical level can be introduced *intrinsically*, in terms of a set of privileged quantities which are treated as 'beables' in the Bell sense. In conclusion, certain difficulties having to do with this attempt are stressed in connection with the conservation rules.

Finally, the interface between classical and quantum theories is dealt with by Enrico Beltrametti and Slawomir Bugajski, from an entirely different viewpoint, by their analysis of the mathematical foundations of both realms in their *On the Relationships between Classical and Quantum Mechanics*. Beginning with the notion of the *state* of a physical system, as a *primitive undefined entity* that is mathematically characterized as being part of a *convex* set, they are able to provide an abstract definition of *observable* that encompasses both the classical and quantum cases. With this, one can abstractly define a *Bell phenomenon*, that reduces to the familiar Bell inequalities for quantum models but subsists for classical ones as well. As a matter of fact, the *classical extension* of a quantum system provides a procedure for building some examples of a *classical EPR*. The definition of the set of states S as *convex* makes known the physical fact that states can be mixed; the *pure states* are then the *extremal elements*, that is those that cannot be thought of as convex combinations of other elements (in classical physics S_C is typically the set $M^+(\Omega)$ of *all probability measures* on a phase space Ω and pure states are *Dirac measures*; in quantum physics S_Q is the set of *density operators* on a *Hilbert space* $\mathcal{H}$, and pure states are one-dimensional projectors). The basic difference between the two constructions is that the decomposition of a *mixed state* in terms of pure states is unique only in the classical case. In other words, S_C is a simplex while S_Q is not. Now, an *observable* is defined as an affine mapping of S into the convex set of probability measures $M^+(\Xi)$ on some measurable 'value' space Ξ describing the distribution of results upon observation. A model based on the set of states S is said to admit an extension if there exists a 'richer' set of states $\bar{S}$ and a way to 'coarse-grain' it into S. Thus, to every observable B on S there corresponds an observable on $\bar{S}$ obtained by first *reducing* the $\bar{S}$-state to S and then applying B. It is remarkable that every *quantum model* (based on a S_Q) admits a *classical extension* built, roughly, by considering a classical system whose admissible pure states correspond to the quantum pure states $S_C = \partial S_Q$. The statistical distributions of a quantum observable and of its classical representative are the same, so that typical quantum features such as uncertainty relations are preserved in the classical extension. This is not paradoxical since the classical observables will not be *sharp* (mapping pure states onto *Dirac measures* on 'values' as those usually considered in standard classical mechanics): instead they will be *fuzzy*. One can find a characterization for a family of observables on a set of states S that embodies the physical content of the uncertainty relations in terms of probability measures. Now, the main point is that if a quantum model admits a *Bell phenomenon*, so will its classical extension. It is clear that this is not possible for families of sharp classical observables, but we already know that the observables of a classical extension are fuzzy, thus preserving some of the original 'quantum flavor' for the classical domain. Some references in the article provide pointers to worked examples.

Now, some words as to the 'origin' of this volume: It contains the revised papers from an International Colloquium held in Parma in May 1995, as the Annual Meeting of the *International Academy of Philosophy of Science*. This event - entitled *Observability, Unobservability and their Impact on the Issue of Scientific Realism* - was organized under the auspices of the University of Parma and of the Italian Ministry of the University and Scientific and Technological Research. The editors are indebted to the Institutions and individuals whose contribution and cooperation made the Parma Colloquium possible. Among the various institutions which have supported the Meeting in the form of generous financial help, the editors would like to mention in particular: the Commission of the European Communities DG 12, the Italian National Research Council, the Italian Ministry of the University and of Scientific and Technological Research, the Parma Municipality, the Foundation of the bank Cassa di Risparmio of Parma & Piacenza, The Industry Union of Parma.

Finally, the editors would like to take this opportunity to express their gratitude to Professor Bob Cohen for accepting this volume in the Boston Studies series. They would also like to thank Kluwer Academic Publishers and, in particular, Annie Kuipers and Stephanie Harmon, for advice and assistance during the preparation of the book. The editors are also grateful to the components of the 'Just in Time' office, Miss Francesca Bandini and Miss Marion Alka, and would like to thank them for the technical preparation of this volume and the realization of the camera ready copy, as well as Professor Mauro Dorato for his valuable assistance with the editorial work.

Evandro Agazzi
President
International Academy
of Philosophy of Science

Massimo Pauri
Vice-President
International Academy
of Philosophy of Science

ERHARD SCHEIBE

THE ORIGIN OF SCIENTIFIC REALISM: BOLTZMANN, PLANCK, EINSTEIN[1]

1. SCIENTIFIC REALISM INVENTED BY PHYSICISTS

Scientific realism today is an issue much debated by philosophers of science.[2] However, to the best of my knowledge it was invented by physicists, and this is a fact that seems to have fallen into oblivion. Moreover, scientific realism emerged in a period where there was a general turn in the attitude of the physicists in matters of the philosophical foundations of their science. In obvious connection with the establishment of theoretical physics as a new discipline at the end of the 19th century, we see the new theoretical physicists becoming engaged in a lively debate on various philosophical questions concerning physics. And not only did this happen but it was also immediately noticed - from without and within. In the years before the First World War the theologian Adolf v. Harnack is said to have said on occasion: "People complain that our generation has no philosophers. Quite unjustly: it is merely that today's philosophers sit in another department, their names are Planck and Einstein."[3] And as early as 1901 Wilhelm Ostwald in his lectures on natural philosophy testified to the change: "The mental operations by which scientific work is organized ... are not essentially different from those that are investigated in philosophy. The awareness of this situation was indeed obscured for some time during the second half of the 19th century; but in our days it has awakened to a most vivid efficacy, and everywhere the spirits are aroused in the scientists' camp to make their contribution to the whole of philosophical knowledge."[4]

Even more extensive commentaries on the entire process can be found without going outside of physics. At the end of the period in question, in a lecture of 1948,[5] Arnold Sommerfeld summarizes the development in the following words: "During the 19th century the relation between physics and philosophy was strained. First philosophy dominated and wanted to prescribe physics its way ... Later the physicists had become suspicious and rejected any philosophy..." The quarrel between physics and philosophy that Sommerfeld refers to had evolved from that unhappy marriage that some natural scientists had contracted with the natural philosophy of Schelling and Hegel in the early 19th century.[6] The divorce that followed was so thorough that it led to the definite methodical emancipation of the *Geisteswissenschaften* as well as to a lasting puristic and partially positivistic attitude of the physicists towards their discipline. However, Sommerfeld goes on to tell us: "In the 20th century the relation between physics and philosophy changed fundamentally. Right at the beginning in 1900 Planck discovered the quantum of action ... He thus gave philosophy a hard nut to crack which it will have

E. Agazzi and M. Pauri (eds.), The Reality of the Unobservable, 31-44.

to deal with quite a while ... The decisive step towards a philosophically deepened physics was taken by Einstein in 1905." Here Sommerfeld alludes to the two new physical theories of our century, relativity and quantum theory, that introduced profound conceptual revisions and set physics in such a contrast to its past and to common sense that philosophical reflection became imperative. He could even have referred to the earlier kinetic theory of heat that had raised basic questions concerning atomism. At all events Sommerfeld can crown his review with the words: "Since Einstein there is no longer any alienation between physicists and philosophers. The physicists became philosophers, and the philosophers are on their guard not to become engaged in a conflict with physics."

That the physicists became philosophers did, of course, not mean that all of a sudden philosophical articles in a professional sense flew from their pen. Gilson has coined the malignant saying: "Nothing equals the ignorance of modern philosophers in matters of science, except the ignorance of modern scientists in matters of philosophy".[7] In fact, one can observe how uneasy the physicists of the first generation of our period felt whenever they were forced by external reasons to philosophize before the public. Ostwald tells us that he had been urged by friends and students to give his lectures on natural philosophy (= *Naturphilosophie*), and he confesses right at the beginning that "[he] may not call philosophy a subject that [he] had studied in the normal sense of the word".[8] Similarly, it delights us to hear what Boltzmann says at the beginning of a series of lectures on natural philosophy[9] that the Vienna ministry had demanded of him. He comments on the large number of people attending the inaugural lecture by saying that he could explain this only by the fact that "[his] present lectures are indeed a curiosity in academic life in a certain respect." And this respect then turns out to be that he as a philosophical layman now has to give these lectures on natural philosophy. He seeks consolation in the most far-fetched explanations for why the ministry had imposed this burden on him of all persons, and he assures the astonished audience that his objections were settled by the ministry with the remark "that any other person would not do better." However, we shall see that what Boltzmann and his followers did when they started philosophizing really was not that bad. There was no doubt a new and creative philosophical spirit among the physicists at the turn of the century.

This being the situation in general, I now turn - still by way of introduction - to the more special question: What is scientific realism? Questions about human access to reality have been asked since the beginning of western philosophy. Already in Plato's "Theaetetus" the equating of knowledge and perception is denied, and Protagoras' doctrine that man be the measure of all things is attacked.[10] In modern times Descartes was the first to give a reformulation of the problem.[11] Starting out from sense illusions, he makes it clear that we can be deceived by almost everything. The epistemic immediacy concerning external things as it characterizes naive realism gets lost, and the monstrous viewpoint of solipsism - an invention of Descartes - occurs. Henceforth all philosophers, including Descartes, try to find *proofs* of the independent existence of bodily things. But one and a half centuries later, Kant[12] has still to notice that it remains "a scandal of philosophy and general human reason to have to accept the existence of external things as a matter of faith ..." Kant added a new proof which, however, did not find general acceptance either. How desperate the situation finally became is certified after another one and a half centuries in a lecture by G.E. Moore, having the title "Proof of an External World".[13] At the end of his lecture, after laborious considerations including Kant, Moore gives the proof in question by proving that, for instance, two human hands exist. And answering the question: "How?" Moore even gives his audience the following details: "By holding up my two hands, and saying, as I make a certain gesture with the right hand, 'Here is one hand', and adding, as I make a certain gesture with the left, 'and here is another'".

I don't mention this proof of an external world in order to demolish the reputation of Moore whom one cannot but appreciate in his way. The point is not the philosophical importance of all these proofs that is different from case to case and certainly different in the cases of Kant and Moore. The point is that all proofs are restricted to the kind of experience as we make it in *daily life* whether this is evident as in Moore or not so evident as in Kant. It is only in recent times that philosophy deals with the problem in question under explicit consideration of the fact that for more than three hundred years there has been a continuously progressing *scientific* experience. And this fact can be made the crucial point of an argument in favour of realism.

Among the physicists, Helmholtz still argued in favour of realism by *physiological* investigations: "Our sensations", he says, "are effects produced in our organs by external causes".[14] By contrast, Planck is the first who clearly bases realism typically on a development of *physics* that leads *away* from all questions of sensation or perception in the usual sense. He supports realism by the continuous empirical success of physical theories that are about more and more remote objects, divested of the subjectivity of human perception, but at the same time fundamental for the construction of physics. It is the realism backed up by this kind of argument from scientific progress that may justifiably be called 'scientific' realism. And at least one line of development of this scientific realism leads us to the physicists around the turn of the century: It leads us to Boltzmann, Planck, and Einstein. Also it leads us, however, to the great anti-realist Ernst Mach whose authority none of the aforementioned could ignore.

2. BOLTZMANN: BETWEEN INSTRUMENTALISM AND REALISM

Ludwig Boltzmann, whom I am going to address first, has passed through a mental development that did not lack a tragic stamp. He stood at the threshold of the final establishment of modern atomism. As a young man he was a pure physicist who had enlarged the kinetic theory of gases, initiated by Clausius and Maxwell, by adding the famous equation that bears his name. Kinetic theory, however, was an atomistic theory, and atoms remained a speculative object during the whole of the 19th century. Many physicists, disappointed by the influences from German idealism, adopted a decided positivistic or, as it was called in those times, phenomenological attitude that was not in favour of atomism.[15] Besides empirical successes there were also physical difficulties for kinetic theory, and Boltzmann, who was convinced of the atomistic approach, felt himself more and more bound to propose *philosophical* arguments to support the theory. Indeed, during the last ten to fifteen years of his life, Boltzmann has done mainly philosophical work. The corresponding publications, however, did not have the intended effect. Searching for philosophical support of atomism, Boltzmann saw himself more and more driven into the camp of his adversaries.

It would be an exaggeration, though, to call this process a conversion, as has recently been done.[16] Boltzmann believed in atoms in the sense that "[the theory of gases] agrees in so many respects with the facts that we can hardly doubt that in gases *certain entities*, the number and size of which can roughly be determined, fly about pell-mell."[17] Yet Boltzmann was an atomist of sorts. He took a position somewhere between the naive belief in atoms and a methodical phenomenalism. This position is perhaps most adequately conveyed by looking at the concept of *physical theory* that Boltzmann developed. On this matter, views were circulating at that time which Boltzmann could by no means be in agreement with. A laconic formulation of Kirchhoff's had risen to fame according to which it was the task of mechanics "to *describe* the motions occuring in nature," where this was meant in the restricted sense: "that the issue can only be to point out *which* are the phenomena occurring, not, however, to discover their *causes*".[18] It goes without saying that kinetic theory hardly satisfied Kirchhoff's condition. Already here we have reached a point where an unambiguous statement on Boltzmann's position can be made: Whether or not he was a realist in matters of the atoms, he pleaded for a concept of theory liberal enough to guarantee their *possible* existence. What concept of theory was this?

Nowadays, one would like to say that Boltzmann suggested the use of theoretical terms. In the language of his time the crucial relevant metatheoretic concept was that of a *picture* (Bild), sometimes also *model*. It is a *ceterum censeo* in Boltzmann's papers that theoretical physics, strictly speaking, does not deal with things themselves but with certain pictures instead that we take of them. Boltzmann gives Maxwell the priority to have introduced the idea of a picture, and he repeatedly mentions Heinrich Hertz as the one "[who] makes physicists properly aware of something philosophers had no doubt long since stated, namely that no theory can be objective, actually coinciding with nature, but rather that each theory is only a mental picture of phenomena, related to them as sign is to designatum."[19] If Boltzmann here quotes that in the theory we make ourselves pictures of the *phenomena*, this conceals a little his proper point that in theory we try to use pictures precisely where the phenomena are *missing*. Such was the case with kinetic theory: It outlined a picture of something that had not appeared to anybody by that time. It is in this sense that Boltzmann says in the same article quite unambiguously: "Phenomenology believed that it could represent nature without in any way going beyond experience, but I think this is an illusion ... [Every] equation ... idealizes [the processes] ... thus going beyond experience." This transcendence belongs to the nature of the mental operation "consisting as it does in *adding something* to experience and creating something that is not experience and therefore can represent many experiences."[20] Accordingly, Boltzmann's presumably strongest argument against the phenomenological position was that it, too, goes beyond the phenomena, for instance, by assuming matter to be a continuum.[21]

Boltzmann was seen by his contemporaries precisely as a representative of this liberal view of theories. At the annual meeting of the "Gesellschaft Deutscher Naturforscher und Ärzte" in 1895 in Lübeck, Ostwald[22] calls out to Boltzmann (in unintended prophecy): "We have ... finally to give up all hope to give a pictorial (*anschaulich*) interpretation of the physical world by reducing the phenomena to the mechanics of atoms." And to the question whether there is a still available mean "to make us a picture of reality", Ostwald answers in the presence of the man whose gas lectures appeared under the motto "Alles Vergängliche ist nur ein Gleichnis"[23]: "In view of such questions, I want to call out to you: You *shouldn't* make yourselves any picture or simile (Gleichnis)! Our task is ... to [look at] the world ... as directly as the constitution of our mind will allow. To relate *realities* ... is the task of science, and we cannot solve it by the underlaying of any hypothetical picture ..." Evidently, speaking of "realities" Ostwald here means "appearances", and these he sees as *rivals* of the theoretical pictures of Boltzmann and Hertz.

Ernst Mach has seen the situation essentially in the same way, though he was even prepared to make an important concession. He would accept an atomistic theory as a mean if it leads to useful results on the level of phenomena and no realistic consequences are drawn from it.[24] He appraises the kinetic theory in view of its successes, and he is not against "the liberty that one takes by assuming invisible hidden motions." But this method has a decisive instrumentalistic proviso. Despite the admissibility of arbitrary ideas as means of research, it is imperative from time to time "to purify the representation of the results of research from the superfluous and inessential ingredients which intervene by the operations with hypotheses." It is precisely because our research work ends with this elimination that everything is permitted in advance for the uninterpretable parts of a theory.[25] Just in case the atoms don't belong to the world of phenomena we are free, God knows what mathematical ideas to relate to them. If the atoms are not perceivable, why then make a picture of them as if they were perceivable or would become so one day?[26]

If we now come back to Boltzmann, we must observe that his statements on the point in question are more ambiguous. He oscillates between realism and instrumentalism. Unambiguous - as we have heard - is his advocacy of a theory concept that leaves room for the introduction of as yet empirically uncertain entities. But if we ask how he intends to fill this place his statements become increasingly cautious. As a physicist he simply believed in atoms and could be very drastic about this. Mach reports that Boltzmann criticized his analysis of sensations by the remark, "no sooner could one analyse the sensations than the paths of the atoms in the brain were not known."[27] Again, in a commemorative address on the occasion of Loschmidt's death, Boltzmann remarked that Loschmidt's body had now decayed into its atoms and added the comment that the deceased himself had put us in the position to know into how many it had decayed - Loschmidt's number being on the blackboard.[28]

This directness of the pragmatically minded physicist contrasts markedly with the philosophical *scepsis* in the man. Boltzmann is most anxious to transform inadequately posed into adequately posed questions. On occasion of his report[29] on Mach's intrusion into a discussion on the value of atomistic theories by the words "I do not believe that atoms exist" Boltzmann starts an argument that here, i.e. in the case of space, time, atoms, etc., as distinct from things like tables, dogs, and human beings, one might not even know "what is meant by asking whether these things exist." In the address on Loschmidt he considers the question of the constitution of matter as being one of the most important questions of the time.[30] It is only that one puts it somewhat differently today as compared with earlier times. "While formerly one was looking for the ultimate elements of ... matter itself, nowadays it is asked from which simple elements the mental *pictures* have to be constructed in order to achieve the *best possible agreement with the phenomena*." By adding that both ways of speaking presumably have the same meaning, Boltzmann seems to suggest: One *ought* to mean the same by the question whether atoms exist as by the question whether the theory in which we have introduced atoms by means of such and such a picture is empirically successful.

Mach has passed judgement on our subject with the impressive words: "If one day the now living physicists will have made their exit from the scene, a future historian ... will easily ... disclose how fearfully serious and terribly naive the mechanical, particularly atomistic ideas have been conceived by a large majority of outstanding scientists of our times, and how few scholars of a peculiar way of thinking belonged to the party in opposition."[31] In the sense of this distinction we certainly find Mach and Boltzmann on the same side. On the other hand, Einstein once said that physicists have to be judged, not on account of their words, but their deeds.[32] And if we take this seriously then the facts are that Boltzmann's major work was a book on the atomistic theory of gases and Mach's one on the analysis of sensations. This opposition of their deeds has been somewhat obscured by their words, however well chosen they might have been.

3. PLANCK: PROGRESS THE CRITERION OF REALITY

Boltzmann had second thoughts also on the development of physics. He can even be seen as the founder of an important tradition of thinking in matters of theory progress.[33] In the preceding section we have seen that Boltzmann viewed the empirical success of a theory as a criterion of the reality of theoretical entities introduced into that theory. It was left to ***Max Planck*** to link this idea to the idea of progress in physics. In his earlier days, Planck was an opponent of atomism who did not believe in Boltzmann's statistical foundation of the second law of thermodynamics.[34] But he changed his mind at the beginning of our century in connection with his own epoch-making contribution to the theory of heat radiation. Henceforth the existence of the atoms was no real problem for Planck: "The atoms", he writes in 1908, "however little we know of their properties in detail, are no more and no less real than the celestial bodies or the terrestrial objects of our environment."[35] Right after his conversion, Planck sees the problem of realism, as far as it concerns physics, in a wider scope within which atomism played an important but not an all-important role. This view already transpires from his first philosophical paper of 1908 coming from a lecture given at the university of Leiden.[36]

The question Planck seeks to clarify in this paper is the question "[whether our physical world view] is merely a convenient but basically arbitrary creation of our mind or ... mirrors real natural processes independent of ourselves." Neither his own position nor that of his opponents is adequately described in this first formulation. But one thing Planck makes entirely clear right at the beginning. He wants to link this question with the other, in what sense and in what direction physics has made *progress* and whether this direction can be determined as a development towards the *unity* of physics. The main thesis then is that this particular development towards unity does not only actually occur but is also an impeccable sign that physics is concerned about a real external world and steadily increases our knowledge of it as being an entity independent of the human mind. In closer detail, the development is described as a *twofold process* characterized by losses and gains. The losing business in this process is the *deanthropomorphization* of our primary world - "the conspicuous elimination of the human-historical elements from all physical definitions."[37] Planck admits outright that this abstraction "is a heavy drawback for the exploitation [of the emerging, purely physical world view] in the reality [of our life]." He speaks of "invaluable advantanges that are worth such a self-renunciation" and asks: "What is the peculiar moment that in spite of these obvious disadvantages provides the future world view with such a decisive precedence?"

The answer is given by the other part of the development. This part consists of the amalgamation of an originally extremely disparate phenomenal world into a *unitary system*. "The signature of the entire former development of theoretical physics is a unification of the system which is obtained by a certain emancipation of the anthropomorphic elements and the specific sense impressions in particular ..." In another place Planck uses an old criterion of wholeness: "The old system of physics did not equal an unique picture but rather a picture collection....one could remove [each picture] without affecting the others. This will be impossible in the future physical world view. There will be not a single feature of it that could be omitted, every one is rather an indispensable constituent of the whole." Physics attempts *intentionally* at "the complete detachment of the physical world view from the individuality of the creative mind." And in doing this it shows itself that physics develops into a conceptual system of ever increasing simplicity and unity. The argument in favour of realism thus springs from a synopsis of that two-fold process: The process cannot be explained but by the assumption that physics is about an external world independent of the mind. This self-understanding of physics was certainly not new at the beginning of our century. But it seems fair to say that in Planck's paper it has found a formulation obligatory upon the whole century. Scientific realism is the conviction that the fundamental epistemological problem can be solved by pointing out the success of modern physics.[38]

Towards the end of Planck's paper something strange happens.[39] Planck supplements the argumentation presented so far by contrasting it with the anti-realistic program of Mach's. And in doing this he cannot avoid becoming polemic. Mach's philosophy of science rests on two pillars: with respect to the subject on his phenomenalism and with respect to the method on his principle of economy. In his critique, Planck attacks both parts of Mach's view. The attack on the principle of economy being more revealing, I confine my presentation to it. Not without pathos Planck conjures the heros of physics from Copernicus to Faraday to assist him in his assault. It was a battle indeed "when the great masters of natural philosophy threw their ideas into science ... Economical viewpoints were the very last ones that these men fortified in their struggle against traditional views and eminent authorities. No - it was their unshaken ... faith in the reality of their world view." On this "incontestable fact" Planck then bases the conjecture that by the principle of economy "progress of science would perhaps be unfortunately obstructed." And he concludes this argument, and with it, his paper with a quotation from the Bible that is to separate the false from the true prophets: By their fruits you shall know them!

It goes without saying that Mach could not leave this attack without an answer. The emerging controversy was sterile concerning the subject but instructive in a psychological respect. In his reply Mach considers himself as mostly misunderstood and tries to play down the differences between his and Planck's views.[40] But sometimes also his pen ran away with him: "After having exhorted us with Christian indulgence to respect the opponent, Planck finally stigmatizes me with the well known verse of the Bible as a false prophet. So we see the physicists are on their shortest way to become a church ..." Planck's first reaction to Mach's article foreshadows his eventual reply. In a letter to von Laue we can read: "Mach's provocation in the current issue of the Physikalische Zeitschrift concerning my Leiden lecture I cannot leave without reply. Until October I wish him all happiness about his article; afterwards he might rather wish not to have written it ..."[41] In fact, the ensuing reply can hardly be called fair.[42] After some argumentation strictly confined to the subject in question, Planck suddenly interrupts the discussion in the middle of his paper and turns to the task to compare Mach's considerations in his books on the theory of heat and the history of mechanics as *physical* achievements with his own or at any rate those of orthodox theoretical physics, the verse of the Bible mentioned being the measure of comparison. The result, of course, is not open to question.

Much later Sommerfeld covered this somewhat embarrassing controversy with the cloak of charity by saying: "The discussion between Planck and Mach showed the contrast between a creative physicist like Planck and a reflecting physicist like Mach."[43] In fact there was a good deal of elitism in Planck's attitude. He saw himself not in the position to see the importance of Mach's physical achievements. Consequently, he was irritated by the fact that Mach proclaimed his philosophy of physics *as a physicist.* The proper core of the matter, however, is that from the behaviour of Planck it becomes obvious how *important* the matter was for him. It was an important *philosophical* concern for him to make it safe beyond doubt that physics is about a kind of reality that does not lie on the surface of sense impressions. Of course, this concern by itself does not explain the *faux pas* he was guilty of. One has to add that Planck *emotionally* was sure of the matter to an extent that to counterbalance *intellectually* he saw himself much less in a position than he knew the positivistic position represented by Mach. But at that time he had the impression that Mach's way of thinking began to establish itself in many heads of physicists and could in this way endanger the new generation. Mach denied that at this time his influence had been large or even worth mentioning. Wherever Planck had his impression from, if he had it, then it seemed to him that time had come to raise his voice and to use his own influence.[44]

4. EINSTEIN: CONSTRUCTIVE REALISM

Although such utterances are without any statistical foundation one would like to say today that the controversy between Planck and Mach has been decided in favour of Planck.[45] Some will then ask: But what about quantum theory? Though it was Planck and not Mach who made an original and highly important contribution to this theory, in retrospect, it rather looks as if from a philosophical point of view Mach was better prepared for the theory than Planck. If anything in physics then it was quantum theory that was a serious challenge to the classical view of reality as being independent of the observer, and Einstein, for one, seemed inclined to view the orthodox interpretation of quantum theory as a new kind of phenomenalism. It is precisely for this reason that he rejected this interpretation. To him it appeared mistaken "to permit theoretical description to directly depend upon acts of empirical assertions, as it seems to me intended ... in Bohr's principle of complementarity."[46] By contrast, Einstein held that "there is such a thing as the 'real' state of a physical system existing independently of any observation or measurement and being describable in principle with the expressive means of physics." Ironically he added: "All men, inclusive of the quantum theorists, stick to this thesis of reality as long as they don't discuss the foundations of quantum theory."[47] In this sense, i.e. leaving out of consideration the questions that came up with quantum theory, the concluding remarks on Einstein are to be understood.[48]

It is better known of Einstein than of Planck that he, too, started out as a Machian. Subsequently Einstein passed through a mental development that led him from Mach's phenomenalism to a further variety of scientific realism. He gratefully acknowledged his origin from Mach as well as he did not deny his later turning away from him. In his 'Autobiography' we read: "[Mach's 'Mechanics'] exercised a profound influence upon me ... while I was a student ... [His] epistemological position also influenced me greatly, a position which today appears to me to be essentially untenable ... For he did not place in the correct light the essentially constructive and speculative nature of... scientific thought."[49] Here, as elsewhere, Einstein likes to emphasize his later rationalistic position. And this is related to a parallel development of his view of reality.

The latter's final version could perhaps best be called a "programmatic or constructive realism". Already Planck had summarized his view in the sentence: "[The real external world] does not appear at the beginning but at the end of physical research."[50] Similarly, we read in Einstein[51] that "the 'real' in physics is to be taken as a type of program", or "'being' is always something which is mentally constructed by us, that is, something which we freely posit." Einstein even thinks he may appeal to Kant in this matter. For he points out: "I ... came to understand the truly valuable which is to be found in his doctrine ... only quite late. It is contained in the sentence: 'The real is not given to us, but put to us (*aufgegeben*) (by way of a riddle).'" It may be left open here whether the appeal to Kant is justified.[52] At any rate it elucidates Einstein's opinion. And this was - to borrow a formulation of Brand Blanshard[53] - that "reality is a system, completely ordered and fully intelligible, with which thought in its advance is more and more identifying itself."

We here meet already with the most conspicuous feature of Einstein's view on reality. At face value we seem to be confronted with a quite robust realism. "I tell you straight out," Einstein writes in 1930 to Schlick, "Physics is the attempt at the conceptual construction of a model of the *real world* and of its lawlike structure."[54] However, as soon as the question is one of the *criteria of reality*, Einstein becomes more cautious, and thoughts arise that we otherwise encounter in the camp of philosophical idealism. In a book review[55] Einstein asks the (neo-kantian) author rhetorically: "Are ... the realists and with them all scientists (in their non-philosophical moments) not right when, by the highly stunning possibility of the integration of our experiences into a system of (time-space-causal) concepts, they are led to believe in real things independently of their thinking and being?" Here Einstein still distinguishes between a reality *hypothesis* itself and a reality *criterion*. But this criterion belongs to the coherence theory of reality, not to the correspondence theory. That Einstein was a hidden coherence theorist can be seen even more clearly from his discussion of a quite concrete scientific question: When have we to expect the next decay of a single radioactive atom? Einstein's answer: "One may not merely ask: 'Does a definite time instant for the transformation of a single atom exist?' but rather: 'Is it, within the framework of our theoretical total construction, *reasonable* to posit the existence of a definite point of time for the transformation of a single atom ?'"[56] The refusal to ask the first question reminds us of the typically idealistic rejection of isolated facts, and the suggestion to replace the first by the second question comes up to the typically idealistic postulate to take the integrability of an event into the already existing (and confirmed) theoretical *system* as a criterion of its reality.[57]

But this is not the end of the story. We find Einstein on coherence theorists' ground not only when it comes to the question of reality criteria. Above this Einstein is quite prepared to *identify* reality with (complete) coherence. In the mentioned book review, he continues the quoted passage with the question: "Is there really a difference between the assumption that the totality of [our] ... experiences admits of a logical, conceptual system connecting them, and the reality hypothesis?" Also from the rhetorical character of this question, it is clear what Einstein's answer is. Moreover, not even from the last step that coherence theorists go in this direction did Einstein get frightened away: If coherence is not only a criterion of reality but reality itself, then there can be only one coherent system. In this sense Einstein said already in 1919 of his general theory of relativity: "The chief attraction of the theory lies in its logical completeness. If a single one of the conclusions drawn from it proves wrong, it must be given up; *to modify it without destroying the whole structure seems to be impossible*."[58] And thirty years later he judged of his unified field theory:" In favour of this theory are ... its logical simplicity and its 'rigidity'. Rigidity means that the theory is either true or false, *but not modifiable*."[59] What is here called 'rigidity' of a theory Einstein usually calls its 'logical unity', and the postulate of logical unity is one of the most often repeated requirements of a theory that we find in Einstein's writings. On occasion he can become enthusiastic about the matter. In a paper of 1929 he formulates the postulate in greater detail[60] by saying "that we do not only want to know *what* nature is like ... but also ... want to achieve the utopian ... goal to know *why nature is as it is*." As a paradigm Einstein refers to kinetic theory, and of theories like this he says almost paradoxically that one there succeeds "in conceiving of the empirical lawlikeness as logical necessity ... Even God could not have determined those connections differently from what they in fact are. "This is", Einstein concludes, "the Promethean element of scientific experience as we try to catch it by the term 'logical unity' ... it is, so to speak, the religious basis of our scientific endeavours."

It is in connection with the decisive rational element of logical unity that also in Einstein's thinking the *development of physics* comes into play. Our search for the greatest possible logical unity of the world view is mirrored in the *level structure* of physics.[61] From level to level the system displays a greater logical simplicity, and this is the way in which physics makes progress. At the same time there is a complementary removal of the anthropomorphic elements of our experience, just as we have found it so much emphasized by Planck: "It must be conceded", says Einstein[62], "that a theory has an important advantage if its basic concepts and fundamental hypotheses are 'close to experience'... Yet more and more, as the depth of our knowledge increases, we must give up this advantage in our quest for logical simplicity and uniformity in the foundations of physical theory."

There is a second sort of complementarity at work here. In principle we are completely free to choose the concepts at the various levels. But as a matter of fact, despite all freedom of reason, there is essentially only one route open to us. "One might suppose that there were any number of possible systems of theoretical physics all equally well justified; and this opinion is no doubt correct, theoretically. But the development of physics has shown that at any given moment, out of all conceivable constructions, a single one has always proved itself decidedly superior to all the rest. Nobody who has really gone deeply into the matter will deny that in practice the world of phenomena uniquely determines the theoretical system, inspite of the fact that there is no logical bridge between phenomena and their theoretical principles."[63]

In conclusion, I think Einstein as well as his forerunners Boltzmann and Planck, in spite of obvious differences in their views on reality, would agree that, unless one holds an essentially realistic position, the success of science would remain a *miracle*.[64] And yet there remains the miracle that the world is conceivable *at all*. Einstein has given it the wording: "The eternal mystery of the world is its comprehensibility."[65] And again in greater detail: " I believe that every true theorist is a tamed metaphysicist, no matter how pure a 'positivist' he may fancy himself. The metaphysicist believes that the logically simple is also the real. The tamed metaphysicist believes that not all that is logically simple is embodied in experienced reality, but that the totality of all sensory experience can be 'comprehended' on the basis of a conceptual system built on premises of great simplicity. The skeptic will say that this is a 'miracle creed'. Admittedly so, but it is a miracle creed which has been borne out to an amazing extent by the development of science."[66]

Heidelberg University
Germany

NOTES

[1] A French version of Scheibe's talk at the 1995 Parma Conference has been already published in *Les savants et l'épistémologie*, M. Panza and J.C. Pont eds., Editions Albert Blanchard, Paris 1995. The French Publisher and the Editors are hereby gratefully acknowledged for the permission of including this paper in its English version in the present Collection.

[2] Leplin 1984.

[3] Seelig 1952, p.45; Sommerfeld, 1951, p.37 (engl. transl. in 1949, p.99).

[4] Ostwald 1902, p.3.

[5] Sommerfeld 1948, pp.640ff.

[6] Helmholtz 1862; Jungnickel/McCormmach 1986, vol.I, pp.23f.

[7] Quoted from Jaki 1966, p.341.

[8] Ostwald 1902, p.1.

[9] Boltzmann 1990, pp.12ff, 152ff (engl. transl. in 1974, pp.153-8); see also Einstein 1934, p.113 (engl. transl. in 1954, p.270f).

[10] Plato, *Theaetetus* 151d-186e.

[11] Descartes, *Meditations*, esp. Med.1 and 6.

[12] Kant, *Critique of Pure Reason*, B XXXIX, B 274ff.

[13] Moore 1939, pp.127ff, 146.

[14] Helmholtz 1879, pp.18f.

[15] There were also, of course, adherents of the corpuscular philosophical tradition: see the discussion in *du Bois-Reymond* 1872.

[16] Blackmore 1982; see also Elkana 1971 and Brush 1990, pp.53ff.

[17] Boltzmann 1974, p.202.

[18] Kirchhoff 1872, *Vorrede*.

[19] Boltzmann 1905, pp.137f (engl. transl. in 1974, pp.90f).

[20] Ibid, p.144 (engl. transl. in 1974, p.96).

[21] Ibid, pp.78ff, 145 (engl. transl. in 1974, pp.41ff and 97).

[22] Ostwald 1895, p.162.

[23] Boltzmann 1896, vol. I, p.4.

[24] Mach 1900, p.362f.

[25] Mach 1912, p.467 and Mach 1872, pp.17ff.

[26] Instrumentalistic views were common at that time, cf. Pearson 1892, pp.114f.

[27] Mach 1922, p.256.

[28] Boltzmann 1905, p.157.

[29] Boltzmann 1990, pp.152f (engl. transl. in Boltzmann 1974, pp.153f).

[30] Boltzmann 1905, pp.152 (italics mine).

[31] Mach 1900, pp.363f.

[32] Einstein 1934, p.113 (engl. transl. in 1954, p.270).

[33] Scheibe 1988.

[34] Cf. Jost 1979.

[35] Planck 1949, p.48.

[36] *Die Einheit des physikalischen Weltbildes*, In Planck 1949, pp. 28-51. The following quotations are from pp. 29-31, 45f and 49. See also the reprint in Heilbron 1988, 301-14, and Planck 1910a, pp.1-9.
[37] Concerning this process see Wiener 1900.
[38] Planck's view has been elaborated in Bavink 1947.
[39] Planck 1949, pp. 47-51. For the Planck-Mach controversy see also Heilbron 1986, Ch.II.1.
[40] Mach 1910; see also Adler 1909.
[41] Thiele 1968, p.90.
[42] Planck 1910b.
[43] Sommerfeld 1936, p.610.
[44] Planck never gave up his position. See his *Positivismus und reale Außenwelt* of 1930 in Planck 1949, pp.228-45, and a later controversy in Müller 1940 and Planck 1940.
[45] Sommerfeld 1929, p.1.
[46] Einstein 1951, p.500 (engl. transl. in 1949, p.674).
[47] Einstein 1955, p.14.
[48] Cf. Scheibe 1992.
[49] Einstein 1951, p.8 (engl. transl. in 1949, p.21).
[50] Planck 1949, p.VI.
[51] The following three quotations are from Einstein 1951, pp. 500, 496 and 505 (engl. transl. in 1949, pp. 674, 668 and 680).
[52] Einstein here paraphrases a Kant's statement in his *Critique of Pure Reason*, B 526f. It says that, given something conditioned, in the domain of appearances the regress in the series of its conditions is only *put to us* (*aufgegeben*) whereas in the domain of things-in-themselves it would already be *given* (gegeben).
[53] Blanshard 1939, vol.II, p.264.
[54] Quoted from Holton 1968, p.660 (German original in Holton 1981, p.233).
[55] Einstein 1924, p.1685f.
[56] Einstein 1951, p.496 (engl. transl. in 1949, p.669).
[57] See for this Blanshard 1939, vol.II, pp.225ff.
[58] Einstein 1934, p.131 (italics mine) (engl. transl. in 1954, p.232).
[59] Einstein 1950, p.15 (italics mine).
[60] Einstein 1929, pp.126f.
[61] Einstein 1984, pp.67ff (engl. transl. in 1954, pp.293ff).
[62] Einstein 1950, p.15.
[63] Einstein 1989, p.109 (engl. transl. in 1954, p.226).
[64] Cf. Putnam 1984, pp.140f.
[65] Einstein 1984, p. 65 (engl. transl. in 1954, p.292).
[66] Einstein 1950, p.13.

REFERENCES

F. Adler, "Die Einheit des physikalischen Weltbildes", *Naturwissenschaftl. Wochenschrift*, Neue Folge VIII, 1909, pp. 817-22.

B. Bavink, "Bedeutung des Konvergenzprinzips für die Erkenntnistheorie der Naturwissenschaften", *Zeitschr. f. philos. Forschg.* **2**, 1947, pp. 111-30.

J.T. Blackmore, "Boltzmann's Concessions to Mach's Philosophy of Science', in *Ludwig Boltzmann*, Gesamtausgabe, vol. 8, Ed. by R.U. Sexl, Vieweg, Braunschweig, 1982, pp. 155-90.

B. Blanshard, *The Nature of Thought*, 2 vols., London, 1939.

E. Bois-Reymond, *Über die Grenzen des Naturerkennens*, Veit, Leipzig, 1872, [11] 1916.

L. Boltzmann, *Populäre Schriften*, J.A. Barth, Leipzig, 1905, quoted edition by E.Broda, Vieweg, Braunschweig 1979.

L. Boltzmann, *Vorlesungen über Gastheorie*, J.A. Barth, Leipzig, 1896, quoted after, *Ludwig Boltzmann*, Gesamtausgabe, vol. 1, Ed. by R.U. Sexl, Vieweg, Braunschweig, 1981.

L. Boltzmann, *Theoretical Physics and Philosophical Problems - Selected Writings*, Ed. by B. McGuinness, Reidel, Dordrecht, 1974.

L. Boltzmann, *Prinzipien der Naturfilosophie*, Ed. by J.M.Fasol-Boltzmann, Springer, Berlin, 1990.

St.G. Brush, *Ludwig Boltzmann and the Foundations of Natural Sciences*, in Boltzmann 1990, pp. 43-61.
A. Einstein, "Elsbachs Buch: Kant und Einstein", *Deutsche Literaturztg*. **45**, 1924, 1685-94.
A. Einstein, "Über den gegenwärtigen Stand der Feldtheorie", *Festschrift für A. Stodola*, Zürich, 1929, pp. 126-32.
A. Einstein, *Mein Weltbild*, Querido, Amsterdam, 1934, quoted after Ullstein Sachbuch, Frankfurt, 1989).
A. Einstein, "Autobiographical Notes and Remarks to the Essays Appearing in this Collective Volume", in *Schilpp 1949*, pp. 1-95 and pp. 663-88.
A. Einstein, "Autobiographisches und Bemerkungen zu den in diesem Bande vereinigten Arbeiten", in *Schilpp 1951*, pp. 1-35 and pp. 493-511.
A. Einstein, "On the Generalized Theory of Gravitation", *Sci. Amer*. 182, April 1950, pp. 13-7.
A. Einstein, *Ideas and Opinions*, Crown Publ., New York, 1954.
A. Einstein, "Einleitende Bemerkungen über Grundbegriffe", in *Louis de Broglie und die Physiker*, ed. by A. George. Claassen, Hamburg, 1955, pp. 13-7.
A. Einstein, *Aus meinen späten Jahren*, DVA, Stuttgart, 1979, ^{3},1984.
Y. Elkana, "Boltzmann's Scientific Research Programme and its Alternatives", in *Some Aspects of the Interaction between Science and Philosophy*, Ed. by Y. Elkana, Humanities Press, Atlantic Highlands, 1971, pp. 243-79.
J.L. Heilbron, "The Dilemmas of an Upright Man", *Max Planck as Spokesman for German Science*, Univ. of California Press, Berkeley, 1986.
J.L. Heilbron, *Max Planck. Ein Leben für die Wissenschaft*, 1858-1947, Hirzel, Stuttgart, 1988.
H.v. Helmholtz, "Über das Verhältnis der Naturwissenschaften zur Gesamtheit der Wissenschaft", *Akad. Festrede* (Heidelberg, 1862), in H.v.Helmholtz *Vorträge und Reden*, Vieweg, Braunschweig ^{5}1903, vol.1, pp. 159-85.
H.v. Helmholtz, *Die Tatsachen der Wahrnehmung*, Hirschwald, Berlin, 1879, quoted Edition: Wiss. Buchgesellsch., Darmstadt 1959.
G. Holton, "Mach, Einstein, and the Search for Reality", *Daedalus 97*, 1968, pp. 636-73.
G. Holton, "Thematische Analyse der Wissenschaft", *Die Physik Einsteins und seiner Zeit*, Suhrkamp, Frankfurt, 1981.
St. Jaki, *The Relevance of Physics*, Univ. of Chicago Press, Chicago, 1966.
R. Jost, "Boltzmann und Planck. Die Krise des Atomismus um die Jahrhundertwende und ihre Überwindung durch Einstein", in *Einstein-Symposion Berlin* (ed. by H. Nelkowski *et al.*), Springer, Berlin, 1979, pp. 128-45.
Ch. Jungnickel, and R.McCormmach, *Intellectual Mastery of Nature. Theoretical Physics from Ohm to Einstein*, 2 vols., Univ. of Chicago Press, Chicago, 1986.
G. Kirchhoff, *Vorlesungen über mathematische Physik*, vol.1 Mechanik, Teubner, Leipzig, 1876.
J. Leplin, *Scientific Realism*, University of California Press, Berkeley, 1984.
E. Mach, *Die Geschichte und die Wurzel des Satzes von der Erhaltung der Arbeit*, Calve, Prag, 1872, quoted Ed.: Bonset, Amsterdam, 1969.
E. Mach, *Die Prinzipien der Wärmelehre*, J.A.Barth, Leipzig, 1896, quoted Ed.: ^{2} 1900.
E. Mach, "Die Leitgedanken meiner naturwissenschaftlichen Erkenntnislehre und ihre Aufnahme durch die Zeitgenossen", *Phys. Zeitschr.* XI (1910), pp. 599-606.
E. Mach, *Die Mechanik in ihrer Entwicklung*, Brockhaus, Leipzig, 1883, quoted ed.: ^{7} 1912.
E. Mach, *Die Analyse der Empfindungen und das Verhältnis des Physischen zum Psychischen*, Fischer, Jena,^{7} 1911, quoted Ed.: Wiss. Buchgesellsch., Darmstadt, 1991.
G.E. Moore, *Proof of an External World*, Proc. Brit. Acad. 1939, pp. 273-300, quoted after : *Philosophical Papers*, Allen & Unwin, London, 1959, pp. 127-150.
A. Müller, "Naturwissenschaft und reale Außenwelt", *Die Naturwissenschaften 28*, 1940, pp. 705-9.
W.Ostwald, "Die Überwindung des wissenschaftlichen Materialismus", *Verhandlg, Gesellsch. Deutscher Naturforscher u. Ärzte 67*, 1895, Erster Teil pp. 155-68.
W.Ostwald, *Vorlesungen über Naturphilosophie*, Veit, Leipzig, 1902.
K. Pearson, *The Grammar of Science*, Scott, London, 1892, reprint Thoemmes, Bristol, 1991.
M. Planck, *Acht Vorlesungen über Theoretische Physik*, Hirzel, Leipzig, 1910.
M. Planck, "Zur Mach'schen Theorie der physikalischen Erkenntnis", *Phys. Zeitschr. XI*, 1910, 1186-90.
M. Planck, "Naturwissenschaft und reale Außenwelt", *Die Naturwissenschaften 28*, 1940, pp. 778-9.
M. Planck, *Vorträge und Erinnerungen* ^{5}, Stuttgart, 1949.
H. Putnam, "What is Realism?", in *Leplin*, 1984. pp. 140-53.
E.Scheibe, "The Physicists' Conception of Progress", *Stud. Hist. Phil. Sci. 19*, 1988, pp. 141-59.
E.Scheibe, "Albert Einstein: Theorie, Erfahrung, Wirklichkeit", *Heidelberger Jahrbücher XXXVI*, 1992, pp. 121-38.
P.A. Schilpp, *Albert Einstein. Philosopher-Scientist*, Library of Living Philosophers, Evanston, Ill., 1949.

P.A. Schilpp, *Albert Einstein als Philosoph und Naturforscher*, Kohlhammer, Stuttgart, 1951.
C. Seelig, *Albert Einstein und die Schweiz*, Zürich, 1952.
A. Sommerfeld, "To Albert Einstein's Seventieth Birthday", in *Schilpp* 1949, pp. 97-106.
A. Sommerfeld, "Albert Einstein", in *Schilpp* 1951. pp. 37-42.
A. Sommerfeld, "Einige grundsätzliche Bemerkungen zur Wellenmechanik", *Phys. Zeitschr. XXX*, 1929, pp. 866-71, quoted from Sommerfeld 1968, 1-6.
A. Sommerfeld, "Wege zur physikalischen Erkenntnis", *Scientia*, April 1936, pp. 181-7, quoted from Sommerfeld 1968, 609-15.
A. Sommerfeld, "Philosophie und Physik seit 1900", *Naturwiss, Rundschau 1*, 1948, pp. 97-100, quoted from Sommerfeld, 1968, 640-3.
A. Sommerfeld, *Gesammelte Schriften*, Vieweg, Braunschweig, 1968, vol. IV.
J. Thiele, "Ein zeitgenössisches Urteil über die Kontroverse zwischen Max Planck und Ernst Mach", *Centaurus 13*, 1968, pp. 85-90.
O.Wiener, "Die Erweiterung unserer Sinne", *Deutsche Revue über das gesamte nationale Leben der Gegenwart*, 25. Jahrg, 1900, vol. 4, pp. 25-41.

EVANDRO AGAZZI

OBSERVABILITY AND REFERENTIALITY

1. INTRODUCTION

In spite of many criticisms levelled against it in both recent and much less recent times, observation remains an indispensable ground for every cognitive enterprise oriented towards the so-called "external world" and, therefore, for the empirical sciences in particular. One could say that observation expresses the basic "realist" attitude of any knowing subject, to the extent that this subject *intends* his knowledge to be knowledge of something different from himself, and also *believes* this knowledge *actually* to be of something other. This ontological distinction between the knowing subject and the known objects is commonly expressed by calling "external" world the target of this knowledge. But the adjective "external" — in its implicit spatial connotation — has actually a pictorial, rather than a conceptual flavour, its exact conceptual meaning being reduced to the *ontological alterity* just mentioned. Yet the pictorial suggestion is not fully misleading, for it can offer us a useful analogy. In fact this ontologically distinct world must nevertheless be accessible to the subject, and the less demanding condition of this *accessibility* seems to be that the subject is "open" to the external world thanks to certain "windows" that are constituted by his *senses*. Observation consists in looking through these windows, letting the external world *appear* to us as it is. In the image of the windows there are some sound elements and others that deserve more careful scrutiny. One of the sound elements is the awareness that the world reasonably contains many more aspects or features than those that can be approached through the finite (and actually very limited) number of windows at our disposal, or, to put it differently, that only a limited amount of those aspects and features can appear to us. An element that must be considered with much care is the idea that the process of observation simply consists in *passively* opening the windows and letting the external appearences flow in untouched and unmodified. This is too rough an idea, and the well-known criticisms levelled against *pure* observation in different contexts (from psychology to epistemology and philosophy of science) have shown that the role of the knowing subject — already at the level of observation — is far from being comparable to that of a mirror situated behind a window. However, even admitting that the subject contributes to the organizing and shaping of the appearences, it remains undeniable that these are appearences *of* something, and that this something is not the subject himself. This is paradigmatically clear in the case of visual observation, but is easily confirmed also in the case of other sense perceptions: when we are in such states, we feel them as expressing the fact that "there is something out there".

The statement just used accounts for the spontaneous expression "external world",

E. Agazzi and M. Pauri (eds.), The Reality of the Unobservable, 45-57.

but — as we have said — does not really entail a spatial connotation. In fact, such a world should be external to what? Perhaps to my skin, to my body? Obviously not: if through surgery, or by means of a sophisticated observation device, I were put in the condition of observing certain *internal* parts of my body, I still would say that their *existence* did not depend on my observation, that they remain ontologically distinct from my act of knowledge. I might be tempted to say that they are internal to my body, but external to my mind. This would be a very bad move, however, since it would amount to an arbitrary *spatialization* of the mind, with the only aim of preserving an improper pictorial way of speaking. Therefore, from now on we shall dismiss the expression "*external world*" and speak instead of "*known reality*". What inevitably occurs with known reality is that it necessarily is a connoted reality, a reality presenting itself with particular *connotations* or features, depending on the particular "way of access" within which it can be present to us, and which corresponds, in our analogy, to the different "windows". Colours, sounds, smells, etc. are the different *qualia* under which reality is present to us, and it is simply impossible to know any item of reality without its being characterized by some kinds of *qualia*.

We have recognized that such *qualia* "depend" on the different senses that offer us an access to reality, and this may be misleadingly taken as an admission of the purely *subjective* character of these features. That this need not be the case may be seen even from the analogy of the windows: that the landscape surrounding my home contains a mountain on its north side cannot be known by someone living in the house unless a window looking to the north is open in the house, so that the "known landscape" contains or does not contain the mountain *depending* on the window from which it is observed. Yet nobody would say that the *existence* of the mountain ontologically *depends* on the window or on the observer.

This analogy helps us in understanding why we are spontaneously convinced that the different *qualia* exist *in the objects* of our knowledge: they are not characteristics of our cognitive organs, but those features of a world distinct from us that can be detected thanks to our particular cognitive organs. However, precisely owing to this circumstance, they do not have an *existence in themselves*: we do not see, for example, the red of a rose, but we *see* the rose *as* red, and this red is a quality that exists *in* the rose as a consequence of the rose being perceived by my eye under a certain light. In the language of traditional philosophy this fact was expressed by saying that colours and other *qualia* are *accidents*, that have certainly a kind of existence, but precisely exist *in a substance*. This old doctrine deserves perhaps more attention than we usually pay it: in fact the word "accidental" means in common language something like "negligible", "inessential", "trivial", while in its classical sense the accident is simply something real that is not endowed with an *independent existence*. However this circumstance by no means trivializes the accidents: on the one hand they are, so to speak, the concrete *way of being* of any substance, and on the other hand a substance can be *known only through its accidents*. This is again well in keeping with our common beliefs: we consider it "normal" to perceive certain *qualia* when a substance is there (that is, something having an existence in itself and being the object of our perception), and we easily understand that we cannot experience a substance without experiencing it as a structured set of *qualia* or *attributes* (properties, relations, functions).

2. THE QUESTION OF REFERENCE

What has been said expresses what we have called the *normal situation*. Yet we know of situations (such as dreams, hallucinations, sense illusions, mirages) in which it occurs that we perceive *qualia* also in the absence of the substances that should constitute their *ontological carriers*. We call such situations abnormal, but this abnormality does not consist in their being unusual (in fact many of them, like dreams, are very ordinary): it consists precisely in the fact that *qualia* are conceived as the way of being present to us of a certain substance, and this is their *proper* nature and role, so that we are surprised when they do not perform this role, when they "are there" without a substance "being there".

But how can it be known that the substance is *not* there? The question is intriguing, for there is no perception of the non-existing, and we cannot make appeal to a perception whose *content* is a non-existing object. The ways in which we usually try to recognize whether or not certain perceptions are "illusory" are basically of two kinds. One consists in opening some other "window" on the substance that should be the carrier of the perceived accident, and checking whether also the accident corresponding to the different window appears (for example, optical illusions may be shown to be such through tactile perception that does not "find" the object, or the feature of the object, that sight seems to present us). The second consists in repeating the same kind of observation several times, and under different conditions: if we do not perceive what we had perceived the first time, if the first perception has vanished, and if we have no reason to believe that the corresponding substance has been destroyed in the meanwhile, we usually conclude that the first perception was an illusion. These two familiar procedures rely upon certain tacit presuppositions. The first is that substances can be known only through their accidents, and that these accidents inevitably appear if the corresponding "window" on the substance is open. The second is that substances, having an existence in themselves, cannot "vanish", so that if an accident vanishes, it could not really be an accident of a substance. What could then be such an illusory perception? The answer is: a mere *representation*.

We are led in such a way to the following portrayal of the cognitive situation. Substances exist in themselves, and are known only through accidents that exist *in the substance* and are accessible to the knowing subject through specific sense organs under suitable conditions. The way of being present of these accidents produces a modification of the subject, and this modification is a representation that exists *in the subject*. Owing to their origin, representations always keep an «orientation» to the substance, but, being at the same time things that have an existence in the subject and not in the substance, they have a kind of autonomous life (in the subject) and can be kept, combined, recalled without their corresponding substances being actually present. Moreover, this dynamics of representations oversteps our conscious control, and there are situations (like dreams and hallucinations) in which these representations are present to us, in the absence of the corresponding substances, and bring us to believe that such substances exist.

The particular way of being present of things to knowing subjects, which is different from the spatial or physical presence, is permitted by a special capacity of this subject, that is not shared by all natural beings, and which has been called *intentionality* in the philosophical tradition. Intentionality is the capacity of opening itself to the world that certain beings possess, and which consists in making representations of the world that are a certain kind of modification of the subject itself. Therefore, representations may be called *intentional entities*, which — though they are in the subject — intrinsically *refer* to the world of substances ontologically distinct from the subject. Using a word taken from linguistics and philosophy of language, we can call *referents* such objects which have an existence independent of the subject, in the sense that they belong to a non-intentional reality. Using this terminology we can now say that what is perceived in dreams, hallucinations, sense illusions, etc. has an *intentional reality*, but without referents. Now we can say that, being aware that the world of representations — or *intentional world* — has its own autonomy, we may legitimately be doubtful as to whether or not certain representations have referents, and therefore try to check them in this regard. However, since the only immediate access we have to referents is offered by *observations* (which in any case can offer us only new representations), what we can do is to reiterate observations and possibly combine them so as to reasonably exclude the possibility of error, according to the two strategies mentioned above. This can be resumed by saying that observation is the ultimate warrant of *referentiality*.

3. SUBJECTIVITY AND OBJECTIVITY

The above remarks allow for a critical appreciation of the dichotomy "objective-subjective". We have seen that a certain attribute or property (for example the colour red) is the way of being present *to us* of a certain *referent* or substance (for example a rose), through a specific sensory apparatus (sight), and under suitable *conditions* (the correct functioning of our visual apparatus and the illumination of the rose by a standard light). It follows therefore that the attribute has a *composite* nature, being the *result* of three basically distinct factors, some of which are on the side of the subject, and some on the side of the object, so that it is not correct to say that this attribute is exclusively objective or exclusively subjective. Of course, a colour is a *perceptual* attribute of a referent, and we can connect its perceptual component to the sphere of subjectivity by stressing that it is present to us under the form of a *visual representation*; but the other components must be linked with the sphere of objectivity, and include the *instruments* of representational activity (that is, the visual apparatus and the light), and the features of the *referent*. In other words: my having a perception of red (representation) depends on conditions ontologically different from my strict intentional capacity, and belonging to the physical world, such as my visual apparatus, the lighting conditions, and the features of the object, so that I could have a different perception not only if the object had different physical features, but also if my visual apparatus were affected by some abnormality, or if the lighting conditions were significantly changed. It is because these levels are interrelated, but also autonomous to a large extent, that certain problems exist. If the colour red is defined in the way proposed above, the fact that a certain subject has a perception of red is not a complete warrant that a red object is there, because this perception could be produced by purely representational mechanisms (like in a dream or hallucination), or by an alteration of the visual apparatus, or by a suitable arrangement of the lighting. Why are we spontaneously ready to say that in such cases we have to do with an error, an illusion, an abnormal situation? The reason is, as we have already stressed, that representations are of an *intentional* nature, that is, they are "pointing" to a referent, to an ontological pole different from the subject, and if this pole fails to exist or to conform to the representation, we feel that the representation falls short of its proper task. A way of expressing this situation could be to say that representations remain, by their intrinsic nature, representations of a *possible world*. That they actually represent the *real world* could be granted, in principle, by checking that the two non-subjective conditions are satisfied (that is, that the functioning of the sensory apparatus is correct, and that the external conditions are standard), but this checking is by no means straightforward nor easy, and this is why we normally rely upon the indirect strategies already mentioned.

4. THE WIDENING OF OBSERVATION

The foregoing analysis has dealt with normal unaided sensory observation, but its elements will now prove useful for the discussion of scientific observation and of the requirement of observability in science. The question that leads to the idea of widening the horizon of usual sensory observation is — using our analogy — the following: is it possible to open new windows on the world? The answer to this question can be affirmative, provided we are aware of certain conditions. From the four elements considered above: representations, cognitive apparatus, instrumental conditions, features of the real world, the opening of new windows would amount to producing new representations, which should be representations of unknown features of the real world, as a consequence of the creation of new instrumental conditions and (perhaps) of the development of our cognitive apparatus.

A first perplexity in this widening of observation can be expressed by saying that, when using a particular instrument, we are not sure that the representations we obtain through it are representations of some real accident of the world or, rather, they are simply produced by the instrument. This perplexity is not sound, since it ignores that *every* attribute of reality is relative to certain conditions of ascertainment. These can be labelled *instruments* in a general sense: for example, when we say that a rose is red, we may overlook that this is said with the understanding that the instrument of observation is the human eye plus solar light, but this is in fact the situation (the same rose might not *be* red under different lighting conditions). Therefore, if instead of the naked eye we use an optical microscope or telescope, we may become able to see features of reality that remained hidden before. We can trust these instruments, roughly speaking, because we have the possibility of comparing the images they offer to our sight on some common *segment* (that is at a scale of magnitude where the naked eye is also still able to perceive certain features), and find that they coincide within that segment. Therefore, we trust the instrument to do the job equally well also outside the segment, and say that we *observe* bacteria or stars through the instrument, which is considered, in such cases, as a kind of *prolongation* of our sense organs: more than opening a new window, we have enlarged an existing window, and remain within the realm of visual representations.

5. INSTRUMENTAL OBSERVATION

However, such situations (in which there is a common segment within which both unaided and instrumental observation can be equally well performed) are rather exceptional and, moreover, they do not help us to understand the reasons for which it is correct to say that we *observe* entities that are accessible *only* by means of instruments that cannot be applied to such a common segment. For example, if we use a radio telescope or an electronic microscope, it would be naive to say that we trust them because we can use them to see certain objects accessible to our naked eye, and find that the two visual representations are identical; nor do we rely on them because, say, there is another common segment of observation which they share with already granted instruments (such as optical telescopes or microscopes), within which they give the same visual representations. The real reason for trusting such instruments is quite different: we confidently use them because we accept a certain *scientific theory* which gives us a *causal account* of what we see through them (in our case of the visual representation) as being the *effect* of a given property of the object that we want to observe. Of course, we still have to do with a visual representation (we see, for instance, certain shapes), but we *immediately* refer it to certain external objects because we can rely on the correct functioning and use of the instruments, and because we have also developed our *cognitive apparatus*, that is, because a certain scientific theory has become an integral part of our *way of knowing* the world.

6. THE NATURE OF PERCEPTION

In order to correctly understand this last statement it is necessary to introduce a digression of a rather general kind, in the course of which certain useful distinctions will be stated.

The doctrine according to which only unaided sensory observation deserves being qualified as observation is normally founded upon two tacit presuppositions: (1) The equating of *observation* and *perception*; (2) Some form of the so-called *causal theory of perception*.

This theory may be summarized as follows: physical objects possess a great deal of physical properties; some of them, under suitable conditions, produce physical effects that may act upon our sense organs and are clearly accounted for by certain physical theories (for example, physical bodies, when they receive a beam of light constituted by electromagnetic waves of different length, absorb a part of such radiation, but reflect another part of it; this reflected part may reach our eyes and, in particular, our retina). Then a second causal chain starts, which is again accounted for by other scientific theories such as physiology (the retina reacts to these radiations in the sense of undergoing certain chemical and electrical phenomena that stimulate in turn the optical nerve; this stimulation is transmitted along the nerve and reaches a particular region of our brain, that we can call the visual region, and there it *produces* the visual *perception*).

What remains not accounted for in this theory is (at least) the jump that takes place when perception comes about. Indeed no physical *predicates* may be employed for describing it: it has no mass, no charge, no location in space, no momentum, no energy in any physically definable sense, and so on. This simply means that there is no way of *explaining* it on the basis of physics (or chemistry, or even pure biology), because every such explanation (in a rigorous sense) must be spelled out as a deduction using laws and conditions expressed through these *specific predicates*. And since causal explanation in science is simply explanation based on scientific *laws*, it follows that the alleged causal theory of perception is no theory at all in a *rigorous* sense. This becomes even more evident if we reflect on the fact that more or less the same physical events are sufficient for explaining how the picture of a physical object may be obtained on the plate of a camera, but nobody would say that the camera *perceives* or sees the object. In other words, the causal theory of perception confuses perception with simple *interaction*, overlooking that a certain interaction may be a *necessary condition* for perception, but not a *sufficient* one. The additional element needed for the coming about of perception resides in some particular feature of the entity which is *able* to perceive.

This point may be clarified by distinguishing a few types of interactions that different kinds of entities can have with their environment. Several entities interact with the environment in a passive way, that is, they may receive and exercise causal effects from and on the environment, but their existence does not depend on those interactions. Physical bodies are examples of this kind. Other entities need the environment for their existence, and have an active interplay with it, that is they destroy certain parts of the environment, they *interiorize* them, and make of them a part of themselves. This is typical of living beings that are all characterized by metabolism. But there are also entities that interact with the environment, are able to interiorize it, and yet leave it intact, do not "eat" it. Animals, at least starting from a certain degree of complexity, are able to perform this astonishing interaction, which consists in *perceiving* the world, and in this sense "interiorizing" it without destroying it (something that plants cannot do), since the world becomes part of them under the form of *representations.* If we want to single out the peculiar features that enable the different entities to manifest these diverse kinds of interactions, we could say that the first is possible thanks to a *physical* constitution, the second is due to *life*, and the third is due to the possession of *intentionality*. Accordingly, we can understand that a camera cannot perceive a house *because* it is only a *physical* entity, nor can a plant have such a perception *because* it is only a *living* entity, while a cat can perceive it *because* it is an entity endowed with *intentionality*.[1]

7. PERCEPTION AND OBSERVATION

We come now to the other point, that is, the tendency (indeed very common) to equate perception and observation. Our claim is that not all perceiving beings are capable of observation. In order to see this we can note that even in common language observing something means more than just seeing or perceiving it. This something more can be located in a sort of "directionality" or selectivity of the perceptual activity, which is concentrated or focussed upon a particular object within the perceptual field. Perception, thus, can be considered a manifestation of a lower and almost *passive* stage of intentionality (reducible to the "openness" to the environment that makes this "internally present" to the perceiving being under the form of a representation). Observation includes, in addition, the manifestation of an *activity* by the observing being, whose most elementary form we can call "attention". We quite often recognize that in certain cases we had actually "seen" something, but had not "noticed" it. This "seeing" is the first stage of observation, of which not only human beings, but also many animals are capable. Humans can do much more, however: they are able to perform *voluntary* (and not just spontaneous) observation. This implies a *conscious* decision to observe, with the precise specific purpose of knowing "how things are". This higher level of observation (which, by the way, is the only one that is significant for *scientific* observation) is possible thanks to an additional capability possessed by the observing being, that of "judging". We have now singled out the typical factor that distinguishes perception from observation (in its fullest sense): this factor is *judgement*. The implicit (but characteristic) goal of an observation is that of attaining a correct judgement of the "external world" (in the sense already explained above), and contains as such the tacit awareness that there are two distinct parts of reality: one which is ourselves, and one which is ontologically different from us, and of which we have certain representations. When we express a judgement regarding a piece of such reality, we manifest our *adhesion* to that representation, meaning that we believe it to manifest reality such as it actually is. A common way of expressing this fact is to say that we attain *true judgements*, and this is by no means an innocent way of speaking: it entails the very important consequence that perceptions are neither true nor false, while observations (conceived in their fullest sense, according to which they culminate in a judgement) are true.

We are fully aware that our discourse has been a little vague and schematic. In particular the proposed distinctions are not always clear-cut, but gradual, and admit of many intermediate cases, but this is something very common (for example, black and white are clearly different colours, but there are many grey objects that may be "almost" black or white, without any objective possibility of tracing an absolute threshold between them). Moreover, the different functions and conditions outlined above are strictly intertwined (physical interactions are necessary for perception, and perception is necessary for observation), so that it may be difficult to locate the point of discontinuity in a process that shows much continuity. However, it is undeniable that objective differences and distinctions exist, and it is these on which we have focussed our attention, since they enable us to clarify the analytic elements we need for the prosecution of our inquiry.

8. EPISTEMOLOGICAL DUALISM

Let us come back to the four constituents of observation we have already mentioned: representations, cognitive apparatus, instrumental conditions, features of the real world.

What is the status of *representations*? Do they belong to the knowing subject or to the "external world"? The answer is apparently straightforward: they obviously belong to the subject. However, this answer is obvious only if we conceive the external world according to the pictorial spatial image we have already criticized. But if the external world is meant to be the reality ontologically distinct from the mind, then it is possible to think that representations are the *object* of our knowing activity, that they are the only things we can know, and in this sense they would be in a way "external" with respect to this activity. This position is not rare in the history of philosophy, and was characteristic of the so-called "modern" philosophy comprised, roughly speaking, between Descartes and Kant. The tacit presupposition of this philosophy was precisely that we know our representations (or "ideas", as they used to be called), but have no warranty that there is an independent reality "behind" these ideas, or at least that this reality has the features that our ideas *represent* to us. This is indeed a purely dogmatic presupposition, for which no evidence nor argument is provided. The common sense belief (duly deepened and sustained by a long classical tradition) is that we know reality *through* our representations: they are not that which is known (*id quod cognoscitur*), but that through which knowledge obtains (*id quo cognoscitur*). What is strange and unjustified in modern *epistemological dualism*[2] is that on the one hand, it is admitted that the aim of our knowledge is to attain reality, but that, on the other hand, this aim is frustrated because we only know our representations and must look for indirect arguments in order to be confident that these representations *correspond* to reality (Kant, as is known, clearly declares that we have no knowledge of reality in itself). Of course, one of the most immediate objections to this strange dualistic presupposition is as follows: if all we know are our representations, how and why could we imagine that there is a reality different from them? Why should our cognitive efforts be directed to attain such a mysterious reality whose existence is even dubious? However, we are not interested here in criticizing the dualistic doctrine, but we want rather to point out that it actually makes of representations something distinct from our minds: if they are the *objects* of knowledge, they cannot be at the same time the *constituents* of knowledge, whereas they did not have such a strange status in traditional epistemology, where representations were *mental* entities *constituting* the particular way things are *present* in the mind (*intentional presence*, different, for example, from the physical presence of an object in a room).

The causal theory of perception is a contemporary version of epistemological dualism, for it maintains that the objects of perception are not things of the external world, but the outcome of a long chain of physical, chemical, biological processes. The proponents of this theory overlook the fact that what we actually *see*, for example, are *not* the neuronal processes which render vision possible (we do not see our neurons when we look at a house), and even if the causal outcomes of the causal chains are called "internal" because they take place in the organs of the perceiving subject and also depend on his constitution, they are not the *object* of perception, and their presence in the subject preserves the features of a *physical* presence and not of an *intentional* one. The result of this critical discussion is the following: representations belong to the knowing subject, are *constituents* of him; they are the way things are intentionally present to him. Because of that, they are "internal" to it in the non pictorial sense of this expression. This does not imply that representations have no ontological status: they possess a reality, and precisely an *intentional reality* quite different from physical reality. Their reality ontologically depends on the constitution of the knowing subject, and it is natural that their features and behavior may be rather different from those of physical reality. This is why representations (though not being disconnected from physical reality because they are primarily the way of being present of such a reality in the knowing subject) may also show some peculiar patterns depending on what we have called the "cognitive apparatus".

9. THE COGNITIVE APPARATUS

The cognitive apparatus is that which constitutes the *specific cause* of perception and observation in those beings that are endowed with it. Though not being physical (in the sense of not being describable by means of the predicates of the physical sciences), possesses its structure and way of functioning, that men have tried to describe from the time of Aristotle up to the investigations of contemporary psychology. As the structure and functioning of the physical world are described in physics by determining properties, relations and interactions of the ontological constituents of this world (that is, of physical bodies), so the structure of the cognitive apparatus may be described by determining the properties, relations and interactions of its ontological constituents, that is, of the intentional entities or *representations*. Already at the level of *perceptions* we know that these representations may have certain properties, that they can be retained in memory, associated, combined in new representations, and that this dynamic has an influence on the formation of new perceptions (the same stimuli may lead to different perceptions according to the different perceptual experience already stored in the cognitive apparatus of a subject). The situation becomes much more complex at the level of *observation*. As we have seen, observation implies judgements, and these in turn presuppose the possession of *concepts*, and the ability to connect them. Concepts and judgements grow up in a nest of correlations of increasing complexity, and all this means again a modification of the cognitive apparatus, that may affect the formation of new observations (the same perceptions may lead to different observations according to the observational experience already stored by the observer). However, observation is not the highest level of cognitive activity. The connection of judgements of which we have spoken includes *intellectual* operations such as inferences, deductions, formation of global representations, construction of new intellectual tools. Taking all this into account, we can say that the cognitive apparatus includes a whole of elements that we can call *cognitive background*. This background may even influence our perceptions, but it certainly influences our observations, in the sense that the judgements we attain in the process of observation cannot fall outside the framework of our cognitive apparatus such as it concretely is (that is, including its cognitive background). This fact has been stressed, but also partially misunderstood, by the advocates of the so-called theory-ladenness of observation. The misunderstanding consists in the fact that they usually mean that an observation is *determined* or at least influenced by the *theory* in which it is allegedly situated, so that it cannot provide the *independent* criterion for evaluating the soundness of this theory, or for comparing different theories. The mistake in this position resides in the fact that (except in very special cases), observations can actually arise independently of the theory *in which* they are performed, though they depend on the *general* cognitive background that, as an overall framework, includes both the observation and the theory concerned.[3]

The influence of this general background can be more clearly found in other steps of human cognitive activity, which go beyond simple observation, that is, in *interpretation* and *explanation*, that are present in common knowledge, but also in science, where they lead to the constructions of *theories* in a proper sense. But we shall not discuss this point here.

10. SENSE AND REFERENCE

It may be noted that the preceding considerations could be summarized by saying that our cognitions always are in a fundamental sense re-cognitions, that is, identifications of some items of reality that *exemplify* certain general and *abstract* patterns we already have in our minds. Plato was the first who explicitly thematized this feature of our knowledge, and he was led to interpret it as "reminiscence", starting in such a way the long history of innatism that has not ceased to stimulate complicated philosophical discussions even today. However, innatism is not the only solution for this puzzle. If we are aware that representations are not the *content*, but rather *constituents* of our cognitive apparatus, we can find the lines of a more balanced understanding by spelling out rigorously the ideas contained in the following analogy: in the same way as pieces of *material* things belonging to the external world become material constituents of our living organism through the biological metabolism that is specific to us (for example, animals cannot metabolize inorganic materials, while plants can), similarly it can be said that the *intentional* items of the external world can become constituents of our cognitive apparatus through a kind of psychic and mental metabolism, that is specific to this very apparatus.

It is not necessary to dwell more deeply on this analysis, since what we have proposed is sufficient for resuming our more general discussion.

Our representations are by their intrinsic nature *representations of something*, which we shall call their *referent*, and this simply because there cannot be a representation of *nothing*. However they have —as we have seen— also an independent form of reality, that is an *intentional*, or *noetic* reality, that we can call the *world of meanings*. This means that we can well have certain meaningful representations without a referent that *exemplifies* them (this is very common in every work of phantasy, where we are conscious that we are constructing a picture of some possible world of referents that may have no counterpart in the material world). In order to ascertain whether such referents of a non intentional nature actually exist we must rely upon some *instrumental operation*, which by its very nature must be proportionated to the nature of the referents we are looking for. In everyday life such instruments (as we have already shown) are our natural sense organs, which we use confidently and spontaneously without having to know any *theory* about them. Yet such a theory is not impossible, and in fact we have given a few ideas of it in the example of vision. This theory enables us to determine the so-called "normal" functioning of our sensory apparatus, and we can use it in order to detect certain "wrong" perceptions we may have (for example, in the case of colour-blindness). However even in everyday life we rely upon other instruments for affirming the existence of certain referents. For example, in order to affirm the existence of a given historical event we must rely upon reliable *documents*, and this is by no means a trivial operation (archaeology, papirology, paleography, numismatics, etc. are quite sophisticated disciplines that allow us to recognize and use correctly the historical documentation, and are not intrinsically different from the theories we need for correctly using a laboratory instrument in physics or chemistry). Even in the case of fictions we have instrumental criteria at our disposal: we can say that it is *true* that Hector is a Trojan warrior in the *Iliades*, because we have non-mental tools for ascertaining this fact (that is, we can find evidence in favor of our claim by *reading* the Homeric poem, which is different from just *thinking* about Hector). Note that in all these cases the non-mental operation we perform has in the last analysis the sense of a *causal explanation* of our instrumental evidence, though in a non strictly physical sense of causation.

11. SCIENTIFIC OBSERVATION AND REFERENCE

Scientific *observation* can consistently be embedded in this general schema. In the case of a papyrus we certainly have visual evidence under our eyes, but we need much interpretation and historical background to grasp the *fact* which is described in the document. And in the case of Hector, we must know an ancient language in order to know what kind of person he was in the *Iliades*. Similarly, when we observe something through a laboratory instrument, we must, for example, visually perceive certain shapes, but we cannot *interpret* them as evidence, say, of the trajectory of an electron without a good expertise in the use of the instrument, and a theoretical background that *causally* relates these shapes with an entity whose features are determined in a specific physical theory. However, since *observation* is *not* identical with *perception*, we can correctly say that we have *observed* the electron, in spite of having *perceived* only certain shapes, for example, on a photoplate.

In the case of the advanced sciences this link with observation as a warranty for referentiality is absolutely necessary, because we really open "new windows" that are intrinsically different from those naturally given us by our sense apparatuses (at variance with historical or literaray objects). In fact the representations of physical reality we propose are not *visual* (or the like), but *conceptual*, and here the autonomy of the noetic world may pose problems. Even with visual representations — we have seen — we are free of combining them in a fictional picture that represents only a *possible* world of referents with no concrete counterpart. With the conceptual representations we have an even greater possibility of constructing "theoretical models" whose degree of abstractness has only certain weak limitations of mathematical and logical character. The "accidents" determined in the construction of such models are far from being visual, acoustic, tactile, and certain difficulties experienced at the first stages of the development of quantum physics were mainly due to the impossibility of *visualising* the subatomic particles, say, as material points or as waves. But there was no need of such visualization. The only serious requirement is that they be in keeping with our cognitive apparatus which, in this case, is not our sensory apparatus, but the whole of accepted physical knowledge, with its concepts, laws, principles, and theories. The condition for being confident that these models do not just represent a possible world, but describe certain properties of the concrete (in this case of the *physical*) world, is that of *instrumentally* linking this model with certain features of the physical world that *causally account* (in a physical sense) for the perception obtained through the instrument.

Of course, this does not imply that our observations are always and necessarily *true*, but this is a very different question. As we have seen, observation is not perception, but judgement, and *every judgement* may be wrong (even a direct visual observation may be wrong). The reasons why an observation may be in need of correction are quite various: the instrument used might be incorrectly employed or have certain defects, or the causal theory of the instrument might be inaccurate, or the theory belonging to the background knowledge incorporated in our cognitive apparatus might be incorrect, or the interpretation based on a correct theory might not be sound, etc. Human knowledge is always fallible (at every stage), but this does not condemn us to skepticism, provided we do not require from knowledge an *absolute certainty*. It is sufficient that our knowledge be established beyond any *reasonable doubt*, as we always require from knowledge in everyday life, being content with what we are able to know at a given historical stage. In the case of science, this reliable knowledge undoubtedly exists, and has the great advantage of being constantly criticizable and improved through the use of our intellectual tools. No less importantly, this knowledge is also achieved through constant observation, provided we conceive it as an intellectual procedure linked to sensory perceptions but not limited to them. By means of this cooperation between *sense* and *logos* we can attain our referents, and increase our knowledge of the real "external world".

Fribourg University - Switzerland
Genoa University - Italy

NOTES

[1] I have developed the above considerations in several papers and in particular in discussing the problem of artificial intelligence, maintaining that a radical difference between minds and machines is represented by the fact that machines lack intentionality (at least in the sense that we do not need to introduce intentionality in order to ***explain*** their behavior, while we must do that in order to explain the behavior of human beings, and even of animals). I presented these ideas already in a paper of 1967 (Agazzi 1967), and revisited them in other papers (Agazzi 1981, 1987, 1991). A few years later, this became a central topic in the discussion on artificial intelligence, following some famous publications of Searle.

[2] This expression is due to the Italian philosopher Gustavo Bontadini, who has devoted penetrating analyses to this doctrine. Contemporary phenomenology has also significantly handled this thematic.

[3] More details on this issue may by found in three articles of mine (Agazzi 1988, 1990,1992).

REFERENCES

E. Agazzi, "Alcune osservazioni sul problema dell'intelligenza artificiale", *Rivista di filosofia neoscolastica* **59**, 1967, pp. 1-34.

E. Agazzi, "Intentionality and Artificial Intelligence", *Epistemologia* **IV**, Special Issue, The Mind-Body Problem, 1981, pp. 195-227.

E. Agazzi, "Intenzionalità ed intelligenza artificiale", in P. Prini ed. *Informatica e pastorale*, Morcelliana, Brescia, 1987, pp. 87-105.

E. Agazzi, "Do Experiments Depend on Theories or Theories on Experiments?", in D.Batens, J.P. van Bendegem eds. *Theory and Experiment*, Reidel, Dordrecht, 1988, pp. 3-13.

E. Agazzi, "Aspects herméneutiques et référentiels de la comparaison des théories", in E. Agazzi ed. *La comparabilité des théories scientifiques*, Editions Universitaires, Fribourg, 1990, pp. 13-25.

E. Agazzi, "Operazionalità e intenzionalità: l'anello mancante dell'intelligenza artificiale", in S. Biolo ed., *Intelligenza naturale e intelligenza artificiale*, Marietti, Genova, 1991, pp. 1-13.

E. Agazzi, "Intelligibility, Understanding and Explanation in Science", in C. Dilworth ed. *Idealization* IV*: Intelligibility in Science*, Rodopi, Amsterdam-Atlanta, 1992, pp. 25-46.

PAUL M. CHURCHLAND

A NEW APPROACH TO HUMAN COGNITION AND ITS SIGNIFICANCE FOR THE PHILOSOPHY OF SCIENCE [1]

I will return to our focal topics of truth, objectivity, and scientific realism toward the last third of this essay. Those issues are usually discussed within a framework of unquestioned beliefs about the basic nature of cognition, and I want to set the stage by outlining an *alternative* conception of what cognition is. It is a conception of cognition - both perceptual *and* theoretical cognition - that derives from recent work in computational neurobiology, connectionist artificial intelligence (AI), and parallel distributed processing research in psychology.

It is of interest because its basic kinematics and dynamics for cognitive activity is radically different from the kinematics and dynamics with which we are so familiar. Common sense and traditional AI and psychology all agree that the basic kinematic element in cognition is the belief, or sentence, or proposition. Our current cognitive state is relevantly represented by the set of beliefs we hold, the set of sentences to which we are committed. And the dynamics of cognition, on this view, concerns the inferences that Reason requires one to draw from accepted sentences to further sentences. Fresh observation sentences can confirm or refute old theories; new theories can be embraced because they explain old puzzles; discovered contradictions can force deletions and adjustments among our beliefs; and so forth. In these ways does the set of accepted sentences, or perhaps the probabilities that we assign to them, change over time.

So far the tradition. But here I want to argue for a new conception of what cognitive activity is, and for a very simple reason. The conception I have just described to you is almost certainly *false*, as a description of what cognitive activity is in creatures generally, and even in humans specifically. At the very least, it is a highly superficial conception of cognitive activity, and we are likely to profit from the perspective that a deeper conception will provide, a neurocom-putational conception, for example.

Obviously, there are many reasons for pursuing a neurocom-putational approach to cognition, but the ones that motivate me here concern the consequences it has for our understanding of *perceptual recognition* and *explanatory understanding*, and more generally still, for our understanding of the notion of *truth*, and for the process of pursuing truth over individual and historical times. But I will come to these consequences after I have put you in the picture. My immediate aim is to prepare a novel perspective from which to address such classical philosophical problems.

I begin by drawing your attention to figure 1, because I have found it extremely helpful in locating people in the relevant conceptual space. What we have here is a cartoon sketch of one of the initial pathways in the brain. Of course, projections go from the back of the brain to hundreds of other places in the brain as well, but if you

E. Agazzi and M. Pauri (eds.), The Reality of the Unobservable, 59-86.

draw too many, it just looks like spaghetti, so I have drawn just three stages here for reasons that will become clear presently.

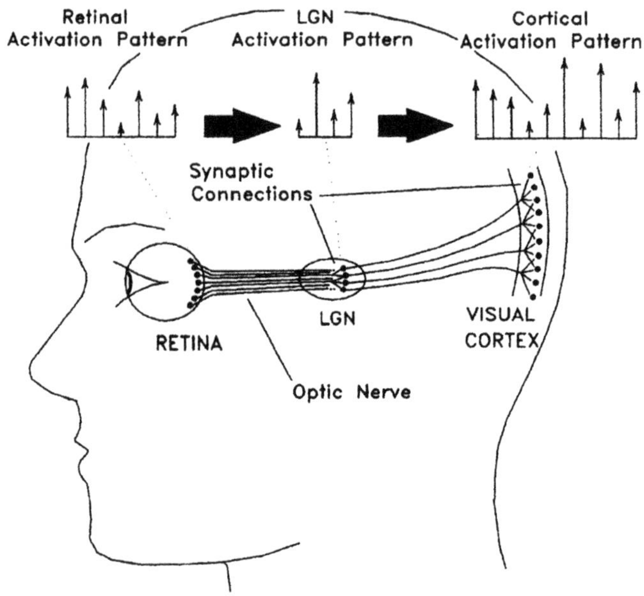

Figure 1

This diagram has two virtues. First, it exploits the fact that you already have no problem appreciating how a *pattern of stimulations* across your retina can *represent* the world. Any particular scene you might be viewing will produce a characteristic pattern of stimulations across the cells of the retina. You have about a hundred million cells on each retina, though I have drawn only eight retinal cells in figure 1. But the point is clear enough: we can specify any pattern of retinal stimulation by a unique list of numbers - decimal numbers between zero and one, say - each one telling us by what fraction of maximum its corresponding cell is stimulated.

Theorists call such a list the *input vector*. A vector is just an ordered list of numbers, and you can see why the notion of a vector is useful here. When I am talking about vectors in this essay, it is simply as a shorthand for a pattern of stimulations across some population of neurons. Your retina is something you know and love, so I am starting out with something that is not a mystery at all.

Your retinal cells *project*, via a large number of axons, their collective pattern of stimulations to another population of neurons at the lateral geniculate nucleus (LGN). Figure 1 shows four cells - although in fact you have something closer to ten or a hundred million neurons at that site - and each one of the projecting retinal axons divides at its tip into a large number of terminal end branches. These make excitatory contact with many different cells at this second or target population. The basic result is that a pattern of stimulations at the retina produces a secondary pattern of stimulations across the cells of the LGN. There is a change induced as the first pattern produces the second pattern: by means of the various excitatory contacts, the brain is processing, filtering, pulling out the special bits of information that it wants.

The LGN projects in turn to the visual cortex, and again the axons divide so that each axon has an impact on a large number of neurons at this target population, typically between 10^3 and 10^4 of them. Each cortical neuron thus ends up wreathed in terminal end branches from many different axons, rather like an old oak tree wreathed in thousands of reaching ivy creepers. What these creeping projections do, collectively, is to produce a *third* pattern of stimulations across the neuronal population at the visual cortex. It is again a distinct pattern, since the pattern arriving from the LGN goes through millions of those little terminal connections, called *synapses*, which vary widely in their conductivity. The arriving pattern is thus transformed as it acts on the target population: it produces a new pattern of stimulations across the cortical neurons. What we have then is a device for transforming a complex pattern at the sensory periphery into a series of different patterns at subsequent layers of neurons.

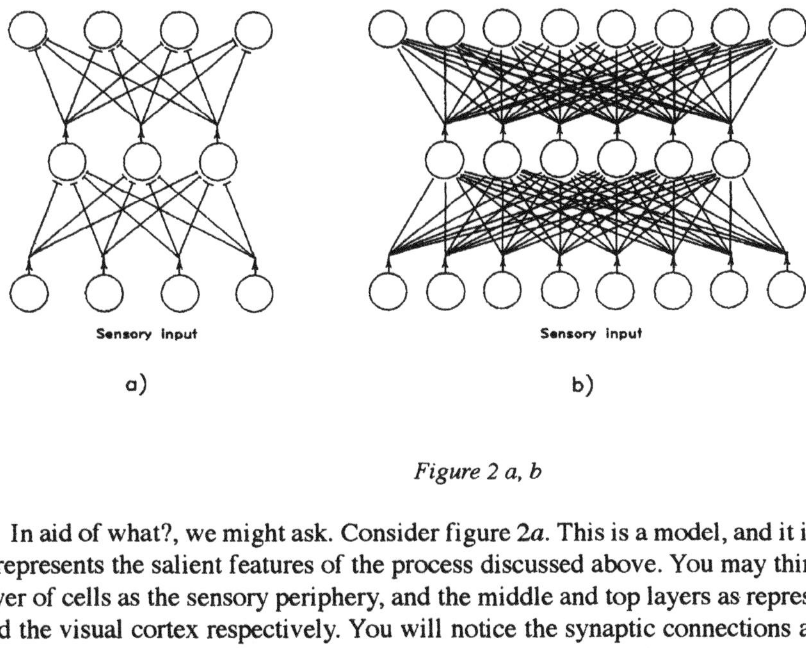

Figure 2 a, b

In aid of what?, we might ask. Consider figure 2*a*. This is a model, and it is useful because it represents the salient features of the process discussed above. You may think of the bottom layer of cells as the sensory periphery, and the middle and top layers as representing the LGN and the visual cortex respectively. You will notice the synaptic connections at the tip of each axonal end branch, represented by small bars, and you will notice that some of them are rather bigger than others. That means that any stimulation arriving at such a synapse will have a large effect on the receiving cell. The same stimulation arriving at a small synapse will have only a small effect on the receiving cell. You may think of these as *gates* of varying sizes. Depending on what overall configuration of gates mediates the incoming pattern of stimulations, we get a specific *transformation* effected between the vector that is conveyed upwards toward the middle layer, and the vector (= pattern of stimulations) that finally results across the middle layer neurons themselves.

You can think of the neurons, by the way, as humming to themselves, and to each other. For they emit pulses along their axons with a frequency varying between zero and perhaps 500 Hz. With a high level of activation a cell is humming a high note; with low activation, a low note. So you have a sort of barber-shop quartet humming away at the bottom layer, and their musical chord is sent upwards towards the middle layer where it is transformed by the many synapses into a different chord struck at the middle layer. The same process occurs at the next layer, where yet a third chord of notes ends up being hummed across the output cells. The parts of the brain, then, are humming to each other in something like the language of complex musical chords.

Of course, the chords struck in the brain are typically not the simple three-to-four element creatures we encounter in music. Networks can be very, very large, and you have already seen from the example of the retina that the musical chord at the human visual periphery has something like a hundred million notes in it! So it is a very complex representation. In fact, we can make neural networks as large as you like - each horizontal layer in figure 2*a* could be extended to the left and right out to twice its original size, as in figure 2*b*, or out to 100 million cells or even larger. And an interesting feature is displayed in pattern-transformers of this kind. This much larger "computer" will compute its more complex function just as swiftly as the tiny network of figure 2*a* computes its much simpler function. Indeed, the speed of computation is *independent* of the cell population at each layer. All that matters is the *distance* between layers, and that can be the same for two networks, even if their layers differ widely in cell population. This means that a network of this kind could be computing simple one-digit sums, or it could be computing second-order differential equations. It would make no difference. It would complete the computation just as quickly.

There is a further virtue of this network configuration. You could go inside the network with a tiny shotgun and start blowing away its connections at random, and it would *continue* computing more or less as it had before the damage. Since each of the millions of synaptic connections contributes such a tiny part of the collective computation, the loss of a random small portion of them affects the network relatively little. What would happen is that the *quality* of the computation would go down. Slowly the network's performance would degrade. Those of us over forty will begin to recognize ourselves in this portrait.

One of the most important features of these networks is that the size or "weight" of each little synaptic connection is *plastic*. You can change the weights, and when you change the weights to a new configuration of values, you get a different transformation effected between a given input vector and the resulting output vector. You can toy with those weights until you get any transformation you want. In fact, a recent result shows that, for any input-output transformation whatsoever (as long as it is an "effectively computable function", for those of you who know or care what that is), there exists a three-layer network of the kind you have just seen, with a suitable configuration of synaptic weights, that will approximate that input-output function to any degree of accuracy desired.

This is a universality result, for neural networks, analogous to the Universal Turing Machine result that so excited people in AI several decades ago. It means, not to put too fine an edge on the point, that they can do *anything*. They just have to be big enough, and have the right configuration of weights.

That problem - how to find the right configuration of weights - was solved by David Rumelhart and others about 1980. But as the story has been related to me, nobody realized at the time that the solution was a genuine solution. Rumelhart invented the "back-propagation algorithm," but Geoffrey Hinton, his colleague, looked at it and said, "That's not going to work; it's going to get stuck in false configurations all the time." So they more or less put it on the shelf for four years. After it had sat there for a good while, Rumelhart pulled it down one day to try to solve a weight-configuration problem that had resisted other methods. It worked beautifully. So they tried it on other problems. And it worked again. They discovered empirically that it only rarely got stuck in unhappy configurations.

The back propagation learning algorithm works as follows. It repeatedly changes the synaptic weights of the student network by tiny increments, so that the weights slowly approach the configuration that makes the network do what you want. This is best illustrated with a real example, to which I now move. See figure 3.

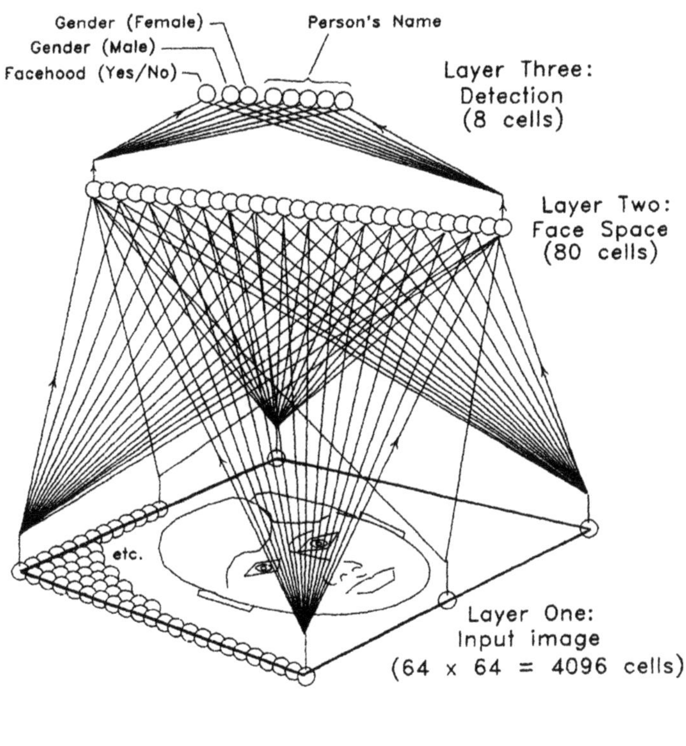

Figure 3

Suppose we want to train a network to do the following kind of transformation. We are going to feed it, as input vectors, vector codings for various two-dimensional images. And we wish to train it so that it invariably produces, at its output layer of neurons, a vector that indicates *1)* whether the current input image is or is not a human face, *2)* whether that face is male or female, and *3)* the identity of the individual person there portrayed. In short, we wish to recreate the human capacity for recognizing other human faces.

This capacity is precisely what some neural-net researchers have recently succeeded in modeling. (Thanks here to the work of Garrison Cottrell's group at the University of California, San Diego.) Their three-stage artificial network is schematically portrayed in fig. 3. Its input layer or "retina" is a 64 X 64-pixel grid whose elements each admit of 256 different levels of activation or "brightness". This allows the network to code recognizable representations of real faces. Fig. 4 shows several of the many input photographs on which the network was actually trained. That "training set" contained 64 different photos of 11 distinct individuals, plus 13 photos of non-face scenes. Cottrell himself is the bespectacled face at the lower left.

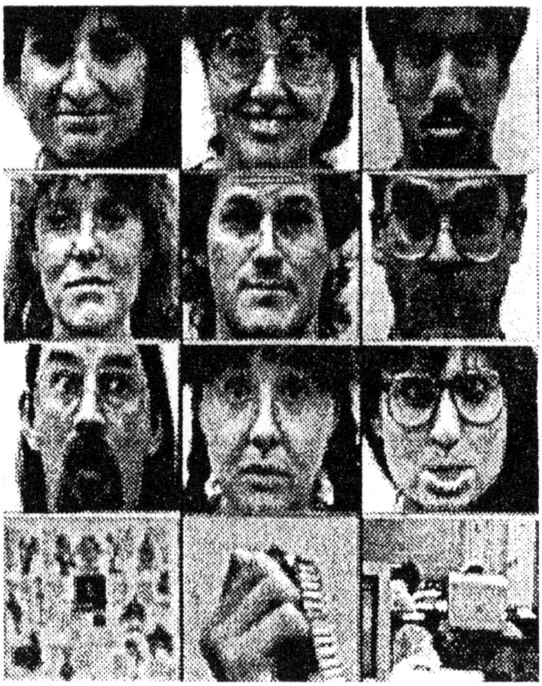

Figure 4

Each input cell projects a radiating set of axonal end-branches so as to make a synaptic contact with each and every one of the 80 cells at the second layer, which layer represents an abstract space of 80 dimensions in which the input faces are variously coded. This second layer projects finally to an output layer of only eight cells. These output cells have had their synaptic connections carefully tuned so as to discriminate, at the second layer, between facial patterns and nonfacial patterns, between male patterns and female patterns, and finally, to respond with the person's "name" (a numerical code arbitrarily assigned to each person) to the re-presentation of any face that the network "got to know" during its training period.

What actually does the work in Cottrell's face-recognition network is the overall configuration of synaptic connections - positive and negative, weak and strong. It is these, and only these, that progressively transform the initial 64 X 64-element pattern or vector into a second and finally a third vector that represents the input photo's facehood, gender, and name explicitly. But it is not at all obvious how to configure the connections, to achieve this end, in such a large network. After all, it contains three distinct layers, 4184 cells, and a grand total of 328,320 synaptic connections!

In fact, Cottrell's group had no idea how they should be configured. That was part of the problem they had to solve. Fortunately, there exists a general technique for finding synaptic solutions to transformational problems, a technique widely used by neuromodelers. It is biologically realistic in the single respect that it involves the steady adjustment of a network's synapses in response to the pressures of its ongoing experience. Sadly, it is biologically unrealistic in just about every other respect. But beggars can't be choosers. Until we sleuth out the brain's actual learning procedures, we have to use whatever is available. This interim technique is called "synaptic adjustment by the successive back-propagation of errors", or just *back-propagation* for short. It has two outstanding virtues. It works very well, at least for smallish networks. And the procedure can be administered by a conventional serial computer, which takes the extraordinary drudgery of the procedure out of the neuromodeler's hands. It works roughly as follows.

The network modeler begins by setting all of the synaptic weights at random values, both positive (excitatory) and negative (inhibitory), not too far from zero on either side. This constitutes an utterly blind "stab in the dark" at the desired configuration of weights. By itself, this gets us nowhere, but have faith for a second.

What this configurational stab permits is the discovery of just how *mistaken* it is, for when we present the network with a specific input image - the upper left female face in fig. 4, for example - the resulting output vector will almost certainly be gibberish. It will fail to code facehood, gender, and name accurately, because the 328,320 synaptic weights that produce the output vector were blindly set at random values. But given the specific input image presented, we know perfectly well what the activation vector across the output neurons *should* have been: ideally, it should have been < 1; 0,1; 1,1,1,1,1 >, where a "1" indicates a neuron's maximal level of excitation and a "0" indicates its minimum. For that vector is the pattern that identifies <yes, it's a face; no on male, yes on female; and, this face is Mary's>.

The crucial point here is the *difference* between the ideal vector we want and the garbled vector that the untrained network actually produced. We want to modify the network's initially random synaptic configuration so as to reduce that difference. This is what the "back-propagation algorithm" does for us. It considers each synapse in the network, one by one and in isolation, and determines whether a tiny increase or decrease in its weight would yield an output vector that is slightly *closer* to the ideal output for the current input image. If so, it makes the appropriate adjustment and then moves on to the next synapse to repeat the process. After all of the weights in the network have thus been nudged to slightly happier values, a new input image - the top right face, for example - is then placed on the network's input layer, and the entire process is repeated for this second image.

In this way, all of the digitized photos in the training set are presented to the network, and its many weights are successively nudged through a series of ever more effective configurations. After tens, hundreds, or even thousands of encounters with the entire set of training images, the network typically settles into a configuration of synaptic weights that produces surprisingly accurate output vectors for any given input image. Once this goal has been achieved, the training ends and the network's weights are frozen at their final values. We then have a network whose cognitive capacities and internal coding strategies we can begin to explore.

Cottrell's face-recognition network topped out at an impressive level of performance. It achieved 100% accuracy on the training set of images, with respect to faceness, gender, and the identity of the individual portrayed. In itself, this is not necessarily impressive, since the network may just have "memorized" the finite set of input/output pairs presented to it during training, rather than acquired some truly general understanding of how to represent faces. A more severe and more relevant test occurs when we present the network with photos it has never seen before, that is, with various photos of the same people drawn from outside the training set. Here again the network performs well. It identified correctly 98% of novel photographs of the people encountered in its training set, missing the name and gender of only one female subject.

A second and even more severe test of the network's ability to generalize to new cases involved asking it to discriminate the faceness and gender of completely novel scenes and people. Here it was 100% correct on whether or not it was confronting a human face, and it was roughly 81% correct in its verdicts on the gender of the novel faces presented. (It got the male faces mostly correct, but showed a decided tendency to misclassify some of the female faces as male.)

A third and very intriguing experiment tested the network's ability to recognize and identify a "familiar" person when one-fifth of the person's face was obscured by a horizontal bar across the input image (see fig. 7*a* below). Surprisingly, the network's performance was impaired hardly at all. Subjects were identified correctly despite the incomplete or degraded character of the input images. We shall discuss this intriguing behavior more deeply in a moment.

How on earth does the trained network do all of this? What is going on inside it to make these surprising skills possible? It is all very well to be told that a zillion synaptic weights have been "suitably configured", but is there any more informative or illuminating description of what has happened to the network during training?

Yes, there is. For starters, let us focus our attention on the 80 cells at the second layer of the network (see again fig. 3). What we now want to know is this: exactly *what* facial features are being coded by the cells at this level of the network? What effective coding strategy was gradually discovered by the network during the relentless pressure of its training period? To put the very same question in a third and final form, what input activation pattern has become the *preferred stimulus* for each of these second-layer cells?

With an artificial network, as opposed to a living brain, we can get an absolutely determinate answer to this question, for each and every one of the 80 cells at issue. This is because the serial computer that conducted the network's training knows the exact value of the synaptic connection between any two cells in the entire network. In particular, it knows the pattern of connection weights onto each of the 80 face-dimension cells. Such connection-weight patterns uniquely determine the retinal input pattern to which a given face cell will give its maximal response. By examining the final configuration of weights for each cell - we can reconstruct the retinal input pattern that now constitutes its preferred stimulus.

The result of doing all this, for each of the 80 cells, yields a bit of a surprise. We might have expected each of these cells to become focussed on some localized facial feature such as nose length, mouth width, eye separation, and so forth. But reconstructing the actual preferred stimuli of the 80 face cells reveals that the network settled into a coding strategy quite different from this.

Figure 5

Fig. 5 reconstructs the preferred stimulus of six typical face cells from layer two of Cottrell's network. One can see immediately that each cell comprehends the entire surface of the input layer, and represents an entire face-like structure, rather than just an isolated facial feature of some sort or other. Neither do these preferred stimuli correspond to individual faces in the original training set. (There are 80 cells, recall, but only 11 distinct faces in the training set.) Rather, they seem to embody a variety of decidedly *holistic* features or dimensions of facehood, dimensions for which ordinary language has no adequate vocabulary. And yet, a given face presented at the input layer will variously activate each of these 80 holistic features, thus producing an activation vector at the second layer, a vector unique to that face. And *different* photographs of the same person presented to the input layer will produce essentially the *same* activation vector across the cells of layer two, thus allowing the output cells at layer three to identify that individual correctly.

Janet Metcalfe, Cottrell's co-worker on this and other networks, has coined the term *holon* for a preferred stimulus or feature of this diffuse, input-spanning sort. It is a coding strategy that networks quite regularly discover during training. This is presumably because it is an efficient and effective way of coding the information needed to solve the problem of finding the right input/output transformation. But it has a vitally important virtue beyond its efficiency: it helps to make the trained network functionally persistent against scattered cell damage and synaptic malfunction.

Since each pixel segment of the input image has some small effect on - that is to say, its information is *distributed* across - every cell at the second layer, and since every cell at the second layer contains at least some important information about the entire input layer, then the scattered loss of cells and connections throughout the network leaves us with a network whose function is somewhat degraded, but still closely similar to its function in its undamaged state. With both the coded representations *and* their transformations being widely distributed across the entire network, there is no "bottleneck", neither representational nor transformational, whose isolated failure would bring the network suddenly to its knees.

Beyond this, an examination of the coding strategy at work at the second layer reveals something familiar and very welcome emerging, namely, a hierarchical structure of recognizable *categories*. Fig. 6 attempts to portray the 80-dimensional space of the cells in layer two of the trained network. It suppresses 77 of them, of course, and settles for three typical dimensions so that we can comprehend the point visually. Notice that, thanks to the network's training, this space now shows a primary partition into two regions: a large region in which all faces tend to be coded, and a smaller one close to the origin in which all nonfaces tend to be coded. This latter region is smaller because the cells at layer two respond hardly at all to any input that is not a face. The connection weights have been configured so that the dynamic range of layer two is spent primarily on discriminating among faces. That is why the facial region is the larger of the two subspaces. If something isn't a face to begin with, it is hardly noticed, it barely even exists, so far as layer two is concerned.

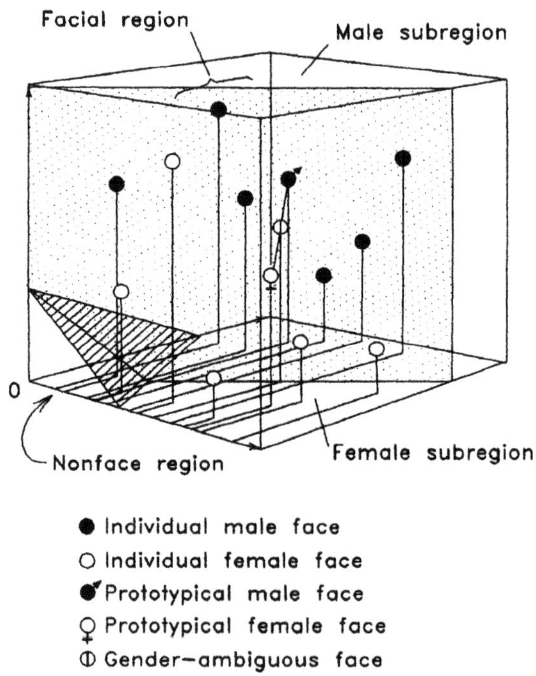

Figure 6

Within that facial region, there is a secondary partition into male faces and female faces. These two regions are roughly equal in volume, reflecting the network's rough parity of discriminatory power within each class of faces. This is no surprise: it was trained on roughly equal numbers of male and female faces. Had the training set been biased in either direction, the resulting partition might well have been lopsided, as we will shortly see.

At the female subregion's "center of gravity" is the coding point for the prototypical female face. The prototypical male face occupies a complementary position next door. Scattered throughout these two subregions are the proprietary coding points for each of the several faces in the training set. These partitions across the activation space of the cells at layer two - these *categories*, for there is nothing else to call them - slowly emerged and stabilized during the course of the network's training. And it was to these partitions, and to the activation patterns that they variously represent, that the cells at the final layer slowly became tuned in turn.

The appearance of these hierarchically divided regions in "activation space" provides us with a further way of describing and explaining the network's acquired skills of recognition and discrimination, a way beyond the austere vocabulary of synaptic connections. Not to put too fine an edge on it, what the network has developed during training is a family of rudimentary *concepts*, concepts that get variously activated by sensory inputs of the appropriate kind. This way of talking has yet to be fully justified, but you can already see the point of introducing it. The suggestion we are exploring is that the appearance of concepts in living cognitive creatures consists in the same sort of learned partitioning of neuronal activation spaces.

Recall now the ability of Cottrell's network to identify a familiar face hidden behind a horizontal bar (see fig. 7*a*). This ability to identify familiar faces correctly, despite the loss of 20% of the information in the original input vector, illustrates a marvelous property of neural networks, a property with far-reaching consequences. This property is their capacity for what is called *vector completion*. Since the output layer identifies the input person correctly as Mary, it must be responding to a facial coding vector at the second layer that is very close to the proper coding point for Mary. At the least, that second-layer coding vector must be closer to Mary's point than to any other person's coding point, else the output layer could not have identified her correctly.

a) **b)**

Figure 7

How is it that something very close to a complete Mary-vector at layer two can result from a 20% *in*complete image vector of Jane at the input layer? The answer, in part, is that the information in the remaining 80% of the input image is still strictly sufficient to distinguish Jane's face, though not from *every* other possible face, at least from each of the other ten faces in the training set. While there might exist other faces identical to Jane's face save for that portion of her face hidden behind the bar, certainly none of the other faces in the training set meets this condition.

The second part of the answer is that the network's second layer, having been trained primarily on faces, tends to "fill in" any empty portion of an input facial image with the appropriate face-like features, features *similar to* the unobscured features of the original and now "familiar" input image. Fig. 7*b* illustrates the informational content of layer two given the obscured image of a face (7*a*) as input. Compare that filled-in face to the face of the original and unobscured subject, the upper-left female face in fig. 4. The reconstruction isn't perfect, but it's not bad either.

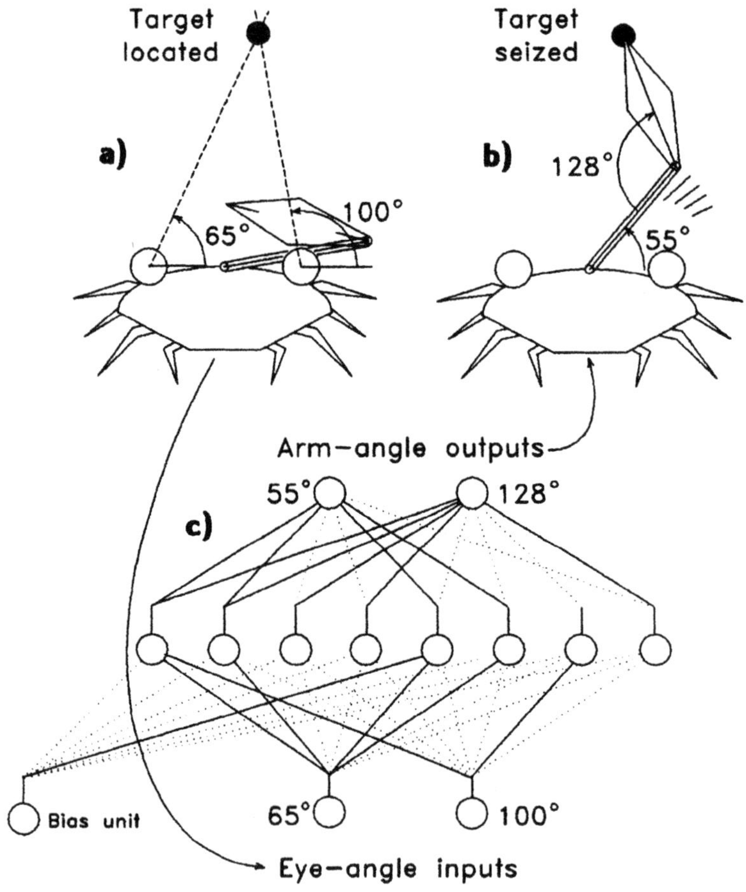

Figure 8 a, b, c

The third and most important part of the answer is that, during its training on the original eleven faces, the network has partitioned its activation space at layer three into a set of eleven "basins of attraction", each one centered on the prototypical coding point generated for each of the eleven individuals in the training set. After all, it was not trained on all possible faces: only on these eleven. It was not asked to discriminate a million faces: just these eleven. The pressures thus exerted on the network during its training period forced it to become extremely sensitive to any and all input features that happen to point, even feebly, toward any one of exactly these eleven faces. This hypersensitivity regarding certain features of the input vector, and this forced concern with eleven specific individuals, means that the trained network has a strong tendency to "jump to conclusions" about the identity of an input face, even when the input is missing some standardly available information. To put the same point in another vocabulary, the network has acquired an especially strong tendency to activate, at layer two, some one or other of the eleven face-vectors to which it was so carefully trained, even when the input vector is partial or degraded. It tends to activate, at layer two, the specific facial vector that is most probable - *among the favored eleven* - given the "evidence" of the input vector received.

What we are observing here, in the phenomenon of vector completion at layers two and three, is a primitive form of inductive inference, or more accurately, of what C.S. Peirce called *abductive inference*, and what we moderns call "inference to the best available explanation". Indeed, it may be that vector completion is *the* basic form that inductive or *ampliative* inference takes in living creatures generally. We shall address this issue in a few moments, when we address the nature of scientific reasoning. For now, let us just note that vector completion or inductive inference appears as a natural and inevitable cognitive phenomenon, even in the very simplest of neural networks.

I am leading you up by stages here, because I want ultimately to get to networks of the complexity of *you*. In this regard I should mention briefly that we can train networks to solve sensorimotor problems as well. Figure 8*a* shows a schematic crab that locates external objects by triangulation with its two eyes, and then reaches for them with a two-jointed arm (8*b*). I have trained up a simple network (8*c*) to take a pair of eye-angles as input and deliver a pair of joint-angles as output, such that the tip of the crab's arm reaches for the triangulation-point of the eyes, wherever that happens to be. In this way it can grab whatever it sees. Such problems are far from trivial, especially with more complex animals, but you can see roughly how networks offer a general account of sensorimotor coordination, and I want now to move to something else. Consider now the familiar but problematic case of *ambiguous perceptual figures* (see figure 9*a*).

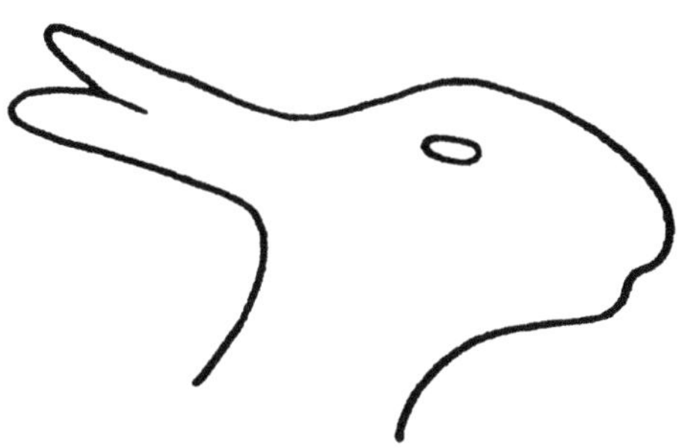

a)

b)

Figure 9 a, b

Here is a case where you can activate one of two different prototype vectors within your brain - prototypes developed during prior learning - and thus see a given object in two quite different ways: as a duck, or as a rabbit.

The old woman / young woman (9*b*) is another familiar case. How do neural networks recreate this sort of phenomenon?

Consider a simple feedforward network as portrayed in figure 2*a*, but this time let us add what are called "recurrent pathways" (see figure 10). Adding recurrency brings many interesting features to a network, but in this context it permits the following. Depending on the network's current "state of mind" at the third layer of figure 10, information will be sent down to the hidden units, information that will shape or shade the way in which the bottom-most input vector impacts the hidden layer. You can thus look at the duck-rabbit figure and direct yourself, "See it as a duck," and you do; or, "See it as a rabbit," and you do. What is going on here (I hereby hypothesize) is that recurrent pathways are doing that job for you.

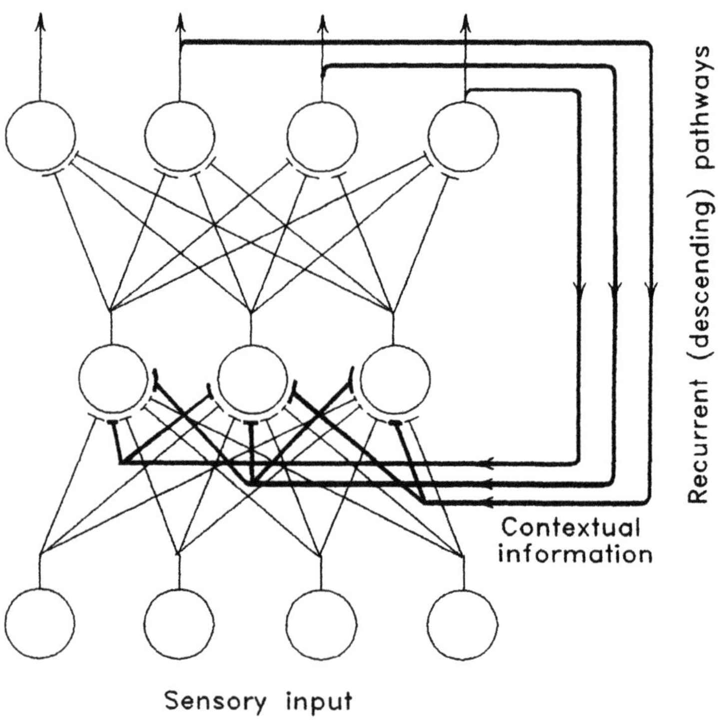

Figure 10

The required wiring is there. Recurrent pathways are not just common, they are the rule throughout the brain. The LGN, for example, which projects to the visual cortex (see again figure 1), receives about ten times as many axons *from* the visual cortex as it sends there. Why is there so much effort spent on these pathways? I think an obvious hypothesis is that their function is to modulate the way in which the earlier layers process the input vectors arriving from the sensory periphery. This makes us, by the way, genuine dynamical systems, because once we kick the network into operation we can shut down the peripheral inputs and the system will wind away on its own. It will show interesting dynamical behaviors such as bifurcation, chaos, and so forth.

The next three ambiguous figures introduce a new dimension of difficulty, for they are not neatly "bivalent" as in the two examples just shown. Probably you do not see anything at all in 11*a*, unless you have seen this famous picture before. It looks like sheer noise to you. Here is a case where you do *not* activate any prototype that prior learning has produced in you. But I will help you to do that. Figure 11*a* is a degraded input which requires me to provide a little bit more information. There is a Dalmatian dog, in the middle of the picture, walking on a brightly sunlit but shadow-mottled field. His head is down, sniffing away at the ground, and he is walking away from you and to the left, toward an ornamental tree with a shadow underneath. You should be able to see the dog now, because I have provided enough contextual data that your recurrent pathways could nudge your hidden units in the direction of your "spotted dog" prototype vector, close enough so that the noisy input of 11*a* finally produced an activation pattern that fell into that waiting attractor at layer two.

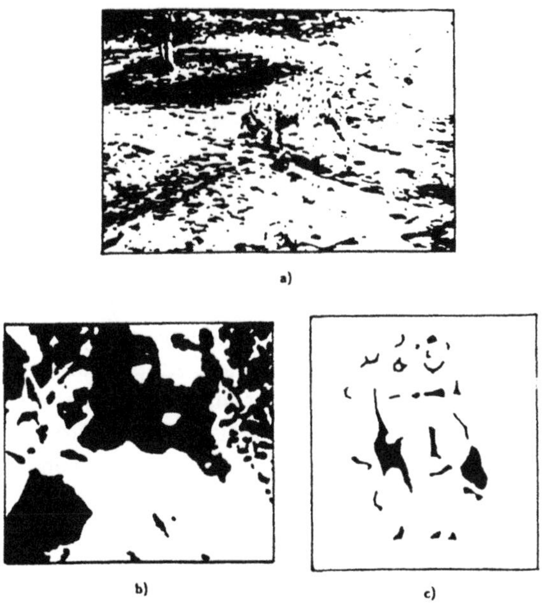

Figure 11 a, b, c

Now observe figure 11*b*. I got this one from a book written a long time ago by N. R. Hanson,[2] and he tells you at the bottom of the page what this is. (But even though he tells you, it was ten *years* before I finally saw what was there. You may find it similarly difficult.) It is the face of Jesus Christ. Or perhaps Charles Manson, as someone once remarked. His forehead is cropped at the top of the picture, but his bearded face can be seen in the upper center of the frame. The light is coming from the upper right, illuminating one cheek and eye-socket, and so forth. the added information, to be delivered to your relevant hidden layer by your recurrent pathways: it is a man on a horse.

The horse is facing left, with its two small ears and long face just visible at the upper left. Below this is its chest and two front legs. Likely, now the rest will pop into recognition: its back legs and tail to the right, the two extended feet and bent arm of the rider, and so forth. (The rider remains indistinct, even for me. Perhaps this one is too hard. But one needs a range of examples.)

What I want to suggest here is that - in creatures like mice, or cats, or humans -what learning does, via some learning algorithm or other (not the back-propagation algorithm) is generate a large library of partitions on your activation-vector space, a *very* large library. These partitions contain central prototypes with similarity gradients falling away from them in many dimensions. You can think of these prototypes as dynamical *attractors* into which various stimuli, even degraded ones, will tend to "fall". At bottom, this is how you recognize the objects and situations in the world.

Obviously this needs a lot of expansion if I am to account for the kind of cognitive complexities we display. Let me make a start on that in the following way. What are the neuronal resources we are playing with? There are at least 10^{11} neurons in the brain. If you think of a standard three-story colonial-style academic's house filled from basement to rafters with grains of sand, then that is the number of neurons you possess. The relative proportions of grain-to-house and neuron-to-head are about the same.

If we suppose that the brain divides itself into (very roughly) 1000 distinct subsystems, each devoted to processing some proprietary type of information, then each such subsystem will have something like 10^8 neurons in it. This means that such a subsystem can represent things with a vector that contains 10^8 (100 million) elements. That vector is big enough to represent the contents of an *entire book* with one fleeting pattern. So the representational resources here are very rich indeed, rich enough, evidently, to represent things a complex as . . . a stellar collapse. It does not have to be something simple like "male face." Such a large vector can represent cell meiosis, or oceanic rift, or harmonic oscillator, or economic depression, or any complex theoretical notion you choose. We have the resources to represent, vectorially, phenomena as complex as these quite easily. We have more than enough resources to do it.

All right, suppose that we do. What then? Then we can give a theory of perceptual recognition, and explanatory understanding, and abductive inference that counts all three as instances of the same thing: *prototype actvation*. In figure 12a we have a well-trained coyote who espies a little tail sticking out from a tuft of desert grass, and what gets activated there is his "desert rat" vector. Notice that this involves an *ampliative* inference: the coyote might be mistaken. The tail could be the tail of a poisonous snake, in which case the coyote is in for a surprise. Networks can make mistakes.

For a human case (figure 12*b*) we have my colleague, the astronomer Margaret Burbidge, looking through a telescope at the planet Jupiter and noticing that its figure is not a perfect sphere, but rather an oblate spheroid. She is a well-trained network, and so the "rotating plastic body" vector gets activated straightaway. Rotating plastic bodies are of course typically deformed in the fashion at issue, so she has *an* understanding of what is going on. It may be incorrect, of course. The activation here is ampliative, since the prototype vector portrays information beyond what the retinal input vector strictly guarantees. But we have seen that this is typical of the behavior of trained networks. They exploit information gleaned from past encounters with the world to "fill out" current inputs in prototypical ways.

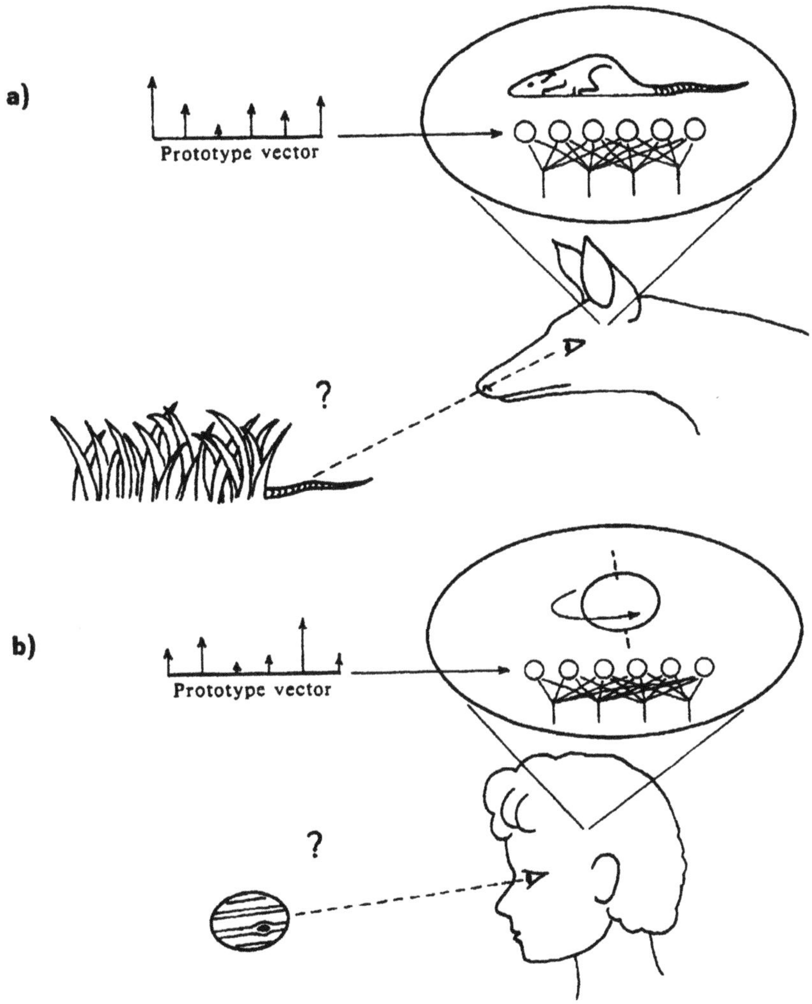

Figure 12 a, b

By now, you can probably see where all of this is leading. I wish to propose the preceding as a general theory of explanation, or more precisely, of *explanatory understanding*. And I will now try to illustrate how one of humankind's crowning glories - the activity of scientific theorizing - can also be understood as just a high-level instance of the cognitive activities already examined.

The central phenomenon to be explored here is the brain's *vector completion* of partial or degraded inputs, a completion often aided by the brain's recurrent manipulation of the relevant population of representing neurons. In plain language, it is the phenomenon of your *recognizing* - perhaps slowly at first, but then suddenly - some unfamiliar, puzzling, or otherwise problematic situation as being an instance or example of something well known to you. The coyote and the astronomer provided two fairly humble examples of this phenomenon. They illustrate how a neural network with recurrent pathways would naturally give rise to both a context-sensitive plasticity in its perceptual processing, and to sudden interpretational "successes" when the continuously cycling recurrent system finally activates some vector close to one of its antecedently-learned prototypes. Let us now look at some rather grander and historically more celebrated examples of the same thing.

Consider looking up at the stars on a clear, moonless night, from some pastoral vantage point free of the occluding haze and background light of the city. Thousands of stars are visible, scattered carelessly both in space and in brightness. Here is a "degraded perceptual input" indeed! For sheer unstructured chaos, it surpasses any of the random-dot figures or splotchy photos presented so far.

And yet, all human cultures have imposed a structural order of some kind or other upon the contents of the night sky, interpreting one group of stars as a dipper, another as a flying swan, a third as a hunter with dogs, a fourth as a scorpion, and so on. Few of these interpretations are very compelling, visually. And certainly none of them yielded any useful predictions of stellar behavior, despite the elaborate mythologies in which they were often embedded. The scorpion never stung anything; the dogs never caught anything; the dipper never poured out any water. In this respect, these interpretations of the visual chaos were not "good theories" about stellar phenomena.

The absence of any such "action" in the sky reflects the fact that the positions of the stars, relative to each other, remained *constant* over time. With the marginal exception of a puzzling handful of *planetes* or "wanderers", every star had a fixed and utterly permanent position relative to every other star in the sky. This constancy allows any nighttime observer to notice that, *collectively*, the stars *do* display a very regular form of behavior.

An hour's peaceful attention reveals that the stars at the eastern horizon are climbing into the sky at the rate of fifteen degrees (thirty moon diameters) every hour. In the same period, the western stars have been sinking below their horizon at the same brisk rate. Indeed, the entire "framework" of stars, the *planetes* included, moves as a single unified object relative to the circle of the earthly horizon, as if the stars were all permanently positioned on the inside surface of a vast sphere surrounding all of *terra firma*, a sphere that rotates in a most magisterial fashion about a gargantuan axis that cuts the sphere at Polaris, the pole star (fig. 13).

Here, in the last clause of the preceding sentence, is the interpretation that suddenly unites the scattered elements of the night sky as visible manifestations of a familiar object: *a rotating sphere*. What is unusual about this particular sphere is its vast size, its utterly regular rotation, and the fact that we are viewing it from the inside, from somewhere close to its center. Beyond these notable novelties, and given a clear view of the night sky's behavior, it is close to being visually *obvious* that we are here dealing with a large sphere. Moreover, and unlike the animistic mythology of the individual constellations, this rotating-sphere interpretation of the whole sky allows us to predict the motions and future positions of all of the stars with great accuracy. Its ultimate truth or falsity aside, this interpretation of the initial chaos was a very *successful* theory. That is partly why it was, in some version or other, the accepted theory of the cosmos in almost every culture from the ancient Greeks until post-Newtonian Europe.

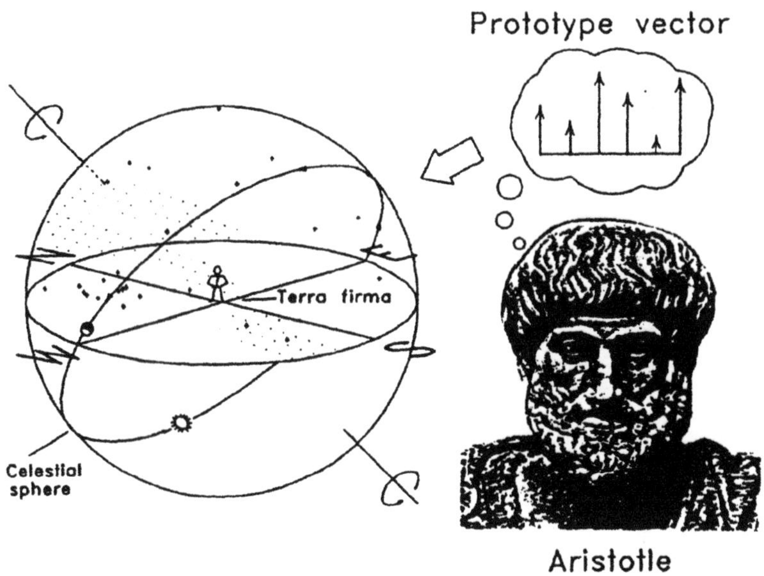

Figure 13

I am here representing the cognitive achievement of an ancient cosmological theory as being similar to the cognitive achievement involved in recognizing a familiar sort of object or process in any other problematic context - any context that involves incomplete or degraded input, for example, or an unusual sensory perspective, or other novelty sufficient to produce the kind of confusion encountered in figs. 11*a-c*. This assimilation of "theoretical insight" to "prototype activation" has the further advantage that prototypes, especially prototypes with a temporal dimension, typically represent far more information than is present in the sensory input that activates them on any given occasion. Those prototypes were originally acquired during training over many and varied instances thereof. This means that a prototype carries a substantial predictive content - about what further or subsequent features will be perceptually discovered in addition to those already observed. This content can go uncorroborated, or be flatly contradicted, by subsequent experience. In this way are "theoretical" interpretations, just like interpretations generally, subject to empirical criticism.

Consider a second historical example of a theoretical insight: Descartes' dynamical account of the motions of the Sun and planets. Why do the planets all circle the Sun? Why do they all revolve in the same direction? Why are the revolutions slower the more distant the body is from the Sun? What *is* the Solar System?

For one who held, as Descartes did, that space must everywhere be continuously occupied by some sort of rarefied and translucent fluid matter, the circling behavior of the planets suggested nothing so much as a giant *vortex* in this universal fluid medium. And that was exactly Descartes' hypothesis. The planets were then seen as being like leaves swept around in a giant whirlpool, with the leaves closer to the center - Mercury and Venus - being carried around the fastest (fig. 14).

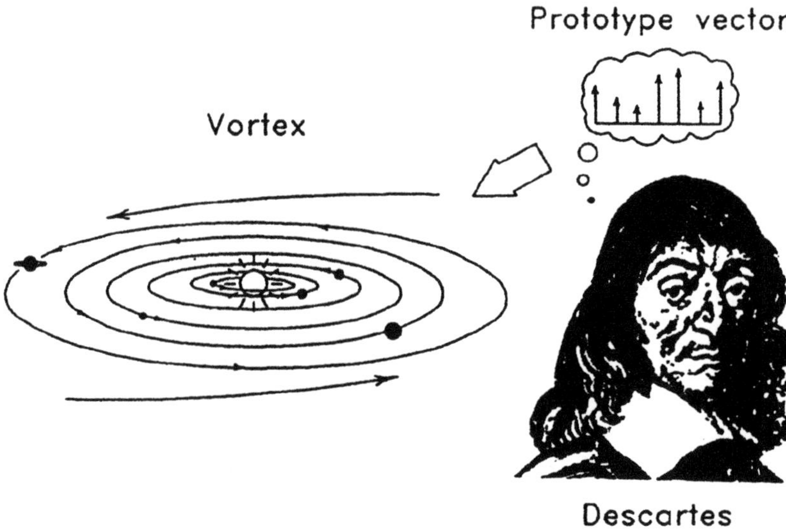

Figure 14

Here is a dynamical interpretation that makes familiar and unitary sense of the many motions at issue. Descartes knew that the Sun was by far the largest of the bodies in the Solar System, and so it was only natural that it be stable at the center of all this rotating fluid. And he also knew (from Galileo's sunspot observations) that the Sun itself rotated, in the same direction as the planets, only faster than any of them, as befits the very center of a whirlpool. The secondary motion of the Moon about the Earth, and of the Jovian satellites about Jupiter, were clearly small subsidiary whirlpools being carried around within the larger. The axial rotations of both the Earth and Jupiter matched the direction of rotation of the tiny moons at issue, and they matched the rotational direction of the Solar Vortex as well. Altogether, it was a compelling interpretation of the substances, forces, and observable motions involved. Once again, a puzzling phenomenon was cognized as an unusual instance of something familiar.

And once again, the interpretation was false. Or, at least, Isaac Newton came up with a much better one. Instead of interpreting the Moon's circular motion around the Earth as the consequence of the Moon's being carried around by a swirling liquid medium, Newton saw the

Moon as being more like a circling stone at the end of a centrally attached string, where the force of the Earth's gravity played the role of the endlessly tugging string. The Moon's motion was therefore an instance of a *body endlessly falling* towards the Earth. The combination of *a)* this steady accelerated motion Earthward, plus *b)* the Moon's tangential, straight-line, inertial motion away from the Earth, produces the roughly circular orbit we observe (fig. 15). What would otherwise be a uniform, straight-line motion for the Moon gets continually deformed into a closed elliptical path by the centrally-directed force of the Earth's gravity.

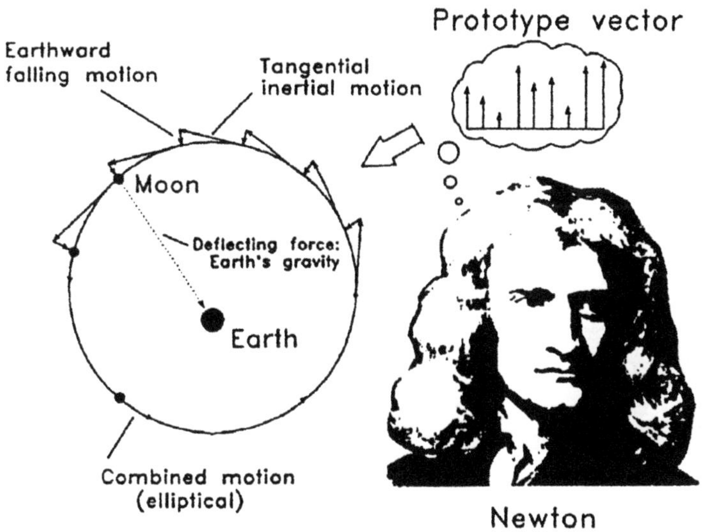

Figure 15

The same interpretation was imposed for the much larger orbits of the six known planets, only this time it was the massive Sun that provided the central attraction. The planets too were endlessly falling, *away* from their natural inertial path outwards, and *toward* the attracting Sun. With the further assumption of an inverse-square law concerning the fading *strength* of gravity as one recedes from its central source, Newton was able to account exactly for the relative revolutionary periods of the six known planets, for their elliptical shape of their orbits, and for their individual variations in orbital velocity to boot. In all, the "central force" interpretation gave us a much more detailed and accurate model of the various lunar and planetary motions. Newton's interpretation of the situation made systematic sense of even the subtlest of planetary behaviors. A body moving under exactly those influences would display exactly the sort of motions observed, at least to the limits of our ability to measure them. Once again, a familiar form of order - a form familiar to Newton, anyhow - can be seen in what is initially a puzzling diversity of planetary behaviors.

Intriguingly, this brilliant interpretation eventually proved false as well. Or at any rate, Albert Einstein came up with a still better one. The so-called "force of gravity" is an illusion, said Einstein. A planet's "curved" path in three-dimensional *space* is in reality a *straight* path (a so-called "geodesic" or "shortest path") within the non-Euclidean geometry of the *four-*

dimensional *spacetime* continuum that surrounds the "attracting" body. Since the planet's path is a straight line in four dimensions, there is no "deflection" in need of a deflecting force for its explanation. What the Sun does is not to exert a *force* on anything; rather, its great mass just deforms the local geometry of spacetime. Given that geometrical deformation, the familiar prototype of an inertial or "straight-line" path takes care of all the rest (fig. 16).

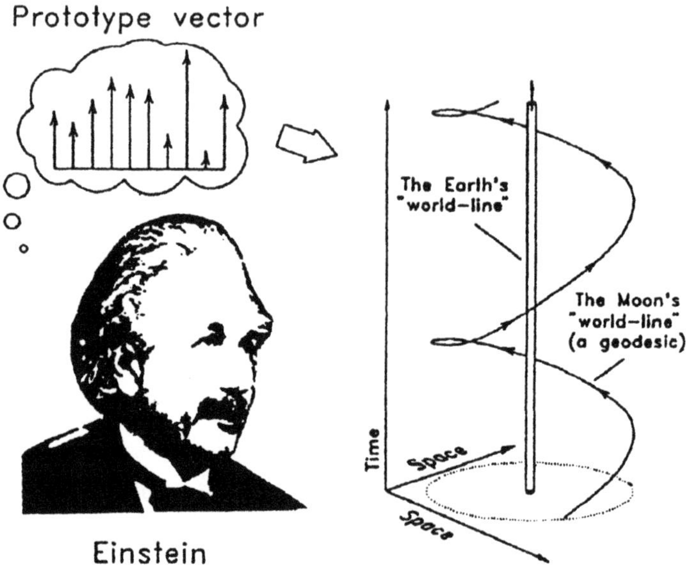

Figure 16

This prototype - geodesic paths in a non-Euclidean spacetime - is admittedly arcane to most of us. But that is beside the point here being made. It *was* a prototype with which *Einstein* was at least moderately familiar. And it was the contextual activation of that very prototype, as an interpretation of the planetary motions, that constituted Einstein's novel insight into the nature of gravitational phenomena.

That prototype involved a family of observable planetary behaviors that were closely similar to the behaviors involved in Newton's earlier prototype. It was therefore and automatically an alternative possible interpretation of the same domain of phenomena. But it also involved some subtle *deviations* from Newtonian behavior - such as the advancing major axis of any highly elliptical orbit - deviations that astronomers had already observed in Mercury's orbital behavior, and quite independently of Einstein's theoretical musings. Further divergences between the two prototypes were also investigated empirically, and it was always the Newtonian expectations that went disappointed. Once again, an ever more penetrating interpretation displaced an earlier one.

Einstein's theory of gravity has some more recent competition of its own, but I will close the series of illustrative examples here. The point of this brief and highly selective excursion into the history of science has been to portray some of the most sophisticated of our intellectual achievements as involving the very same activities of vector processing, recurrent manipulation, prototype activation, and prototype evaluation as can be found in some of the simplest of our cognitive activities, such as recognizing a dog in a low-grade photograph. What distinguishes scientific cognition is just the unusual ambition of its interpretive enterprise, the sophistication of many of the prototypes deployed, and the institutional procedures that govern the evaluation of the competing interpretations proposed. At its core, scientific cognition involves the very same cognitive mechanisms as define cognition generally. And those mechanisms, according to the present story, are precisely the mechanisms embodied in a large and highly-trained recurrent neural network.

If this really is how scientific cognition is embodied in human brains, what are the consequences for the Philosophy of Science? They are many and far-reaching, I believe, but I shall briefly explore only a handful on this occasion. The first important consequence is that there are at least two major and quite distinct kinds of learning. The first is a relatively slow process that involves many small structural changes in the brain. This is the process in which the brain's many synaptic connections are slowly modified in response to repeated experience. As those weights are modified, the activation spaces of various neuronal populations are gradually sculpted into a hierarchy of categories with prototypical cores. This is the process by which the brain's cognitive categories are formed in the first place.

The second kind of learning can happen quite quickly, on the relatively rare occasions when it does happen. It can happen on a time scale of seconds or less, and it involves little or no synaptic change at all. It is a consequence not of structural changes in the brain, but of *dynamical* changes in the internal behavior of a recurrent network. This second kind of learning is the process wherein already-learned categories are variously redeployed in increasingly esoteric domains of experience, domains distinct from the domains in which those categories were initially learned.

What drives such redeployment is the recurrent modulation of one's default or "feedforward" cognitive skills, and the occasional vector completion that finds a familiar kind of order in a novel or puzzling domain of sensory inputs. This is the process, I suggest, that accounts for what we think of as "conceptual breakthroughs" in science. It is typically, followed, of course, by an extended period of evaluation, in which the predictive, explanatory, and manipulative virtues of the newly-deployed prototype are carefully explored. This is essential, because a theoretical interpretation that seems compelling and useful at first glance may show itself to be hopelessly problematic when systematically applied to an expanded perceptual domain.

A second consequence is that, if human learning is essentially the same as that observed in the recent artificial neural network models, then the external world itself plays a robust and central role in steering cognitive development in general, and scientific development in particular. In other words, relativist worries that the development of science is or must be steered *entirely* by social factors is clearly mistaken. Neural networks can be fiercely efficient at pulling information from their sensory environment.

On the other hand, a third and cautionary consequence is that all cognitive *responses* (to raw stimulation patterns at the sensory periphery) must always be *theory-laden*. This is because any cognitive response to peripheral stimuli is always the result of passing the peripheral input activation pattern through a specific matrix or configuration of weighted synaptic connections, a configuration that carves the adjacent neuronal activation space into one set of presumptive categories to the exclusion of millions of other possible sets, a configuration that makes the network highly sensitive to certain kinds of structures and patterns in the environment at the expense of millions of other possible structures and patterns. Theory "loading" thus turns out to be, not an unwelcome blemish on an otherwise innocent process, but rather the essence of cognitive activity in the first place.

This suggests that a cognitive *pluralism* is a wise methodology for us to follow, since certain quite genuine patterns in the environment may well be utterly *invisible* to a network trained to employ one conceptual framework, while perceptually obvious to a different network trained to use a different framework. Paul Feyerabend[3] has made roughly this point in a classical setting. It is a consequence as well of the neurocomputational perspective. Which brings me to my fourth and final consequence.

I return at last to the question of truth. It connects with the preceding fairly quickly. On the neurocomputational view here presented, the principal mode of *representation* that humans and all animals use is not the sentence or the propositional attitude. It is the activation vector across a large population of neurons. And the principal mode of *computation* is not the inference from one sentence or belief to another; it is the swift transformation of one activation vector into another by means of a matrix of synaptic connections. Gaining knowledge consists in configuring your synaptic weights so that your network performs the transformations that you want, the ones that help you find your toothbrush in the morning, find your way to school, pass your algebra test, make yourself dinner, play the flute, find yourself a mate, and all of the things that make for a functional life.

Truth may be a virtue of sentences, but it is not obviously a virtue of activation vectors, or of weight configurations. These latter invite a more pragmatic dimension of evaluation, since their cognitive role is to facilitate useful anticipations of the future and useful sensorimotor coordinations. One is tempted to think of the hidden-unit partitions as "corresponding" in some way to the structure of the world, or to some part of it. But there are many ways of "corresponding" to the world, and it is not obvious that we can here demand the uniqueness that the classical notion of truth implies.

A conceptual framework is a partition across your activation spaces, or rather a set of partitions, and these can be dramatically complex, especially in the case of a brain as large as yours. Let us pursue the possibilities here. How many neurons in the brain? 10^{11}. How many connections per neuron? At least 10^3. So the total number of synaptic connections in the brain is at least 10^{14}. That is a very large number, and it is nice because we get to exploit the combinatorics of 10^{14} weighted elements, each of which can be set at . . . what? Let us say ten different values. There have to be at least two: the connection weight has to be either big or small. And probably there are no more than 100 functionally distinct values for any given weight. I guess 10 here in order to do some easy arithmetic.

How many different synaptic-weight configurations are possible for the human brain? Remember, it is the weight configurations that dictate the activation-space partitions and the resulting input-output function, so we are asking after the number of possible conceptual frameworks or functional accommodations with the world possible for the human brain. The relevant calculation is simple. There are ten possible weights for the first connection, times ten for the second connection, times ten for the third, and so on through all of the 10^{14} connections. The total number of distinct configurations, therefore, is ten to the 10^{14}th power, or $10^{100,000,000,000,000}$.

How big is this number? For comparison, recall that the total number of elementary particles in the physical universe, photons included, is only about 10^{87}. Alternatively, pick a unit of space, a cubic meter, say. That is a size you can get your arms around. How many cubic meters in the astronomical cosmos? Only about 10^{80}. Physical space may be large: it is astronomically large. But conceptual space is larger still. It is superastronomical.

But as soon as I have said this, an objection arises One might say, "Look, Paul. It is not true that changing the value of one synaptic connection will produce in me a (significantly) different conceptual scheme. In fact, you told us earlier that one could go into a network with a shotgun and blow away large numbers of its connections entirely, and yet have only a minor or a negligible effect on its activity." This is correct. The alternatives calculated above count only as "just barely discriminable" alternatives, not *significantly* different conceptual alternatives.

Very well, how many *significantly* different conceptual alternatives are possible for a human-sized network? That is hard to say, given the vagueness of the term "significant." But we can do a very rough estimate in fixing a lower bound. Consider the baby artificial networks like the one I discussed at the outset. Such research networks typically contain fewer than 1000 units, usually much fewer, closer to 100. (The reason they do not contain more is that it takes too long to train them if they are much larger.)

However, if we limit ourselves to networks of 1000 units and roughly 10^6 connections (the number of connections tends to be roughly the square of the number of units), it must be said that these easily admit of thousands of "significantly different" conceptual partitions and functional properties. There are people on both sides of the Atlantic who are training up small nets to perform all sorts of wonderful things, such as compose music, or convert graphemes into phonemes, or distinguish mines from rocks, or scrutinize loan applications to ferret out the deadbeats, or discriminate plastique explosives in airplane luggage, or play Backgammon, or coordinate jointed limbs, or recognize the curvature of surfaces, or discriminate valid inferences in the propositional calculus, or perform stereo vision, and so on. The number of significantly different partitions imposed on essentially the same physical network resource is already into four figures and is still climbing. And this variety is achieved in a network with no more than 10 to the 10^6 distinct possible weight configurations.

How then do *our* combinatorial resources compare with those of a baby network? Ours are greater by a factor of (10 to the 10^{14}) over (10 to the 10^6), which is equal to a factor of $10^{99,999,999,000,000}$. That is still a spectacularly large number, and we have to multiply it by at least 1000, which was our lower bound measure of the significant conceptual variety possible for a baby network. The resulting lower limit for our own conceptual diversity is therefore $10^{99,999,999,000,003}$. Rounding down, for the sake of being conservative, we still get $10^{10,000,000,000,000}$ significantly different conceptual alternatives possible for the human brain. Correcting for our initially inflated estimate still leaves us with a superastronomical space of conceptual possibilities.

Consider further. Suppose you had the talent to change every one of your synaptic weights to some desired new value, that is, to adopt an interestingly different conceptual scheme, every 100 msec. That is, ten times a second. So, bingo, you are Mozart. Bingo, you are Attila the Hun. Bingo, you are Virginia Wolfe. Bingo, you are Einstein. And so forth. And you did this for your whole life. How many different frameworks would you go through? You have about 10^8 seconds in a lifetime, and you would streak through 10 schemes every second. So you would traverse a total of 10^9 different schemes during your lifetime.

But recall from the last paragraph just how much space you are trying to explore. Even at the breakneck pace described, you will have explored only an infinitesimal portion of the vast conceptual space available. Evidently, exploring conceptual space with the resources we have, even if accelerated in implausible ways, is going to be about as effective as exploring the astronomical universe with nothing more than one's own two feet. It is just too large to hope to explore all of it. (Of course we can still *explore* it. And we do.)

Notice further that conceptual space will be even larger for any creature with more neurons and more synaptic connections than we have. And since there is no theoretical limit to how large a brain might be - we can always add more units - there is therefore no limit, in principle, to the size of conceptual space. It is infinite. For a given network, of course, there is always a finite conceptual space. And for a larger network, there is a yet larger, yet still finite, space for it to explore. And so on. But this progression goes on indefinitely.

From which I draw the following conclusion. It is not a demonstration, strictly, but rather an interesting possibility. Perhaps there is no unique point in human weight space, nor perhaps in any other weight space, that yields the One True Theory of Everything. In fact, when we are contemplating conceptual variety and conceptual spaces on this scale - spaces, moreover, that are indefinitely extendible - then the idea that there is a single, Uniquely Correct account of the universe starts to look . . . well, silly. Perhaps theories or conceptual frameworks just get better and better, ad infinitum. . . . That would be nice. It means we need never stop having fun.

And it also means that the cognitive adventure may be rather more like the evolutionary adventure - the *biological* adventure - than we first imagined. Rather than a convergence towards a Unique Truth, perhaps we confront an indefinitely fecund radiation. Recall that the idea of a convergence towards perfection, or towards God, was common in early discussions of biological evolution also. But this is inconsistent with Darwin's message that there is nothing teleological about the process. It is not converging on any point. It is a process of blind *radiation*, nudged by endless contingencies. And it is being snipped here and there, by sundry teeth and claws. So selection is going on. But it is a process of blind radiation to explore the many niches that are there. From the model of this paper, it rather looks as if the same thing is true of our individual and collective cognitive activity. This theme is not novel with me. Thomas Kuhn[4] and Donald Campbell[5] have been pushing it for years. But it is evident that the recent developments in computational neurobiology provide a new dimension of support for their general point of view.

Now all of this tends to put me into the same bag with the various people who are beating on the realists and trashing objectivity and so forth, and I fear that the preceding reflections may distress some of my colleagues at the conference. For I fear that I was invited here in the hope that I would Stand Up for Truth. As you see from the preceding, I feel somewhat chary of doing so. But I would like to stand up for something that may be close to it.

We have learned something that we did not know in our early history: that there is no *largest number*. The sequence of natural numbers forms an infinite set. There is no Biggest Number. Children often have a difficult time getting hold of this, at first. They think there must be a biggest number, and they are a bit surprised to learn that there is no such number. But no one has ever exclaimed, with alarm, "Oh my god, there's no biggest number! And so we can no longer tell whether any number is bigger than another number, because we cannot tell whether it is *closer* to the Biggest Number. How terrible this is!"

Such a worry would be confused. We can still make accurate comparative judgments as to whether any number is bigger than another number, even though there is no Biggest Number from which to measure their relative distance. Similarly, we can still tell whether one theory is better than another, without appeal to the respective relations they might bear to a Unique Truth. In fact, the difference in performance is *usually* dramatic, even though there is no single Best Theory. And the differences between them, in their anticipatory and coordinative properties, will be as objective as anything can be. What is free to vary is our interest in which particular anticipatory and coordinative properties to pursue.

Altogether then, the conception of cognition portrayed here today invites, I think, the following view. First, it invites a humility with regard to our current conceptual achievements. Considering the space we are exploring and the little journey we have already had, we cannot have achieved *very* much, relative to what is possible. Second, what is *possible* starts to look pretty good. There must be a lot of cognitive fun left to be had, and lots of rewards to be discovered. And finally, the idea that, cognitively speaking, things can just get better and better and better, ad infinitum, may be a more effective draw to continued intellectual activity than the idea of a Final Resting Place could ever be.

University of California at San Diego
U.S.A.

NOTES

[1] This 1995 presentation, at Parma, outlines some research already published, and some research still in progress. My thanks to the editors at the MIT Press for permission to use some material from ***The Engine of Reason, The Seat of the Soul***, by Paul M. Churchland (Cambridge, Mass., The MIT Press), 1995.

[2] Cf. Hanson 1958.

[3] Cf. Feyerabend 1978.

[4] Cf. Kuhn 1962.

[5] Cf. Campbell 1974.

REFERENCES

D. Campbell, "Evolutionary Epistemology", in P. Schilpp ed., *The Philosophy of Karl Popper*, Open Court, La Salle, 1974.

P.K. Feyerabend, "How to Be a Good Empiricist: A Plea for Tolerance in Matters Epistemological", in B. Baumrin, ed., *Philosophy of Science*, Vol. 2, pp. 3-19, Interscience Publications, New York, 1975.

N.R. Hanson, *Patterns of Discovery*, Cambridge University Press, Cambridge, 1958.

T.S. Kuhn, *The Structure of Scientific Revolutions*, University of Chicago Press, Chicago, 1962.

JEAN-PIERRE DESCLÉS

ABDUCTION AND NON-OBSERVABILITY

Some Examples from Language Science and the Cognitive Science[1]

1. INTRODUCTION

The notion of abduction, developed more than a century ago by the logician, semiotician and philosopher Charles Sanders Peirce (1839-1914)[2], is essential in epistemology and in the formation of new knowledge. However, this notion has not received, we think, enough attention from philosophers of science. Besides, this notion is still misunderstood because it is too often confused either with the deductive inference of *modus tollens* or with inductive inference. Abduction has not yet been subjected to an in-depth critical analysis in the same way as deductive inference, which has been analyzed by logic, or as inductive inference, which has been analyzed by means of statistics and probability calculus. Yet, abduction is employed in spontaneous reasoning and also in a large number of scientific reasoning (including mathematics). Man in this way has made it possible for himself to state hypotheses that refer to some not directly observable entities.[3]

These hypotheses, however, are related to some other hypotheses already accepted or directly observable facts and can be rejected or modified later on. The search for evidence reinforces the *heuristic plausibility* of the formulated hypotheses. Numerous examples of abduction can be given.

> "(...) Numberless documents and monuments refer to a conqueror called Napoleon Bonaparte. Though we have not seen the man, yet we cannot explain what we have seen, namely, all these documents and monuments, without supposing that he really existed. Hypothesis again".
>
> (Peirce, *Collected Papers*, II. 625)

A policeman's reasoning is often abductive. Taking in the observed clues, he sometimes will be led to formulate a hypothesis ("it is X that is (perhaps) guilty"); he will then try to look for some further evidence that will enable him either to reinforce or to eliminate his hypothesis. A physician proceeds often by abductive reasoning as well. Having observed certain symptoms, he formulates a hypothesis : "it is the disease M or the disease M′ that has caused the observed symptoms". To eliminate this indetermination, he will then have to rely on the recognition of some complementary evidence (for example, results of the laboratory analyses); this complementary evidence will favor the determination of the hypothesis H ("it is the disease M") by eliminating the hypothesis H′ (" it is the other disease M′") or *vice versa* or else another hypothesis

E. Agazzi and M. Pauri (eds.), The Reality of the Unobservable, 87-112.

H″ will be formulated, which will possibly need justification by still further evidence. The personal conviction in the French law is the result of abductive reasoning : on the basis of a series of consistent indices, the judges are led (perhaps in an abusive manner, according to some) to turn a hypothesis that is simply plausible into an assertive judgement; the role of the lawyer consists, among other things, in the identification of the relevant indices which would be liable to conflict with the accepted plausible hypothesis, so as to render this last less plausible in the eyes of the judges.

Independently of C.S. Peirce, the mathematician Georges Polya[4] called the mode of reasoning by abduction "heuristic syllogism" or "plausible reasoning".

> 'The indices that convince the inventor that his idea is valid, the indications that guide our actions in our daily life. (...) all these types of proofs have two essential points in common. First, they do not offer the certitude by a rigorous demonstration. Next, they are useful in the acquisition of new knowledge that is not purely mathematical or logical and for all knowledge that concerns the physical universe".
>
> (Polya, *Comment poser et résoudre un problème*: 107)

> "The heuristic reasoning (...), in one word the plausible reasoning, (...) does not try to demonstrate (...) [its] nature seems to be confusing and to be difficult to define when one looks at it from a purely demonstrative logical point of view."
>
> (Polya, *Comment poser et résoudre un problème*: 109)

When Christopher Columbus, explains G. Polya again[5], was sailing towards the West across an unknown ocean, he and his crew felt reassured when they saw some birds because, for them, they seemed to be signs that land was not far. However, these clues were not sufficient as land did not appear; other indications appeared later on: "the crew of the caravel Nina also saw signs indicating the closeness of land. One of these was a little branch covered with berries. Everyone breathed more freely and was delighted in the presence of these signs". Thus, abductive reasoning takes the following form :

When we approach land, we often see birds.	(common knowledge)
Or, we see birds	(verified fact)
So, it is more likely that we are approaching land	(plausible hypothesis)

In this paper, we will first introduce the concepts of *deduction*, *induction* and *abduction* as defined by C. S. Peirce. We will emphasize the significance of *abductive reasoning* by comparing it with both deductive reasoning, (demonstration) which aims at logical conclusions, as well as inductive reasoning, which aims at the formation of laws. After this introduction, we will explicitly distinguish the *abductive act* from *abductive reasoning*. The abductive act is the real source of new inventions. In addition, the abductive act allows one to embark on a series of partially ordered but increasingly refined hypotheses. In part three, we will present a succint analysis of *two examples* of abductive reasoning in linguistics : the discovery by Jean-François Champollion of the hieroglyphic system in Egyptian writing and the postulation by Ferdinand de Saussure of the existence of a *phoneme* in Indo-European languages without any direct trace of it in historically attested languages. These are two paradigmatic examples of abductive reasoning, and they show that certain methodological procedures in the human sciences are in fact solid.[6] In part four, we leave the history of linguistics behind, in order to study the typology of languages. *Language invariants* are abstract and non directly observable entities, but they are formed by abduction from their observable traces in different languages. The Applicative Universal Grammar model of Sebastian Konstantinevitch Shaumyan illustrates well this abductive approach. Part five will review the position occupied by abduction in the *model of cognition* which involves "internal" (symbolic and figurative) representations. These representations are not observable, but they are formed abductively by observable behaviors and by phenomenological traces - linguistic for example - of cognitive processes. The abductive procedure is not simply theoretical and uniquely constitutive of scientific theories. In fact, as we will see in the last part, natural languages encode and 'grammaticalize' certain abductive reasoning: certain linguistic configurations - simple grammatical *morphèmes* in certain languages - activate in the listener the abductive reconstruction of a fact which has not been directly verbalized by the speaker.

2. DEDUCTION, INDUCTION AND ABDUCTION ACCORDING TO PEIRCE

C. S. Peirce distinguishes three classes of inferential reasoning: *deduction*, *induction* and *retroduction*. This latter is also called *abduction* or hypothesis:

> "There are in science three fundmentally different kinds of reasoning. Deduction (called by Aristotle *συναγωγη* or *αναγωγη*), Induction (Aristotle's and Plato's *επωγωγη*) and Retroduction (Aristotle's *απαγωγη*, but misunderstood because of corrupt text and, as misunderstood, usually translated abduction). Besides these three, Analogy (Aristotle's *παραδειγμα*) combines the characters of Induction and Retroduction.
>
> (Peirce, *Collected Papers*, I 65).

To explain the differences between these three modes of reasoning, C. S. Peirce gave an example as follows (*Collected Papers*, II 623). There is a bag of beans. We take out a handful of beans and show them to a subject. There are three kinds of possible reasoning:

Deduction:

Rule:	*All the beans in this bag are white*
Observed fact:	*These beans are taken from the bag*
Deductive conclusion:	*These beans are white*

Induction:

Knowledge:	*These beans are taken from this bag*
Observed fact:	*These beans are white*
Rule: (inductive generalization)	*All of the beans in this bag are white*

Abduction:

Rule:	*All of the beans in this bag are white*
Observed fact:	*These beans are white*
Plausible hypothetis:	*These beans are taken from the bag*

Deduction is clearly expressed by the classic schema of *modus ponens* : if "p" is an accepted premise and if one also accepts the implication that "q is deducible from p" then, by *modus ponens*, "q" is deduced. The procedure of *modus tollens* takes the following inference schema : if the negation of "q" is a premise and if one has the implication "q is deducible from p", then the "negation of p" is deduced by *modus tollens*. The inference mode of *modus tollens* is derivable from the inference mode of *modus ponens* by the law of contraposition: $p \Rightarrow q \Leftrightarrow$ non(q) $\Rightarrow$ non(p). Thus, the following two schemas :

modus ponens	***modus tollens***
p is true	q is false
if p then q	if p then q
q *is true*	p *is false*

Induction generalizes: if one property is true of a sample that is judged as significant of a class, then one generalizes that the same property is true of the whole class.

> "Induction is where we generalize from a number of cases of which something is true, and infer that the same thing is true of a whole class. Or, where we find a certain thing to be true of a certain proportion of cases and infer that it is true of the same proportion of the whole class".
>
> (Peirce, *Collected Papers*, II. 624)

Induction has been taken into account mainly within the domain of statistics which must specify principles that, on the one hand, define a "significant" size of a sample and, on the other hand, generalize, for the entire population, some feature that has been observed to be present in each element of this sample. For Peirce, induction is very different from abduction (Hypothesis).

> (...) By induction, we conclude that the facts, similar to observed facts, are true in cases not examined. By hypothesis, we conclude the existence of a fact quite different from anything observed, from which, according to known laws, something observed would necessarily result.

The former, is reasoning from particulars to the general law; the latter, from effect to cause. The former classifies, the latter explains. The great difference between induction and hypothesis is, that the former infers the existence of phenomena such as we have observed in cases which are similar, while hypothesis supposes something of a different kind from what we have directly observed, and frequently something which it would be impossible for us to observe directly".

(Peirce, *Collected Papers*, II. 636)

"Hypotheses are sometimes regarded as provisional resorts, which in the progress of science are to be replaced by induction. But this is a false view of the subject. Hypothetic reasoning infers very frequently a fact not capable of direct observation. It is an hypothesis that Napoleon Bonaparte once existed. How is that hypothesis ever to be replaced by an induction?"

(Peirce, *Collected Papers*, II. 642)

"A third merit of the distinction is, that it is associated with an important psychological or rather physiological difference in the mode of apprehending facts. Induction infers a rule. Now, the belief of a rule is a habit. That a habit is a rule active in us, is evident.(...). Induction, therefore, is the logical formula which expresses the physiological process of formation of a habit. Hypothesis substitutes, for a complicated tangle of predicates attached to one subject, a single conception. Now, there is a peculiar sensation belonging to the act of thinking that each of these predicates inheres in the subject. In hypothetic inference, this complicated feeling so produced is replaced by a single feeling of a greater intensity, that belonging to the act of thinking the hypothetic conclusion (..) [The] emotion is essentially the same thing as an hypothetic inference, and every hypothetic inference involves the formation of such an emotion. We may say, therefore, that hypothesis produces the sensuous element thought, and induction the habitual element."

(Peirce, *Collected Papers*, II.643)

Abduction leads one back to the hypothesis "p" ("p is plausible") because of the proof that one has the clue "q" and knowing that one also has the implication "q is deducible from p".

"Hypothesis is where we find some very curious circumstance, which would be explained by the supposition that it was a case of a certain general rule, and there upon adopt that supposition".

(Peirce, *Collected Papers*, II. 624)

Let's compare the abductive schema ("hypothetic reasoning" according to Polya[7]) with a *modus ponens* scheme[8]:

abduction	***modus ponens***
if p then q	if p then q
q is true	p is true
p *is plausible*	q *is true*

Take notice right away that while the schemes of *modus ponens* and of *modus tollens* are logically valid, the abduction schema does not have the same *status*. In fact, *modus ponens* and *modus tollens* are two theorems (tautologies) of classical propositional logic. On the contrary, the corresponding proposition of the abductive schema is neither a theorem (tautology) of classical logic nor a theorem in any softer logic such as intuitionistic logic or minimal logic. In *modus ponens*, if "p" is asserted and if the implication is equally asserted, then the proposition "q" is also asserted, whereas in abduction, if the implication "if p then q" is asserted and if "q" is asserted, the proposition "p" is no longer deducible from these two premises; the abduction schema allows one only to think that p is *simply plausible*. So, deduction aims at the assertion of a conclusion "q" on the basis of (i), the knowledge that one possesses of a rule or of a law ("if p then q") and of (ii), a hypothesis that is supposed to be true; induction aims at the formation of a rule or of a general law on the basis of the proof that each element of the sample (supposed, incidentally, to be significant) possesses a same set of properties. Abduction aims simply at the reinforcement of a plausible hypothesis on the basis of (i), the knowledge that one has of a rule or of a law ("if p then q") and (ii), the evidence that the awaited conclusion "q" is indeed achieved. We can thus summarize that while *deduction demonstrates* that something *must be*, and while induction *shows that some law is indeed operational* and is generalizable, *abduction*, on the other hand, *simply suggests that something is a possibility*.

> "Abduction is the process of forming an explanatory hypothesis. It is the only logical operation which introduces any new idea; for induction does nothing but determine a value, and deduction merely evolves the necessary consequences of a pure hypothesis. Deduction proves that something must be; Induction shows that something is actually operative; Abduction merely suggests that something may be."
>
> (Peirce, *Collected Papers*, V· 171).

Deduction reaches a conclusion "q" from a rule and a hypothetic premise "p" that is accepted as true. Induction forms by means of generalization a rule or a law that is applicable to a population. This is done on the basis of an observed correlation between each *element* of a sample of a population and one same set of properties possessed by each one of these elements of the sample. Abduction tries to reinforce the plausibility of a hypothesis "p" on the basis of the evidence that a conclusion "q" that is deducible from this hypothesis is indeed achieved or observed.

For C. S. Peirce, abduction allows one to form new ideas by relying on the observation of evidence:

> "Presumption [abduction] is the only kind of reasoning which supplies new ideas, the only kind which is, in this sense, synthetic. Induction is justfied as a method which must in the long run lead up to the truth, and that, by a gradual modification of the actual conclusion. There is no such warrant for presumption. The hypothesis which it problematically concludes is frequently utterly wrong itself, and even the method need not ever lead to the truth (...) Its only justification is that its method is the only way in which there can be any hope of attaining a rational explanation".
>
> (Peirce, *Collected Papers*, II. 777).

Classical logic has developed the mathematical analysis of formal deductions. In an axiomatic version of propositional logic, for example that of B. Russell, the elementary schema of deduction is expressed by the achievement of the conclusion. In Gentzen's version of "natural deduction", deduction is expressed by the rule of elimination of the symbol and the implication. Classical logic enables the demonstration of theorems (that is to say, the assertable propositions demonstrated as true following a proof) which are deducible from a certain number of propositions that are provided as being true and so have been previously asserted. For its part, abduction emphasizes the conditions which enable the reinforcement of an hypothesis on the basis of a certain number of semiological clues. These clues cause us to accept the hypothesis as being increasingly more plausible.

> "Retroduction is the provisional adoption of an hypothesis, because every possible consequence of it is capable of experimental verification, so that the persevering application of the same method may be expected to reveal its disagreements with facts, if it does so disagree.
>
> (Peirce, *Collected Papers*, I. 65).

3. THE ABDUCTIVE ACT AND ABDUCTIVE REASONING

We will now show how, with a series of clues, abduction functions to increase the plausibility of an hypothesis, to reject a plausible hypothesis, and to establish a series of hypotheses with increasing plausibility.

3.1. *Evidence in favor of the Plausibility of an Hypothesis*

Think about the following abductive reasoning. Suppose we have a law that allows us to relate a conclusion C_1 to a certain hypothesis H: the conclusion C_1 must be stated, by a more or less direct observation, for example, or because the conclusion must be further testified as the direct consequence of the hypothesis H and of higher theoretical principles. We now have the following propositional implication: "if H then C_1". Furthermore, one does observe the conclusion C_1: either this conclusion is the result of a more or less direct observation, or it is an already well-known fact that has been incorporated into the set of theoretical propositions. Therefore, one infers by means of abduction from this evidence and from the knowledge of the law that the hypothesis H becomes plausible to a certain degree, "degree i", with the following schema :

if H then C_1
C_1 is observed

H *is plausible "to the degree i"*

Now suppose that with the same hypothesis H one can expect to obtain another conclusion C_2 (observed fact or a fact theoretically deducible from H). When one actually observes C_2, we are likely to think, by abduction, that the hypothesis H is plausible to a degree j :

if H then C_2
C_2 is observed

H *is plausible "to the degree j"*

Thus, with the two successive observed facts C_1 and C_2, the plausibility of the hypothesis H increases. The degree j of the plausibility of H becomes higher (in a broad sense) than the previous degree i. As the number of observed results expected from the hypothesis H increase, the plausibility of the hypothesis H increases as well. Generally speaking, with a series of clues $C_1, C_2, ..., C_n$, all of which could be deduced from the same hypothesis H, the plausibility of this hypothesis H can be increased to the extent that, sometimes, it may be accepted as true, when these clues are actually observed. We state therefore that the sequence of clues $\{C_1, C_2, ...\}$ leads by abduction to the plausibility of H.

Abduction is often being conducted in mathematics when forming conjectures. In fact, a conjecture is merely a plausible proposition, it has not (yet) been demonstrated. In the eyes of the mathematician, a conjecture can seem to be plausible when certain clues (which on the contrary are demonstrated mathematical facts) are present. These clues reinforce the belief of the mathematician that the conjecture is highly plausible. And very often, when these clues are displayed systematically, they are so "significant"[9] that they lead to the conviction of other specialists. When a conjecture E is proposed, one shows that if E is accepted as being true, one can then also deduce by explicit demonstrations that the propositions $T_1,..., T_p$ are true (these are theorems). So, we have the following implications: if E then T_1, ..., if E then T_p. Since the propositions $T_1, ..., T_p$ are theorems which have been demonstrated without using the hypothesis E, *these theorems are the clues* which reinforce the plausibility of the conjecture E. Nevertheless, nothing prevents mathematicians from trying to demonstrate, by a demonstrative proof, that the conjecture E is indeed a theorem. Thus this conjecture becomes a true theorem and is no longer a highly plausible proposition. The plausible conjecture can often assist the organization of a set of mathematical propositions into a coherent whole.

3.2. *Refutation of a Plausible Hypothesis*

Supposing now that one has formulated the hypothesis H in the form of a proposition, and that one knows that if the hypothesis is verified then one must necessarily deduce certain conclusions from this hypothesis, for example in the form of observable facts (obtained with the protocol statements destined to manifest these observable facts) or in the form of theoretic propositions. We therefore have a tendency to formulate a proposition C of the following kind: "the expected result has indeed been realized in the protocol statement that was supposed to manifest it" or "the theoretic fact that has, furthermore, been testified". We define the relationship between H and C using the following implication: "if H then C". When the expected result has not been realized, one can reject the hypothesis H, according to *modus tollens*.

Suppose that from a hypothesis H, one should compulsorily obtain the results $C_1, C_2, ..., C_n$ and C_{n+1}. With that, one indeed obtains $C_1, C_2, ..., C_n$ but, on the other hand, (i) the expected result C_{n+1} is absent (as a result of, for example, an experimental setting to produce this result C_{n+1}) and (ii) the result C'_{n+1} is obtained without necessarily the prior implication "if H then C'_{n+1}. In this situation, the hypothesis H must be abandoned according to the schema (of refutation):

if H then C_1

----

if H then C_n
if H then C_{n+1}
(if H then C'_{n+1}) is not true
C_1, ...C_n are effectively obtained
BUT C_{n+1} is not obtained
C'_{n+1} is obtained

---

the hypothesis H is no longer plausible
it should be reformulated

In this case, one must either suddenly reject the hypothesis H, or better still, try to modify it, so that one formulates *another hypothesis* H' from which one can deduce the results C_1, C_2, ..., C_n, C'_{n+1} but not necessarily C_{n+1}.

It is evident that the rejection of an hypothesis should not be made blindly. Here is an example[10]: Clairaut was studying lunar movement. After having made in his calculations some simplifications that he believed to be permissible, he came to a conclusion which was different from what had been observed for the lunar movement at apogee (observation of an inconsistent consequence, contrary to what we could expect). The speed of this movement was two times smaller than the observed speed. As a result, he started having doubts about Newton's law of gravitation (rejection of an hypothesis) and he wanted to find a term with which he could remedy (refined reformulation of the hypothesis) by adding a supplementary term k'/r^4 to the term k/r^2. However, this modification attracted quite a lot of criticism. In order to account for this additional term, he reexamined the approximations that he had accepted in his calculations and he had to conclude that his term of remedy was not valid (1749) : Newton's law should not be modified...

Generally speaking, if one obtains C'_{n+1} instead of the expected result C_{n+1}, then one sets a new abductive schema with the hypothesis H' which adjusts the initial hypothesis H :

if H' then C_1

----

if H' then C_n
if H' then C_{n+1}
(if H' then C_{n+1}) is not true
C_1, ..., C_n and C'_{n+1} are in fact obtained

---

the hypothesis H' is plausible

The new hypothesis H' now becomes more plausible than the initial hypothesis H, not only because H' can account for C_1, ..., C_n but also because it does not imply the result C_{n+1} which has not been observed. In this way, abductive reasoning enables one to formulate a partially ordered series of increasingly refined hypotheses $\{H_0, H_1, ..., H_q\}$, the plausibility of which increases. This is achieved thanks to a sequence of clues $\{C_1, C_2, ..., C_n, ...\}$.

A well-known example is the reformulation of an initial hypothesis in John Kepler's work in astronomy. Kepler had a mystical conception of the organization of the solar system: he believed that the words of the Creator of the universe were meant to be expressed in terms of exact numeric relations, in the same way as Plato's regular polyhedra[11] were expressed. Thus, Kepler was engaged in an analysis of the observation made by Tycho Brahe in order to prove that his hypothesis on the harmony of the solar system was verifiable: "the different distances between the orbits are in a numerically exact relationship" (We will call this hypothesis "hypothesis H_0"). According to Kepler, the distances between the spherical surfaces on which the six planets Saturn, Jupiter, Mars, Earth, Venus and Mercury revolved around the sun were such that between two orbits one could harmonically build one of the five polyhedra of Plato. However, benefitting from the very precise observations of Tycho Brahe, Kepler recalculated the orbit of Mars which did not correspond with the expected regular orbit ("C_{n+1}"): the new observations thus no longer fit in the hypothesis H_0 of regular orbits. Kepler then had to change his judgement to account for the new data ("C'_{n+1}") by renouncing his initial hypothesis and by formulating other hypotheses[12] in his *Nouvelle Astronomie* (1609): "Mars follows an ellipse which is almost circular around the sun which is situated at one of the centres".

3.3. *The Problem of Bootstrapping: the Abductive Act*

The great difficulty in abductive reasoning is obviously *the problem of bootstrapping*: how does one *find, formulate, and set up the hypothesis H_0*, the initial hypothesis in an increasingly plausible chain of hypotheses? Consequently, it is the bootstrap that contains the true innovation and the "good idea", "the sparkling light" in Polya's words. The refinement of different hypotheses can afterwards become a systematic process and can be handled methodically.[13] This bootstrapping is quite well-known among computer scientists[14] and artificial intelligence specialists[15], especially those concerned with the problem of knowledge acquisition.

Hence a difference is to be made between the *abductive act* and *abductive reasoning*. The abductive act includes *inventing and bootstrapping the abductive process* by setting up and formulating with precision an initial hypothesis H_0 (or a limited set of mutually compatible initial hypotheses). This is normally done by relying on clues and various sporadic symptoms. Following this, *abductive reasoning* includes rationally organizing a set of plausible hypotheses (often already partially ordered), eliminating by refutation those hypotheses judged as incompatible with the observed facts, so that a set of coherent hypotheses with a very high plausibility can be established by successive approximations of the initial and rejected hypotheses. In order to discover the way of functioning and the unknown characteristics of a complex system (for example the system of planetary orbits of the solar system or the system of the hieroglyphic writing of ancient Egyptian), we must at least begin with an hypothesis that is not too far removed from its actual function. The hypothesis should be such that, by successive steps one can come closer and closer to this "discovery". Some hypotheses are fertile and indeed they bootstrap the abductive process, whereas others can block the process and divert the researcher from the "right" direction which would otherwise direct him to some very plausible hypotheses.[16]

Often, a large number of facts can be organized starting from an explanatory hypothesis H_0, but some facts, despite being similar, are "resistant" to the explanation provided by H_0. For example, a researcher can have the idea to postulate a non-observed entity *E in such a way as to formulate a new explanatory hypothesis H'_1, which would enable one to organize, within the same paradigm, not only the already explained facts but also previously irreducible facts.[17] In this way, the postulated entity *E formed by abduction, together with other entities which possess, on the contrary, a well-established *status*, reveals its role by its consequences but not in a direct way. Abduction justifies the majority of theoretical projects which rely on empirical materials and which postulate non-observable hypothetical entities. Abduction helps to link up, by reason, the more or less directly observable empirical to the non-directly observable and organizable, theoretical, hypothetical and sometimes explanatory constructs of the empirical project under research.

4. EXAMPLES OF ABDUCTIVE REASONING IN LINGUISTICS

Two examples of "abductive reconstructions" will be given here. The soundness of these two reconstructions has been confirmed by the later discoveries of linguistic materials, and this kind of discovery is often accidental.

4.1. *Jean-François Champollion*

The work of deciphering Egyptian hieroglyphics carried out by Jean-François Champollion in 1822 is an example of how abductive method was employed starting from the linguistic data available on the famous Rosetta stone. The discovery of Champollion resides in the recognition that the hieroglyphic writing system is composed of three overlapping systems : a system of signs of ideograms, a system of phonetic signs and a system of "determinative" signs.

> "This is a complex system of writing which is simultaneously figurative, symbolic and phonetic in a same text, a same sentence, and I may say, almost in a same word."
>
> (Champollion : *Livre du système des anciens hiéroglyphes Ègyptiens*, 1824).

This analysis of the complex system of ancient Egyptian hieroglyphic writing has been achieved by a series of successive and increasingly refined and precise hypotheses, some of which had already been confirmed by the "forerunners" (among them was the famous physicist Thomas Young). Let's look at some of the stages in this discovery procedure[18]:

- Sylvestre de Sacy isolates the demotic inscriptions on the Rosetta stone into groups of signs, each corresponding to the Greek proper names and to the Greek text. The Swede Akerblad assigns a phonetic value to these groups of signs.
- Champollion, by studying a copy of the demotic inscription of the Rosetta stone, arrives at the hypothesis that the hieroglyphics must have the capacity of representing sounds (1810). Later on he has the idea that the Egyptians, as is common in the semitic languages, neglect vowels to a large extent, and very often don't write the vowels in their writing system (1814).
- Thomas Young publishes a table of 220 "words" or groups of hieroglyphic signs. He senses that phonetical hieroglyphs and the interdependence of the three Egyptian writing systems exist, yet he has not succeeded in discovering the inner structure of this writing system (1818/19).

- For Champollion, the demotic characters are only the final deterioration of the original signs. The three Egyptian writing systems originate from a single system (1821).

- In counting the hieroglyphical signs and the corresponding Greek words on the Rosetta stone, Champollion finds 1.419 hieroglyphs for 486 Greek words: so it is impossible that each hieroglyph represents an idea (1821) ; furthermore, as these 1.419 hieroglyphs can be reduced to 166 basic figures, it is also difficult to suggest that each hieroglyphical sign is a sound (1821).

- Like Young, Champollion undertook the work of phonetic analysis of the cartouche of Ptolemy on the Rosetta stone and succeeded in assigning the phonetic value of seven letters from the Coptic name P.T.O.L.M.I.S to seven hieroglyphical signs. He continues this work on a copy of the name of Cleopatra, K.L.E.O.P.A.T.R.A in coptic. He noticed that the three hieroglyphical signs O, P, L and the corresponding sounds are the same in the two cartouches. It seems then that *the Greek names have been transcribed into the phonetical hieroglyphs.*

- On the 14 September 1822, Champollion received from Nicolas Huyot two cartouches from the Egyptian royal dynasty, one of Ramsès and the other of Thoutmès. He arrived at the hypothesis that the phonetical hieroglyphs are related to ideographical signs. He proposed a principle of Egyptian writing : The same writing can sometimes represent ideas and sometimes represent sounds.

> "One can in no way suppose that the Egyptian phonetic writing is a system as fixed and invariable as our alphabets. The Egyptians were used to directly representing their ideas. The expression by means of sounds is only an auxiliary tool in their ideographical writing. And when the need to use the vocal expression arose more frequently, they thought of expanding their means to express sounds, but they did not, for this reason, renounce their ideographic systems to which they were dedicated by religion and the tradition of their continuous use during many centuries..."
>
> (Champollion: *lettre à Mr. Dacier*, 27 september 1822)

We stop for a second to take a closer look at this progression of 14 September 1822, a critical step in the decipherment and the understanding of the Egyptian writing, because it exemplifies a perfect abductive approach. The biographer Jean Lacouture has described for us this famous morning:

> "He [Jean-François Champollion] leans over the first of the Huyot documents. A cartouche obviously signifying the name of a sovereign appears. Jean-François stares at it with eyes wide open : what he sees is a group of signs, the first of which is a type of solar disc, in reddish ocher, followed by a sort of letter with three downstrokes mounted on a rounded T-shaped ear (like the one that completes the cartouche of Ptolemy, taking the value of an s there). (...) It is not difficult for him to assign the value s or se or ses to the double sign at the right (the hieroglyphs, he knows, can be read either from left to right or from right to left). And how about this disc in reddish ocher looking like a sun situated on the left? A sun? This is pronounced as Ra in Coptic, but what does it mean? If the double final signs read as s or sès, and if the first signifies Ra, it would suffice for the middle character, the tripod, to take on the value of m, and he, Jean-François - he holds his breath - would be in front of the cartouche of the most renowned of the Pharaohs, Ramsès, perhaps even the most eminent of all, Ramsès II... (...) There, before his eyes, (...) is the demonstration that his approach, which he believed to apply only to the names of foreign and late sovereigns, would be operative also in reference to the greatest period, the great dynasties of Egyptian antiquity. (...)".
>
> (Jean Lacouture, *Champollion, une vie de lumière*, pp. 453-54)

It is necessary that this hypothesis, which took its form in the mind of Jean-François, be confirmed by further data.

> "He takes up a second piece of the Huyot documents. Here we have another cartouche containing a not yet known group of signs, but again different from the late names that by now have become familiar to him (...). On the left, there is an ibis. Then, the same tripod sign and bristling, recognized in the cartouche that he just attributed to Ramsès, and because of this he knows of its value M. Finally, there is a sort of curved T-shape which, upon reflection, would read as S in the cartouches of the Greek-Latin sovereigns and which hypothetically reads S in that of Ramsès just scrutinized. But the ibis; this is Thot, the God of Thot, the sage with foresight, the very inventor of the hieroglyphs ! And if Thot is a possible reading, and if the second sign would mean m, and if the last is understood as s or ès, it is then the great Thoutmès (or Thoutmosis), perhaps the third, the founder, the conqueror (...). What would make the separate reading of Ramsès and Thoutmès uncertain and haphazard would be reduced to a minimum if one put together the series of hypotheses, each one lending weight to the other (...). He is sure to be in front of a triple but not double system, with the hieroglyphs signifying at the same time and in turns both sounds and ideas, these latter being either properly or figuratively represented. Ra and Thot, sun and ibis, are looking at each other whereas M and S, alphabetic signs, are listening to each other. Other than that, the MS group covers a meaning : "have delivered". The whole system is already there."
>
> (Jean Lacouture, *Champollion, une vie de lumière*, pp. 454-55)

Jean-François Champollion rushed to his brother Jacques-Joseph's home and cried out: "Je tiens l'affaire !"[19]. Later he presented his discovery in a lecture on Egyptian hieroglyphs in front of the Académie royale des Belles Lettres, the 27 September 1822. The reconstruction, in the final step of the discovery, of the decipherment shows perfectly the abductive method of the discovery. In fact, after a succession of increasingly refined hypotheses (and of rejected wrong hypotheses[20]). Champollion reached a decisive moment. From the new documents (the Huyot document), he extended the phonetic principle which functions in the writing of the names of Greek kings to the names of sovereigns of the high Egyptian empire. With this abductive act, he discovered, by applying his hypothesis to other names, the triple system of writing (phonetic, ideographic and symbolic) of the Egyptian hieroglyphs. There is a big difference[21] between the method of Young and Champollion's really abductive method. The former procedes with a series of lucky conjectural identifications, the latter establishes the linguistic system of writing by hieroglyphs, generalizable to the entirety of data[22]. The correctness of the decipherment carried out by Champollion was confirmed permanently in 1866 when his disciple, P. Lepsius, discovered in San, Egypte, a stone in three languages (the decree Kanope): by transposing the Egyptian text into Greek, according to Champollion's model, he obtained a translation which was in agreement with the Greek original.

4.2. *Ferdinand de Saussure*

With his *Thesis on the Primitive System of vowels in Indo-european languages* (Leipzig, 1879), F. de Saussure, at the age of 21, first basked in glory. His second great feat was his famous *Cours de Linguistique Générale*, which he did not write himself as it was drawn up by three of his students. On the basis of his observations on the vocalic alternations in the observed Indo-European languages (Greek, Sanskrit, Latin, Germanic languages...), Saussure was able to formulate an abstract hypothesis ("a certain phoneme, not attested in any language, exists in the Indo-European languages") which makes it possible to arrange the known and observed facts in different attested languages[23]. One can note these observations in the following examples from the Greek language (table 1)[24]:

PRESENT	PERFECT	AORIST
e	*o*	-
petomai ("I fly")	*pepotemai*	*eptomen*
ei	*oi*	*i*
peitho- ("I convince")	*pepoitha*	*epiton*
er	*or*	*r*
derkomai ("I look")	*déedorka*	*edrakon*

Table 1

In these three cases, one finds the Indo-European alternation *e/o/-*; besides, the vowel can be complicated by the sounds *i, u, l, m, n, r*. The alternation is realized in two ways: 1) qualitative alternation: front vowel - back vowel (e/o); 2) quantitative alternation: full vowel / empty vowel. Yet, this excellent table cannot stand in presence of the following alternations (less frequent) in which the long vowels appear and the "vowel - empty" alternation disappears (table 2).

in Greek :		
a	*o̲*	*a*
phami ("I speak")	*phonè ("voice")*	*phatos ("said")*
histami ("I put")		*statos ("put")*
In Latin :		
dare ("to give")	*do̲num ("gift")*	*datus ("given")*
stare ("to stand")		*status ("standing out")*

Table 2

To account for this irregularity, more or less similar principles have been proposed. On the contrary, F. de Saussure maintained that the alternation *e/o/-* was the most common initial form. Having asserted this, he then admitted that in the group *a/o/a*, the long vowels were formed as the result of a contraction of a short radical vowel *with an unknown sound*, unknown because no historically attested languages showed any trace of it (at least not at the time). This sound, which we represent by the symbol *A, must possess, according to Saussure, the properties of a voiced glottal sound (Saussure called it a "voiced coefficient"); its function was to transform the radical vowel into one of the back vowels of the type [a] or [o]. If found in a non-stressed position, the radical vowel would disappear (empty position in the alternation), while the glottal, which represents a voiced sound as a hypothesis, has the trace of a or i . One can present this process of transformation in a very simplified version as follows :

phe+*A+mi	→	pham	ph+*A+tos	→	phatos
histe+*A+mi	→	histami	st+*A+tos	→	statos
pho+*A+ne	→	phone	st+*A+tus	→	status
do+*A+num	→	donum	d+*A+tus	→	datus

F. de Saussure wanted to establish an hypothesis: "In Indo-European languages there exists a certain phoneme, *A, which is a voiced glottal, with an empty position in actual languages". The phoneme *A has been postulated with an abductive act. Its relationship to empirically observable elements is justified by an abductive reasoning which reaches a high plausibility for this voiced glottal. Thus, the "unexpected" observations have been accounted for by the hypothetical existence of *A which serves to rationalize the observation. According to the judgement of the great Danish linguist L. Hjelmslev, "the advantage of such an analysis over the classical analysis resides, in the first place, in the fact that it provides a most simple solution to the problem by eliminating the so-called long-vowels from the system; also, this analysis enables the achievement of a complete analogy with the vocal alternations which have been considered up to now as something fundamentally different...". This analysis of Saussure was quickly accepted by the community of Indo-European linguists (despite some reservations) and it won him a famous reputation. However, that's not the end of the story. In 1927, after the decipherment of Hittite, the hypothetic entity *A was empirically attested. F. Kurilowicz has shown that a Hittite phoneme /h/ is the vestige (a retroflex) of Saussure's voiced glottal *A in the Indo-European languages. He presented the following parallelism which confirmed exactly Saussure's theoretical point of view :

Latin:	*pascunt* ("they defend")
west Tocharian:	*pasken* ("they defend")
Hittite:	*pahsanzi* ("they defend")
Latin:	*novare*
Hittite:	*newahh* ("innovate")
Greek:	*laos* ("army")
Hittite:	*lahha* ("war")

Saussure's abductive analysis is obviously not a prediction, but its correctness has been confirmed in that this abductively postulated entity has been identified by one of the phonemes in Hittite. Here we have an extraordinary moment in the epistemology and the philosophy of Human Science, a victory as glorious as that of the astronomer Le Verrier when he pointed his telescope at the planet Neptune which had not yet been discovered other than by his own calculations. Saussure's analysis shows the absolute maturity of the reconstruction methods employed by the Indo-European comparatists.

5. ABDUCTIVE CONSTRUCTIONS OF THE NON-DIRECTLY OBSERVABLE LANGUAGE CONSTANTS

The abductive method used by the comparatists in their diachronic reconstruction of past linguistic states (such as Proto-Indo-European or Proto-Semitic) which are often hypothetic as no documents that attest them exist, is now being continuously used by contemporary linguists in their synchronic reconstruction of abstract and unobservable language constants on the basis of variations observed in the attested languages. Structural linguistics has shown us how to construct and manipulate non-directly observable significative entities. For example, the well-known distinction between phonology and phonetics is made because of a difference between the non-directly observable phonic entities and the directly observable sounds[25]. A phoneme, a unit of phonology, is an abstract unit and is physically realized by different sounds according to the speakers; the sounds are the objects of studies in phonetics. A phoneme[26] cannot be established as pertinent unless it is in distinctive opposition with other phonemes of the "system" in a same language :

> "By (directly or indirectly) phonological or distinctive opposition we thus understand any phonic opposition capable of differentiating lexical meaning in a given language. Each member of such an opposition is a phonological (or distinctive) unit".
>
> (Trubetzkoy, 1969 : 33-34 , quoted in Shaumyan, 1987: 63)

Thus, in French, the phonemes /b/ and /p/ are in opposition because either can make the difference between two different linguistically significant units /bas/ and /pas/. In addition, there are many different realizations of the phoneme /b/ or of the phoneme /p/, according to whether the speaker is an adult or a child, a man or a woman; one's dialectical origin matters as well (Parisian French is different from the French in Toulouse or in Montréal).

This distinction between phonology and phonetics has inspired the linguist S. K. Shaumyan to differentiate two levels of analysis in languages. The *phenotype level* on which the language structures (phonic, morphological, syntactic and grammatical) are more directly observable. Each phenotype language (Chinese, English, Portuguese, French, Bulgarian, Caxinawa, Waianpi...) is described by a particular phenotype grammar, which describes the more specific features of these languages. The *genotype language*, according to the Shaumyanís model, is realized or encoded[27] in the different linguistic structures (morphological and syntactic) of the phenotype languages, it is described by a genotype grammar.

> Genotype grammar comprises functional units - predicates, terms, modifiers - and abstract operand-operator relations between these units. Phenotype grammar comprises syntagmatic units - morphemes and words- and connections between them in terms of linear order and their morphological properties. (...) The rules of genotype grammar are invariant with respect to various possibilities of their realization by phenotype grammar. (...) The distinction between the genotype and phenotype levels is of paramount importance for linguistic theory, because this distinction puts us on the road to the solution of the basic question of linguistic theory (...): What factors contribute to the similarities and differences between natural languages?"
>
> (Shaumyan, *A Semiotic Theory of Language*, 1987: 97)

One of the objectives of the linguistic typology is to exhibit the constants in terms of observed variations across different languages. Such constants are the abstract abductive constructions designed by the linguist. They are the foundations of an organization of an entire set of different constructions according to languages. For instance, there is the group of *accusative languages* (such as Indo-European languages), the group of *ergative languages* (such as Basque or Georgian...) and the group of *active/inactive* (such as the Guarani, an Amerindian language). These languages have quite different grammatical organizations in diatheses. Yet, one can argue that certain invariant structures transcend the diversity of the types of the observed languages[28]. These invariant structures are abductively constructed and belong to the genotype language. The actual realizations of these invariant structures depend on the different typological constructions of a phenotype language group.

Let us now look briefly at an example of how a language constant is expressed in the phenotype language. It is known that the copula "BE" (expressed in English by is) in the Indo-European languages assumes several different semantic values : (1) value of identification (*Beijing is the capital of continental China*); (2) value of location (*Beijing is in China; China is in Asia*); (3) value of attribution (*Beijing is a gigantic city*); (4) value of inclusion (*The Chinese are Orientals*); (5) value of membership (*Beijing is an urban metropolis*). These five values taken by the copula "BE" are not constants according to languages. Some languages have completely different linguistic forms[29] to express identification on one hand and location and inclusion on the other. Besides, other semantic values (like possession, the mereologic relations between the parts and the whole...) are not expressed by the copula "BE" (*est* in French, *is* in English) but by other expressions like "HAVE" (*John has a car*) "is a part of" (*the finger is a part of the hand*), "is composed of" (*water is composed of hydrogen and oxygen*), "include" (*a meal includes a meat dish and a dessert plus a drink*). Finally, other ways of using the copulas "BE" and "HAVE" are not related to these values because in Indo-European languages, "BE" as well as "HAVE" are used as operators of grammatical auxiliaries (in the constructions of passive and in the temporal resulting constructions of the perfect). Faced with the diversity of linguistic expressions of semantic values (1), (2), (3), (4), and (5) and the differently categorized related values in different languages, an abstract schema can be legitimately given by an abductive procedure. This schema is certainly not directly observable in any particuliar natural language, but it is capable of organizing all these semantic values in one sole network in which these different values of the schema are expressed by the specific labels of each language. The linguist A. Culioli[30] has been doing research on such an abstract schema which he calls the "schema of localization" ("the unit X is located in its relation to the unit Y"). On the basis of his own observation of the analyses of E. Benveniste[31] and that of other linguists[32], Culioli has formulated his schema, by means of abduction, according to the functions of the copula "BE" and its "equivalents" in numerous languages. This abstract schema of localization is in fact anchored in the perception of more or less salient forms where there exists an asymmetry between "a located unit", therefore less referentially determined, and "a locator", therefore referentially better determined. One can thus study the formal properties of such an asymmetric schema of marking and from this can deduce the different specifications associated with the observed semantic values, in a way so as to organize these semantic values in a network of more or less specific values[33]. This schema of localization is an hypothetic construal derived by an abductive approach. It has the status of a plausible language constant that all natural languages must necessarily realize, i.e encode by means of more or less specific categorizations and syntactic constructions. Its function is obviously not directly observable in any particular language. According to our knowledge, no natural language directly encodes the abstract potentiality of the schema of localization by a unique linguistic label which directly expresses all the semantic values specifying the localization[34]. In designing a schema like this, one can however hope to be able to compare, by means of this same abstract standard and of the network that the schema creates, the directly incomaprable and heterogeneous linguistic expressions which are the directly observable constructions. It is in analyzing the linguistic expressions by means of the structured network of semantic values which specify the abstract schema of localization in which one can locate exactly (in the network) the value of each expression related to "BE" and "HAVE". Without this schema of common reference, it would remain impossible to reduce one expression to another and each would be limited to the phenotype languages of which they are the components.

6. ABDUCTIVE CONSTRUCTIONS OF MENTAL REPRESENTATIONS IN COGNITIVE SCIENCE

Behaviorist psychology, in abandoning introspection, turned to the sole observation of behaviors, apprehended by the associations between stimuli and directly observable responses. The observer, as the theorist, in analyzing the behavior of men or of animals, forbade himself to provide any explanation which might imply any hypotheses invoking internal mental structures. The only possibilities were then the stimuli and the responses obtained by their associations in a "black box" which is impossible to open; all that counted were the observable associations between "input data" and "output data" of the black box. Many important criticisms have been raised against the behaviorist psychology. As a consequence, cognitive psychology was developed in order to "explain" and to "account for" certain complex behaviors like perception, action action in regard to the environment, speaking, memory, reasoning, planning...; the psychologists were authorized not to open the "black box" but to postulate "intermediary representations" between the input stimuli and the output responses. More generally, cognitive science has become an interdisciplinary subject, whose goal is to study simultaneously the aptitude and the cognitive capacity of human beings to, for example, detect visual and acoustic signals, recognize faces, talk in a language or reason, and also to all "intelligent systems", be they natural (human beings, animals...) or artificial (cognitive machines[35]). A large number of experiments carried out in the analysis of, for example, the comprehension of linguistic utterances, in the analysis of gestures, the analysis of perspective identification, necessitate a real postulate of "internal symbolic representations[36]" (which means the contents of the "black box"); some of these representations have been qualified as "mental".

Cognitivism, at least in its most classical form, presupposes the existence of unobservable internal representations[37]. Since the psychologist or the linguist cannot enter "into the heads" of subjects, they are therefore to postulate, by an abductive approach, the internal symbolic representations which can to some extent account for, if not explain, or at least model, the observable behavior.

In such a framework, the vocabulary can be represented in the form of a network of intentional concepts (a semantic network) to which a subject has more or less rapid access (Quillian's model). This network is organized, according to the work of E. Rösch, with the "base levels" and the instances more or less representative (typical) of a concept. This network of concepts has the ability to organize the empirical instances which are more or less well identified into extensional classes. This organization procedes in chains of ressemblance (in the sense of Wittgenstein's "family ressemblance") to prototypes. Each prototype is an element which represents "well enough", the entire category as a particular empirical instance. Some of the experimental techniques, like semantic triggering, lead to building and structuring this network. Note well that this network is a model of conceptual knowledge and that it is a construal in a real abductive approach.

In a general way, cognitivism aims at constructing the internal representations (qualified as mental) and the internal procedures which transform one representation into another. These representations and these procedures, which are not directly observable, are designed on the basis of the observable behaviors. However, in postulating such representations, it becomes possible to organize and to model the observable behaviors which manifest them; without these representations, the observed behaviors have "no sense". Thus, abductive abstraction lends meaning to the observable behaviors, at least for psychologists.

Acquisition often procedes by abductive reasoning. Take the example of the categorizations obtained by constructing embedded extensional classes. A certain number of objects $O_1, ..., O_n$ are classified into one same class K_1, because they share the same intentional property P_1 - or concept[38] - which is constitutive of the class K_1 (in this case, K_1 is the extension of the property P_1). After having done this, one can try to "add" other "objects which resemble" but which " do not exactly verify" the property P_1. On the basis of these new instances, one constructs a new property P_2, "slighty different" from P_1: P_2 is constitutive of another class K_2 which is broader than K_1. This can be illustrated by the following two-steps[39] schema:

Step 1 : Let the already constituted $K_1 = \{O_1, O_2, ..., O_n\}$ be an extensional class, so that the property P_1 applies to all the objects $O_1, O_2, ..., O_n$, in which case: $P_1(O_1) = P_1(O_2) = ... = P_n(O_n) =$ "true". Assume that one has the objects $O'_1, ..., O'_p$ which "resemble" the objects of K_1 but such that the property P_1 no longer applies exactly to them, i.e $P_1(O'_1) = ... = P_1(O'_n) =$ "false".

Step 2 : Formulation, by abduction, of the new property P_2 - new concept - which is different from P_1 and such that: $P_2(O_1) = ... = P_2(O_n) = P_2(O'_1) = ... = P_2(O'_p) =$ "true", hence the new extensional class $K_2 = \{O_1, ..., O_n, O'_1, ...O'_p\}$ which comprises K_1.

7. LANGUAGES ENCODE AND GRAMMATICALIZE ABDUCTIVE REASONING

It is accepted that natural languages are systems of highly structured symbolic representations. Natural languages serve the purpose of expressing thought (old conception of the language), communicating and representing by symbolic configurations ó phonic or graphic ó the perceived external environment. Natural languages enable the establishement of real dialogues between speaking subjects, in which the speaking subjects construct progressively and dynamically common referential spaces. To carry out these representative, communicative and conversational functions, the linguistic units, especially the grammatical units, encode not only stable entities, normally expressed by nouns, but also operations (like the operations of predication, of determination, of topicalization, of wording...) and the oppositions of semantic categories for which these linguistic units are the labels. Natural languages are more significant than their encoding by symbolic complex arrangements of stable entities, operations and abstract semantic categorizations: *languages are the semiotic base of verbalized reasoning*. In fact, certain grammatical units are the true triggers of abductive reasoning. In other words, some grammatical units (grammatical morphemes) are not only observable symbolic traces of operations and of abstract categorizations but also the clues which are catalyst to abductive reasoning in the process of comprehension. Take for example the following utterances :

(a) *Well, you have been crying!*
(b) *You must have been crying*
(c) *You have been crying.*

The utterance (a) can be paraphrased by the modal utterance (b). It is therefore not the simple statement of a resulting fact expressed by (c) for which (b) is no longer a paraphrase. When (a) is uttered by a subject S^0 to his listener, subject S^1, an abductive reasoning in S^1 is triggered off at the reception of the utterance, which can be illustrated as follows:

1. it has been observed by S^0 that I, his interlocutor S^1, have red eyes;
2. both of the two interlocutors know that if someone cries,then he has red eyes (shared and memorized common knowledge);
3. S^0 has formulated a plausible hypothesis: the event "S^1 has been crying" has happened;
4. S^0 utters in the form of a resulting state: you are in the state which results from the event "you have been crying".

This reasoning is a direct example of abductive reasoning :

1. presence of an observed fact "q";
2. appeal to memorized general knowledge: "if p then q";
3. on the basis of the observed fact "q" and the knowledge of the general law "if p then q", the claim that "p" (an event has happened) is plausible;
4. the utterance of the resulting state of the event denoted by "p".

Upon hearing the utterance of S^0 "You have been crying!", in order to "understand" this utterance, S^1 has to "reconstruct by an abductive reasoning", the event which has caused the resulting state that has been verbalized by S^0. In fact, S^0 does not utter "*you have red eyes*" which would be a simple descriptive statement, it does not utter "*you have cried*" either, which would be a simple assertive statement. S^0 expresses an hypothesis in verbalizing his thought. This hypothesis is plausible in his eyes, as it is based on a statement of observation : "I see that my interlocutor S^1 has red eyes". As one of the most plausible reasons for this empirical observation is that S^1 has just cried, I can verbalize the state which is the immediate result of this event by using the form of the present perfect continuous. So the utterance of S^0 has a modal value, and this is why the paraphrase of (a) by (b) (*You must have been crying.*) is possible and by (c) is impossible because (c) is a descriptive statement. Moreover, S^1 can respond either by confirming what has been intended as only plausible by S^0 : "*Yes, I have indeed been crying*", or by giving another reason, therefore eliminating the plausible hypothesis provided by S^0 : "*No, I have not been crying, if my eyes are red, it's because a speck of dust has gotten into them*" or again, by asking S^0 for the reasons which have led him to his utterance: "*Why do you think that I have been crying?*" With these explanations, the following dialogue is completely natural :

Look at Mary's red eyes!
She has been crying!
No. If she has red eyes, it's because she must have dust in them.

Many typologically different languages[40] possess more or less specific grammatical processes (for example, the "perfect" forms in Indo-European languages; suffixes and less frequently prefixes in Amerindian languages; or particles in certain Tibetan-Burmese languages) which enable the speaking subject to signal distances of different degrees that he takes in respect to the described situations because he perceives these situations in an intermediate way. The subject suggests, with such grammatical forms, that he is not the first-hand source of the information represented by his discourse because the facts constitute general knowledge, admitted or transmitted by tradition; or they have been brought to his knowledge by a third person or by hearsay; they might also have been inferred from observed clues; or else are the result of reasoning. The subject obtains from this a set of semantic values[41] which leads to a general grammatical category that Zlatka Guentcheva[42] suggests calling "la catégorie du médiatif"[43]. It seems very much to us that this category is an important language function which would encode certain types of reasoning indirectly based on an abductive schema by constructing an hypothesis on the basis of clues (observed facts, signs that are accessible to the senses of perception - vision, audition, touch - hearsay, reported discours). English and French have not developed, unlike certain Amerindian languages[44] for example, systems of grammatical labels specific to this category of the intermediary but they have the capacity to express the semantic values of the intermediary in several ways.[45]... Here are some examples in French[46]:

(a) X serait fort ennuyé par la tournure que prend l'affaire Y.

(b) Touvier aurait quitté la France et se serait rendu au Canada.

(c) Les résultats des examens réalisés par le professeur (...) font état de la présence dans le sang (...) de la morphine en particulier. La cause de la mort serait ainsi une crise cardiaque déclenchée dans un contexte de (...).

(*Le Monde*, jeudi 17 juin 1993)

In this last example (c)[47], the journalist wants to inform the public of the still unclear circumstances of this death. As the test did not provide any definitive answer, he formulates an hypothesis using abductive reasoning and based on the evidence provided by the experts: the conditional here encodes this reasoning. However, the conditional is not the only grammatical means in French to encode such abductive reasoning.

The French "passé composé" can also in some contexts represent this type of inference. We have already had an example. Let us now look at the following dialogue :

- *Regarde! Les bagages de François ne sont plus dans sa chambre!*

- *Il est parti!*

- *Non, il doit avoir déménagé.*[48]

The value of the "passé composé" in the second line encodes an abductive reasoning: noticing the absence of the luggage in the room of François, the subject is led to formulate an hypothesis ("François has left") which could account for the observed fact. However, this hypothesis is only a plausible one, and it can be refuted. The subject verbalizes not the observed state but the resulting state of a reconstructed event.

> "This verbalization is made in three steps:
>
> ***Step I*** : statement of a state, called observed state, situated in a certain realized state and concomitant with the act of utterance: the concierge has red eyes or the luggage of François is no longer in his room.
>
> ***Step II*** : procedure of reconstruction by abduction consisting in looking for a possible hypothesis which accounts for the observed state:
>
> a) One has the general knowledge (expressed in the form of a law): "a given process is the cause of a given observed state" and the resulting state of the process is concomitant to the observed state;
>
> b) As "the resulting state of the process is concomitant to the observed state" and one sees "the observed state" well, it follows by abduction that "the process which has caused the resulting state" is a plausible hypothesis; it is at least a possibility. The abductive procedure requires the construction of an expressed intermediary referential which is distinct from the properly expressed referential.
>
> ***Step III*** : projection of the possible process with its resulting state into the expressed referential. Thus, the process and its resulting state, both of which are inferentially constructed, are in the realized state of the subject."
>
> Zlatka Guentcheva (1994 : 20)

One would have the same type of abductive reasoning for the "effet de surprise" before an unexpected statement[49] as in the following examples:

How this boy is grown up!

Look! It has rained.

As a matter of fact she is beautiful!

In a natural language, the grammatical labels - like the grammatical tenses - are often taken to be associated with a certain number of different semantic values. The identified relevant indications in the context often help to eliminate ambiguity by establishing an hypothesis. Take, for example, the following sentence: *The next day he was taken in.* Without any specific context, this utterance has two semantic values, either a value of unreal, or what we call the value of "new state"[50] ... This will be made obvious when we insert the utterance into two different contexts showing the two values:

(a) *Without his friends' intervention, the next day he was taken in.*
(b) *Thanks to the intervention of police officers, the next day he was taken in.*

In the context (a), the value of the utterance is unreal and the interlocutor can infer that he was not arrested; in the context (b), the same sequence of *the next day he was taken in* takes on a different semantic value because one can infer that he has been arrested. To resolve this uncertainty, the interlocutor processes abductively by identifying in the context the relevant linguistic indications (the symptoms) (the linguistic index *without* is relevant in (a); the linguistic index *thanks to* is relevant in (b)). Having recognized these relevant indications, he is led to infer his conclusion: either the event has indeed taken place, or the event has not taken place but could have taken place.

8. CONCLUSION

Abduction proposes hypotheses or general ideas which deduction later on develops and which induction projects into the future. *Abduction* "provides us with all of our ideas concerning real things, beyond the data of perception, but it is a simple conjecture without convincing power".

> "An abduction is a method to formulate a general prediction without positive assurance that it will succeed in a particular or ordinary case. Its justification is that it is the only possible hope in settling rationally our future conduct, and the induction based on the past experience greatly encourages us that abduction will succeed in the future."
>
> (Peirce: *Collected Papers* : 2.270)

These remarks show that it is highly likely that abduction is one of the essential cognitive aptitudes of human beings because it enables man to continuously construct new knowledge. In this way, man becomes capable of constructing heuristic and explanatory hypotheses and of justifying them with a series of indices as well as protocol statements so that the actual facts which these last represent, might not be (even if only temporarily) put into question, for theoretical reasons for example. Man in this way acquires an enormous power of invention with which he can communicate his creative imagination, reasoning and experimental control.

University of Paris-Sorbonne
France

NOTES

1 The author thanks Wei Hsiung for the English translation of the French text of this paper.

2 See Peirce 1965.

3 Which means "frequently something which would be impossible for us to observe directly", see Peirce, o. c: 2.640.

4 G. Polya o. c., pp. 106-107.

5 G. Polya o. c., p. 99.

6 Other equally instructive examples could be cited in history, archeology or sociology.

7 G. Polya, o. c., p. 108.

8 In the schema of *modus ponens*, one could use the following phrase : *q because of p* according to the law *"if p then q"*; in the schema of abduction, we have: *p because of q on knowing* (shared knowledge) that *"if p then q"*.

9 According to the expression of G. Polya, o. c., p. 102.

10 See Jean-Pierre Verdet 1990, pp. 207-208. G. Darmois who recalls this anecdote (preface to the book of G. Polya, o. c.) concludes : "The second problem was well raised, the first one was not".

11 "My design", writes Kepler in the preface to *The Secret of the World*, "Reader, is to demontrate in this little book that the Creator, the Very Good and Very Great, for the creation of this mobile world and the layout of the skies, has referred to the five regular bodies which, from Pythagoras and Plato to our days, have acquired such great celebrity, and that he has put them in order according to their number, their proportion and the relation of their movements" (Jean Kepler, *The Secret of the World*).

12 Kepler imagined several systems before formulating what have become his "laws". "Coming back to the physics of movements and thinking that the planets are the magnetic bodies, of which one pole is attracted by the sun and the other is pushed back by it, he imagines a libration of Mars along the vector light passing through the sun. But this results in yet another failure. Finally it is the geometric considerations without physical implications, despite what Kepler thinks, that are to lead him to success. Studying the trajectory of Mars when the part which joins the planet in the centre of the orb is perpendicular to the line of absides, Kepler discovers a remarkable relationship between the "lunar" and the optic angle (...). Now, this relationship defines an ellipse which Kepler will accept a few months later". see Verdet 1990, p. 155.

13 "It is without doubt an important matter to invent, but it is an even greater matter to progress in the art of invention" (G. Darmois: preface to the book of G. Polya, o. c.).

14 To start a computer program, a computer program must be loaded by means of a loading program which reads and stores the programs in memory, in such a way that they are ready to be executed at all times. But how is one to read this loading program and put the computer in a normal working condition? In computing technics, one also uses the bootstrap: to carry out a compiler Comp in language L, one hopes to use this new language L to write the compiler Comp. Therefore one writes a first version comp' in another language L' (generally of a lower level) for which there is already a compiler. The compiler Comp' is a program which is written in L' and can "translate" the programs written in L. When the computer Comp' is compiled, it can compile the compiler Comp which is a program written in L. We will then use this compiled compiler Comp to compile the programs written in L and the compiler Comp' will then disappear.

15 "To bootstrap means that one creates an object with the help of this object. This seems strange, for how can one use a non-existant object to create itself? In reality time intervenes : one creates a succession of versions of an object. These versions are increasingly perfected and one uses the version N to create the version N+1", see Pitrat 1993. So, to create a hammer it is necessary to create a hammer. Modern computers are designed on the basis of other computers. "The bootstrap is very useful to resolve very difficult problems, as in the AI. Each progress made in AI must make the ensuing steps easier. What one develops has been decided on the basis of its utility to what we want to do next. AI is very well adapted to a bootstrap because the metaknowledges which process knowledge are also knowledges" (Pitrat: o. c. : 73).

16 The controversy between Young and Champollion on the deciphering of Egyptian hieroglyphics, that we will evoke later, illustrates perfectly the difference between an hypothesis (that of Champollion) which triggers a truly abductive process and an hypothesis (that of Young) which, although good, has not been able to open the road to a discovery of the writing system in which three sub-systems overlap each other.

17 We will see this later in an example based on the "voiced coefficient" of Ferdinand de Saussure.

18 We follow the excellent analysis which, we think, is typically abductive, that has been retraced by Jean Lacouture (see Lacouture 1988).

19 According to the legend, Jean-François Champollion collapsed into a "cataleptic" state during several hours after this discovery.

[20] For a long time, Silvestre de Sacy, and Champollion following him, insisted on the almost exclusive ideographic character of the hieroglyphs: "I persisted in this wrong direction until the moment when the factual evidence [forced] me to recognize a phonetic value in masses of hierographic groups, comprised in inscriptions on the Egyptian monuments of all ages" (Champollion: *Livre..*).

[21] See one of the above notes.

[22] "J.F Champollion's method has (...) not too much to do with that of Dr. Young. His writings have the form of a mathematical demonstration (...). The nature of the work of Thomas Young is not a decipherment but a series of conjectural identifications. The reasons for his success and for his failures are very clear. He worked mechanically, like a school boy who, in discovering in a translation that *Arma virumque* means "the weapons and the man", reads: arma: the weapons; virum: and; que: the man... (...) For lack of method, Young did not have any notion of a phonetic system (...) What he had was nothing but hypothesis, conjecture, and supposition concerning the hieroglyphic writing (...). The great discoveries of 1822 were followed by even greater ones. The phonetic alphabet was the key (of the deciphering) of not only the Greek and Roman names but also all of the periods" (Sir Peter Le Page-Renouf, quoted by Jean Lacouture, o. c., p. 484).

[23] Saussure's procedure does not absolutely end up as an inductive approach, at least not in the sense that we have understood "inductive inference" according to Peirce's classification.

[24] Here we use the analysis of J. D. Apresjan (see Apresjan 1973, pp. 98-100).

[25] The founding of phonology has been difficult and has met with a lot of reservations from those linguists who refuse to accept phonemes that they considered to be fictitions entities.

[26] So, as reminded by S. K. Shaumyan (see Shaumyan 1987, pp. 32-93), a phoneme can not be defined as "a class of sounds which: (1) are phonetically similar and (2) show certain characteristic patterns of distribution in the language or dialect under consideration".

[27] S.K. Shaumyan's genotype language.

[28] Not only the languages of the accusative type and of the ergative type but also the languages called active/inactive (like Guarani, South Dakotan or some Paleo-Siberian languages).

[29] See the wonderful article of E. Benveniste 1966, pp. 187-207.

[30] See Antoine Culioli 1981, pp. 88-117; 1982, pp. 93-126; 1992, pp. 74-77.

[31] E. Benveniste, o. c.

[32] For example: J. Lyons 1974. Also, see the series of studies in Lyons 1967.

[33] See Desclés 1987, pp. 57-78.

[34] The Indo-European languages are, however, very close; the polysemic "BE" in these languages make them very different from other languages. This fact would explain the importance that the western philosophers have given to the notions of "be" and of "being".

[35] See Desclés 1990(a), pp. 121-147; 1995, pp. 145-186.

[36] See Desclés 1990(b), pp. 33-56.

[37] "The rapid development in cognitive science obliges us to add cognitive states of the "ordinary psychology" (...). Other than that they are radically inaccessible to the ordinary introspection, they play an explanatory or causal role according to cognitive science in the formation of beliefs (perceptual) and in their propositional attitudes without, however, the normal interaction with them" (see Jacob 1992).

[38] The concept, in the sense intended by G. Frege, is an application of a domain in the set {"true", "false"}. A concept classifies objects: "those who fall into the concept" or "those to whom the concept applies" on the one hand and "those who do not fall into the concept" or "those to which the concept doesn't apply" on the other.

[39] This concept is completely caricatural. However, one can reconsider in more detail Andrew Woodfield's analysis of "the experience of thought" and show how abduction is involved in the formation of concepts (see Woodfield 1990).

[40] See Zlatka Guentcheva 1994. She quotes the following languages: the Indo-European group: Albanian, Armenian, Bulgarian, Persian; the Altaic group: Turkish, Azerbaïjan, Gagagouz, Ouzbec, Ouygour,...; the Fino-Ougrian group: Lappish, Nenetz, Selkoup, ...; Tibeto-Burmese: Tibetan, Akha, Sherpa; the Amerindian group: Makah, Tuyuca, Wintu, Pawne...

[41] See, for example, the "cognitive maps" or "mental maps" of L.B. Anderson 1986, pp. 261-312.

[42] Zlatka Guentcheva 1990(a), 1990(b), pp. 179-196.

[43] Other names have been given for this category: "evidential", "unwitnessed", "testimonial" (R. Jakobson), "non seen"...

[44] For example, an Amerindian language of the Yuman branch, Maricopa, grammaticalizes directly the intermediary: *Pam-sh Bonnie tpuy-k - 'ish- 'a* "Pam SUBJECT Bonnie kill-k -say+sh-hearsay" = Pam killed Bonnie (I hear tell) = Pam might have killed Bonnie, according to what I have heard; quoted by Zlatka Guentcheva 1994, p. 13: the morpheme

'ish- 'a is the label of intermediary.

[45] See Zlatka Guentcheva 1994, pp. 8-23.

[46] Translations in English:

(a) *It may well be that X is quite bothered by the way the Y affaire is developing.*

(b) *It could be that Touvier has left France and has gone to Canada.*

(c) *The result of the tests carried out by the professor (...) take into consideration of the presence (...) of the morphine in particular in the blood. It could be that the cause of the death is therefore a heart attack caused in a context of (...)* (Translation from *Le Monde*, Thursday, 17 June 1993).

[47] We again use Z. Guentcheva's analysis in *Langue française*, 102, 1994, pp. 8-23.

[48] *- Look! François' luggage is not in his room anymore!*
- He has left!
- No, he must have moved.

[49] This value is sometimes called the "admirative" value, for example, in Bulgarian or in Albanian.

[50] For a more complete analysis of this problem, see J. P. Desclés 1994, pp. 57-88.

REFERENCES

L. B. Anderson, *Evidentials, Paths of Change and Mental Maps: Typologically Regular Asymmetries*, in Chafe, W & Nichols J. eds, *Evidentiality: The Linguistic Coding of Epistemology*, Advances in Discourses Processes, vol. XX, Norwood & New Jersey, Ablex Publishing Corporation, 1986.

J.D. Apresjan, *Eléments sur les concepts et les méthodes de la linguistique structurale contemporaine*, Dunod, Paris, 1973.

E. Benveniste, *Être et Avoir dans leurs fonctions linguistiques* , Bulletin de la Société de Linguistique LV, 1960, requoted in Problèmes de linguistique générale, Gallimard, Paris, 1966.

A. Culioli and J.P. Desclés, *Systèmes de représentations linguistiques et métalinguistiques*, Collection ERA 642, Laboratory of Formal linguistics, University Paris 7, 1981.

A. Culioli and J.P. Desclés, "Traitement formel des langues naturelles. Mise en place de concepts à partir d'exemples", *Mathématiques et Sciences Humaines* **77**, 1982.

A. Culioli, "The Concept of Notional Domain", in *Pour une théorie de l'énonciation*, Ophrys, Paris, 1992.

J.P. Desclés, *Réseau sémantique: nature logique et linguistique des relateurs*, Langage **87**, 1987.

J.P. Desclés, "Architectures, représentations et langues naturelles", in *Les sciences cognitives en débat*, Editions du CNRS, Paris, 1990(a).

J.P. Desclés, "The Intermediary Representations", *Review of Synthesis* IV, 1-2, International de Synthese, Albin Michel, Paris, January-June 1990(b).

J.P. Desclés, "Quelques concepts relatifs au temps et à l'aspect pour l'analyse textuelle", in *Etudes cognitives 1, Sémantique du temps et de l'aspect*, Poland Academy of Science, Institute of Slavic Studies, Warszawa, 1994.

J.P. Desclés, "Intelligenza artificiale e linguaggio naturale (Riflessioni epistemologiche)", in *Interpretazioni attuali dell'uomo: filosofia, scienza, religione*, Guida Editori, Napoli, 1995.

Z. Guentcheva, *Temps et aspect: l'exemple du bulgare littéraire contemporain*, Sciences du langage du CNRS, Paris, 1990(a).

Z. Guentcheva, *L'énonciation médiatisée en bulgare*, Revue des Etudes slaves, LXII/1-2, 1990(b).

Z. Guentcheva, "Manifestation de la catégorie du médiatif dans les temps du français", *Langue française* **102**, 1994.

P. Jacob, *Introduction aux Sciences Cognitives*, Gallimard, 1992.

J. Lacouture, *Champollion, une vie lumière*, Edition Grasset & Fasquelle, 1988.

J. Lyons, *The Verb "BE" and Its Synonyms, Philosophical and Grammatical Studies*, Edited by John W. M. Verhaar, *Foundations of Language*, Supplementary series, D. Reidel Publishing Company, 1967.

J. Lyons, *A Note on Possessive, Existential and Locative Sentences, Foundations of Language* **3**, French translation in *Languages* **34**, Paris, 1974.

C.S. Peirce, *Collected Papers of Charles Sanders Peirce*, vol. I and II, Havard University Press, Cambridge, 1965.

J. Pitrat, *Penser l'informatique autrement*, Hermès, Paris, 1993.

G. Polya, *Comment poser et résoudre un problème*, Edition Jacques Gabay, Paris 1989.

S.K. Shaumyan, *A Semiotic Theory of Language*, Indiana University Press, Bloomington and Indianapolis 1987.

J.P. Verdet, *Une histoire de l'astronomie*, Edition du Seuil, Paris, 1990.

A. Woodfield, "Un modèle à deux étapes de la formation de concepts", in *Introduction aux sciences cognitives*, under the direction of Daniel Andler, Gallimard, Paris 1990.

ROBERTO TORRETTI

'SCIENTIFIC REALISM' AND SCIENTIFIC PRACTICE

The adjective 'scientific' as used in the title of this lecture refers in the first place to modern experimental and mathematical physics; and to the other natural sciences insofar as they follow the lead of physics and thrive in its light. I do not place this restriction on the word in the context of this lecture because I hold other forms of learning in contempt as non-scientific. Far from it. However, the issues of "scientific realism" which are the subject of this conference do not rise outside physics and its near relatives. Thus, I have never heard a musicologist or a philosopher of musicology question the reality of Bach's "Well-Tempered Clavier", or of the Fugue in C minor in Book I, or of the main theme of that fugue. Nor do linguists and philosophers of language have any doubts regarding the pervasive presence of the genitive absolute in classical Greek.

The main question concerning realism in science is whether the nouns and pronouns used in scientific discourse name real entities or refer to 'constructs', i.e. to figments of the scientific imagination. But our self-styled "scientific realists" will not be satisfied by a straightforward acceptance of the former alternative, in the plain naive sense of 'real' applied to ordinary things, which the Romans called 'res' - whence 'reality' and 'realism'- and the Greeks called *pragmata*. They expect the objects of scientific knowledge to be real in a stronger sense than the things of everyday life, such as my two hands, or a red flavorful apple, which they take to exist, so to speak, on loan, as mind-dependent appearances of the really real. Scientific realists fondly trace their ancestry to Democritus, who wrote: "By custom, sweet; by custom, bitter. By custom, hot; by custom, cold. By custom, color. In truth: atoms and void."[1] The dependence on human life of the main features of common sense reality was eloquently proclaimed by Galileo at the dawn of modern science:

> I do not believe that anything is required in external bodies besides their size, shape, multitude, and motions, fast or slow, in order to excite in us tastes, odors, and sounds; and I think that if ears, tongues, and noses are removed, the shapes and numbers and motions will remain, but not the tastes nor the odors nor the sounds. Apart from the living animal, the latter-I believe-are nothing but names, just as tickling and titillation are mere names if the armpit and the skin lining of the nose are removed.
>
> (Galileo, *Il Saggiatore*, §48 [EN, VI, 350])

Galileo gave this Democritean doctrine a Pythagorean twist which remains central to the practice of modern physics: The Book of Nature, he said, "is written in mathematical language and its characters are triangles, circles, and other geometric figures without which it is humanly impossible to understand a single word of it" (*Il Saggiatore*, §6 [EN, VI, 232]). Therefore, as one would say today, the really real can be adequately represented as a mathematical structure. Similar views were held by Descartes, who neatly imbedded them in his mind-body dualism. These early modern

E. Agazzi and M. Pauri (eds.), The Reality of the Unobservable, 113-122.

scientists were also devout Christians and they believed that God, like a Renaissance architect, had designed a geometric plan after which he had created the universe. Their science was bent on retrieving that plan. Pascal complained that the reduction of physical objects to "number, shape and motion" was "uncertain, useless and depressing", but that reduction was taken up by Locke in his widely read *Essay concerning Human Understanding*, and became a paradigmatic trait of scientific realism. Locke's book was the main butt of Berkeley, the founding father of antirealism, who vindicated everyday reality by showing that the abstract idea of colorless, insipid matter favored by the scientific philosophers was founded on and nourished by our ordinary experience of colorful and tasteful things. By conceding that the latter were indeed dependent on mind Berkeley turned Locke's "way of ideas" into grist for his spiritualist mill, but the very cleverness of this move made his vindication of common sense both equivocal and unpopular.

Today's "scientific realists" are, more often than not, materialists, and therefore closer to Hobbes than to Descartes, but, all the same, they remain frozen in a 17th century frame of mind. Paradoxically, they speak persistently of the reality of the *external* world, as if they were disembodied spirits contemplating it from the outside, and, for all their godlessness, they put forward a view of it that is only conceivable from the standpoint of an omniscient God. The cultural distance from "scientific realism" to late 20th century philosophy is so big that someone nurtured, say, in the writings of Heidegger and Wittgenstein and in the tradition of American pragmatism may well despair of converting a "scientific realist" to more sensible views, by reasoning. And yet in our own generation Professor Hilary Putnam has undergone such a conversion, moving - no doubt, on his own initiative and through his own ripening and deepening reflection on how matters stand - from a garden-variety version of "scientific realism" to what he calls, with emphatic pleonasm, 'pragmatic realism',[2] which agrees well with the position I take in this lecture. Still, I am afraid that what I can offer in the guise of argument will be seen by "scientific realists" more like a confession or manifesto, edifying and perhaps enjoyable for those who share my dubious background but decidedly allergenic and not at all cogent for those who don't.

"Scientific realists" believe that reality is well-defined, once and for all, independently of human action and human thought, in a way that can be adequately articulated in human discourse. They also believe that the primary aim of science is to develop just the sort of discourse which adequately articulates reality - which, as Plato said, "cuts it at its joints"[3] -, and that modern science is visibly approaching the fulfilment of this aim. I find it very hard to accept any of these statements or even to make sense of them. The existence of a well-defined or - as Putnam called it - ready-made reality is no doubt implied by the standard monotheistic conception of God, but I have not the slightest ground for thinking that God's worldview can be articulated in human discourse. To entertain the notion that we could convey that view in words is a symptom of acute provincialism. I contend, moreover, that scientific discourse is just the verbal aspect of scientific practice and has no serious justification apart from it; that this verbal aspect is no closer to the aim of science than, say, its manipulative aspects; that science does not have a primary aim but that with her, as with any other form of human activity, the distinction between means and goals is continually shifting from one context to another, so that any goal attained will, while in our possession, be used sooner or later as a means, while most means have at some stage been goals. From this perspective, science is, if I may say so, the continuation of common sense by other means, as required by the overall pragmatic situation. When I search for my car in a parking lot I bear in mind its visual appearance but do not once think about the photon exchange going on between the sun, the car's surface and my eyes. When I reflect about the origin and evolution of the universe I mix the car and myself and everything else up in a single homogeneous worldwide purée. And these two ways of articulating reality - as well as many others - are quite appropriate for finding our own way in it on different occasions and with different purposes.

Enough of generalities. A few examples drawn from past and present scientific practice will, I hope, clarify these ideas and substantiate my contention that it is pragmatic realism, not the nostalgic kryptotheology of "scientific realism", that best expresses the real facts of human knowledge and the working scientist's understanding of reality.

A familiar argument against realism in science runs as follows: *Premise One*: Any set of empirical data can be accounted for by many different physical theories (i.e. by embedding it in different mathematical structures). *Premise Two*: We may prefer the one that *we* judge simpler, or prettier, or easier to calculate with, but we have no grounds for believing that our preferences are shared by the Creator of the universe or, worse still, that they were followed when the universe was born by chance. *Conclusion*: If science aims at establishing the true structure of reality from empirical data it faces an impossible task. Premise One can be readily demonstrated in the case of the simplest theoretical task, namely, to fit a curve to a set of data about two correlated quantities: the data can then be represented by small rectangles on the plane and infinitely many different curves will go through any finite set of them, no matter how small and how densely packed they are. Presumably, the variety of admissible theoretical accounts will only increase as the data to be accounted for grow more complex. However, in real life such variety is somewhat utopian: even if it is a logical certainty that many incompatible theories might account for all the available data, as a matter of fact we do not know a single one which does. What we have are theories that account *in part* for some *specific* family of phenomena. And whenever there are two or more rivals in one field, effects soon appear which give the victory to one. Still, such choices, though in no way uncertain, are often based on considerations of beauty and expediency. For example, data cannot easily clash with a theory involving several adjustable parameters, such as Ptolemaic astronomy or some of the recent rivals of Einstein's theory of gravity. But the very fact that they have such flexibility is enough to damn them if there happens to be a rival theory without adjustable parameters which agrees with the data. This manner of choice is natural if we aim at understanding the reality at hand in the context of our life and practices, yet I do not see how one can be sure that absolute reality, as it is supposedly structured apart from human history and human interests, is not best described by one of those theories with adjustable parameters we so cheerfully discard.

Clark Glymour (1977) studied an interesting case in which the vacillation between divergent theoretical accounts of the same set of data does not result from the fuzziness of the latter and cannot be resolved by criteria of beauty or expedience. Though purely academic, this example is quite instructive. Suppose for a moment that Einstein's General Relativity is the true theory of the world. Glymour was able to prove that Einstein's field equations have alternative, *topologically incompatible*, solutions which are nevertheless *observationally indistiguishable* in the following precise sense: Let $<\mathcal{M}_1, g_1>$ and $<\mathcal{M}_2, g_2>$ be two models of General Relativity, where $\mathcal{M}_1$ and $\mathcal{M}_2$ are four-dimensional real differentiable manifolds and g_i $(i = 1,2)$ is a solution of the Einstein field equations defined on $\mathcal{M}_i$ (i = 1,2). $<\mathcal{M}_1, g_1>$ and $<\mathcal{M}_2, g_2>$ are (weakly) observationally indistinguishable if, for each point $p_1 \in \mathcal{M}_1$ there is a point $p_2 \in \mathcal{M}_2$ such that the past of p_1 is isometric with the past of p_2.[4] Glymour proved that in some such pairs of observationally indistinguishable solutions of Einstein's field equations, the underlying manifolds are topologically distinct. Glymour's proof could be carried out rigorously and abstractly, without regard to a particular hypothetical distribution of matter, because there is a precise topological relation between the two manifolds involved, viz., either one of them is a covering space for the other, or they both have a common covering space.[5] Now, in a relativistic world all empirical data for the corroboration of scientific hypotheses must be fished out of the scientists' pasts.[6] Hence, if the world they happen to live is topologically distinct yet observationally indistinguishable from another one, they must remain forever undecided as to the overall shape of their world. If they are "scientific realists", bent on getting an overall, "God's eye" view of what they consider "the external world", such indecision is fatal to their epistemic ambitions. But it is a matter of complete indifference to the pragmatic realist who, as being-*in*-the-world, seeks to understand the reality at hand. As time goes on, some of his expectations, based on data culled from an earlier past, will be seen as deceptive, but only in the light of data obtainable from the now current past. Thus, for him, two observationally indistinguishable relativistic world models, though topologically inequivalent, could well constitute epistemically equivalent descriptions of one and the same reality. A difference, to be a difference, must make a difference.[7]

The antirealist tendencies among recent historians and sociologists of science draw their sustenance not from a *conceivable* disjunction of *imaginary* theories between which one could not decide because they would all be equally good, but from the *actual* succession of *real* theories most of which one already has decided against because they are not so good as the last. Inevitably the point is made that even the currently accepted theory is not good enough to be final and will in all likelihood be superseded later. The accurate theoretical representation of reality would thus be deferred for ever. Against this conclusion it has been argued that, although science never quite hits the truth about reality, it comes closer and closer to it. This idea of truth by approximation can be held in two versions: either we assume that reality stands there ready made outside the process of scientific research, which aims at finding the truth about it; or we conceive the true articulation of reality as the limit to which the succession of scientific theories converges, and which is constituted by this succession in the sense in which a real number is constituted by a Cauchy sequence of rationals. The second version is perfectly acceptable to the pragmatist - in fact it was put forward by Charles Saunders Peirce - but the "scientific realist" can only countenance the first. I shall not rehearse the arguments that have been made for and against either version.[8] Let us just consider an example which is as good a case of approximation as any one can think of and see what comfort the "scientific realist" can derive from it. Nobody will question that 19th century chemists got very good values for the atomic weights of many elements. Our own values are only slightly better. However, the quantities they measured are now understood very differently. They thought they were reaching for the mass of each indivisible part of each one of the ultimate elements of matter. We no longer think that the parts in question are really indivisible, even though, oblivious of Greek, we still call them 'atoms'. But, more importantly, we now understand that an atomic weight measured by 19th century methods is not usually the mass of a particular kind of atom, but the weighted average of the masses of several kinds of atoms with the same chemical properties, in the proportions in which those kinds (isotopes) are normally mixed in our environment. So, in our view, 19th century analysts measured very nearly *the right numbers*, but *not of the quantity they had in mind.*[9]

Thus the way that scientific theories succeed and displace one another provides strong grounds for doubting the scientific realist's belief that science is getting ever closer to achieving an adequate grasp of the self-subsisting, uniquely defined structure of reality. Moreover, it seems to me that - even if, for the sake of the argument, one grants that there is such an absolute structure- science as it is actually practiced is not in the business of looking for it. I gather this from the manner in which conceptually disparate theories are jointly brought to bear on the understanding of specific phenomena and the solution of particular problems. In order to make this clear I shall propose two examples. I take the first one from Hawking's celebrated letter to *Nature* on black hole explosions (1974).[10] Roger Penrose, Stephen Hawking and Robert Geroch proved in the 60's a series of remarkable mathematical theorems which imply that, under some physically very plausible assumptions, a relativistic spacetime will contain black holes, that is, regions in which the gravitational field is so strong that no matter or radiation can ever come out of them and all incident matter and radiation gets trapped. In the said paper Hawking studied what would happen if matter inside a black hole obeys the laws of (classical, non-relativistic) Quantum Mechanics. As is well known, these laws imply that a particle trapped inside a potential well generally has some finite chance of tunnelling through the barrier surrounding it. Such chancy tunnelling through potential barriers is the key to the quantum-theoretical explanation of radioactivity, which is conceived as the sudden causeless ejection of certain quanta of matter and radiation from the atomic nucleus to which they are normally bound. Hawking applied the same general idea to black holes and calculated the time it would take for a black hole of given mass *m* to evaporate completely. (If *m* is the mass of the Sun, the time in years is of the order of 10^{63}). It must be emphasized (i) that the notion of black holes arises in General Relativity and in accordance with this theory they do not exist in the flat Minkowski spacetime underlying Quantum Field Theory, let alone in the Newtonian spacetime underlying non-relativistic Quantum Mechanics; and (ii) that the Einstein field equations do not set up a quantized field, so that Quantum Mechanics cannot be regarded as a local approximation to General Relativity. Hence, if Hawking's move is judged from the standpoint of scientific realism, one must say that it was made on credit, using both General Relativity and Quantum Mechanics as inaccurate provisional partial substitutes for a future Quantum Theory of Gravity, which would consistently account for the existence of black holes and the way they evaporate. Of course science needs credit no less than other trades; indeed, some methodologists *require* new theories to issue promissory notes redeemable by forthcoming observations and experiments. But to grant credit to a future, as yet unstated theory, which two mutually inconsistent and, according to scientific realist criteria, plainly false theories are expected to approximate, is more than a sensible person can do. I think, therefore, that Hawking's practice cannot be understood in terms of scientific realism, but should be seen simply as what it claimed to be, namely, the solution, based on the most suitable theories at hand, of a problem posed by black hole research (specifically, by the attempt to assign a temperature to black holes).[11]

Hawking's work on evaporating black holes makes my point very clear because it combines disparate theories, I dare say, outrageously. But scientific realists might question its relevance because the very assumption that Hawking's study has a referent - i.e. that there are black holes - rests even today on somewhat flimsy evidence. I shall therefore turn to another example which, though less extreme than the former, concerns an object whose reality is beyond doubt, namely, the planet Mercury and its motion. According to Newton's theory of gravity if Mercury were alone with the Sun it would trace over and over an ellipse of given size and eccentricity having the center of gravity of the Sun-Mercury system at one of its foci. With some abuse of language, I shall refer to this ellipse as Mercury's Keplerian orbit. As a matter of fact, Mercury's trajectory deviates slightly but consistently from its Keplerian orbit. Relatively to the geocentric system of polar coordinates used in observational astronomy, Mercury's perihelion advances each year by somewhat less than 1 minute of arc (close to 56″). Almost 90% of this advance can be explained by the general precession of the equinoxes and is thus due to the choice of a geocentric coordinate system. But, according to Newton's theory, the remaining 10% must be due to gravity. The greatest part of this - some 5.3″ per year or 530″ per century - was accounted for by the perturbational methods of classical celestial mechanics as a result of the gravitational interaction between Mercury and the other planets. However, there still remained a small yet inescapable balance of approximately 43″ of arc per century which never was satisfactorily accounted by Newtonian theory. On December 24, 1907, Einstein wrote to Conrad Habicht that he was working on a relativistic study of the law of gravity through which he expected "to explain the still unexplained secular variations in Mercury's perihelion".[12] The main motivation for this study was, of course, not this small astronomical anomaly, but the open clash between the theory - now known as Special Relavity - that Einstein introduced in 1905 to account for the electrodynamics of moving bodies, and Newton's theory of gravity, which, for good reason, was then acknowledged by everyone, including Einstein, as the most successful theory of physics.[13] As is well known, in Newton's theory gravitational influence propagates instantaneously throughout infinite space, whereas in Special Relativity no information can travel faster than light. Einstein's quest for a theory of gravity ended in November 1915, when in four hectic weeks he communicated to the Prussian Academy three different theories. The weekly succession of theories was interrupted on the third week of November when he derived the unexplained part of Mercury's perihelion advance from an approximate solution of the field equations proposed in the second week.[14] What we now know as the Einstein Field Equations of General Relativity were communicated in the fourth week, but, though they differ both in their mathematical shape and their physical meaning from those of the second week, they agree with them on empty space, through which planets are supposed to travel, and so they inherit from them the third week's solution of Mercury's anomaly.[15] Indeed, the first exact solution of the Einstein Field Equations, independently reached by Schwarzschild and by Droste early in 1916, fully confirmed the solution of the anomaly and has thereafter been regarded as one of the "three classic tests" of General Relativity. In order to see how this test works, we must recall that in General Relativity gravity is conceived as a property of spacetime analogous to the Gaussian curvature of ordinary surfaces, governed by the spacetime Riemann tensor. Test particles - that is material particles so insignificant that they contribute nothing to the gravitational field - follow spacetime geodesics. The Einstein Field Equations sport on one side the Ricci tensor, formed by contraction of the Riemann tensor on two pairs of indexes, and on the other side a tensor built from the stress-energy tensor representing the spacetime distribution of matter. In the absence of matter, Ricci = 0. Mercury's anomaly is overcome by solving this equation, which, as it happens, also occurs in the otherwise very different theory of gravity put forward in Einstein's second paper of November 1915. Suppose then that the spacetime metric is spherically symmetric in space and that Ricci = 0 everywhere outside a small region surrounding the axis of symmetry. The Einstein Field Equations can then be solved exactly. The solution, which involves one constant of integration usually denoted by $2m$, is defined everywhere except on the field's axis of symmetry.[16] If one equates m with the mass of the Sun, a test particle moving through empty space with Mercury's speed at Mercury's distance from the axis of symmetry will have a spatial trajectory very similar to Mercury's Keplerian orbit (with the axis of symmetry at one of its foci), but which will differ from that orbit insofar as the point where it comes closest to the field's axis of symmetry advances on every turn at a rate of 43″ per century. Thus, the secular precession of Mercury's perihelion, unaccounted for in Newton's theory of gravity, is seen to follow from Einstein's. But please note how this result is achieved. Mercury, a chunk of matter about as dense as the Earth, bounded by a surface as least as large as that of Asia and Africa put together, is conceived as a particle of negligible mass, whose presence and motion do not in the least affect the perfect symmetry of space. This streamlined Mercury, suitably placed in the gravitational field of the lonely Sun, will, in accordance with Einstein's theory, display the unexplained perihelion precession of 43″ per century, but not, of course, the precession of 530″ per century that Newtonian celestial mechanics derived from Mercury's gravitational interaction with the other planets. In the symmetric Schwarzschild field, besides the source at the axis of symmetry, we meet only test particles which, by definition, cannot act on each other. And of course, neither Einstein, nor Schwarzschild, nor Droste, were in a position to even approximately solve the Einstein Field Equations for a 10-body system.[17] In their solution of Mercury's problem the relativists simply ignored the 530″ secular precession because they took for granted that, under the circumstances, this can be accounted for by Newtonian theory, and they could safely assume that a relativistic theory of the whole solar system would, if available, yield predictions agreeing with the Newtonian predictions within an acceptable margin of error. Now, a scientific realist can feel perfectly comfortable with such an arrangement if he assumes that General Relativity is the final physical theory of everything. It is then permissible, from his point of view, to describe Mercury's motion by using first Schwarzschild's idealized model because Mercury is negligibly small, then the Newtonian approximation because the contribution of the other planets to the gravitational field is sufficiently small (though, obviously, not negligible), and adding up the results. But if General Relativity is not the final theory the example of Mercury's motion does not differ essentially from my previous example of evaporating black holes: an otherwise unmanageable problem is solved within a margin of imprecision consistent with observational error by bringing to bear on it two conceptually very different theories, irreconcilable in God's eye, none of which can be true of reality in the strong scientific realist's sense.

I have given two examples from the one area of physics to which at one time I devoted careful attention, but I feel sure that similar examples can be found in other areas. In particular, experimental results notoriously depend for their interpretation on the theories involved in the design of the several pieces of equipment employed in the experiment; these are usually old theories, different from and ultimately incompatible with the frontline theory which the experiment is meant to test. Indeed, some methodologists believe that it is viciously circular to test a theory with instruments designed in the light of it. I don't think they are right, but as a matter of fact new theories are being continually tested with equipment built according to older theories that those new theories undercut. The scientific realist may indeed assume that the entire experiment can be satisfactorily reinterpreted in terms of the new theory, but once again this is to live on credit. The pragmatists, of course, do not need to feed on fanciful expectations and can make do from day to day with what is really there, because they readily accept that physics, like every other major human enterprise, is a patchy, makeshift affair.

University of Puerto Rico
Puerto Rico

NOTES

1 Democritus, fr. 9, fr. 125. Translated by David Furley 1987, p. 177.

2 Cf. Putnam 1987, p. 17: "The key to working out the program of preserving commonsense realism while avoiding the absurdities and antinomies of metaphysical realism in all its familiar varieties (Brand X: Materialism; Brand Y: Subjective Idealism; Brand Z: Dualism...) is something I have called *internal realism*. (I should have called it pragmatic realism!) Internal realism is, at bottom, just the insistence that realism is *not* incompatible with conceptual relativity."

3 *Phaedrus* 265e1.

4 This definition of weak observational indistinguishability is due to Malament (1977). The relation thus defined is not symmetric. To ensure symmetry, Glymour's original definition of observational indistinguishability is stronger. For observers who must decide on the basis of available data which of two topologically incompatible models of General Relativity represents their world, it makes no difference whether the relation is symmetric or asymmetric: the decision is not possible if the universe they happen to be in is weakly observationally indistiguishable from the other.

5 Let A and B be two topological spaces. A is a *covering space* for B if there exists a *covering map* $f: A \rightarrow B$, i.e., a *continuous mapping of A onto B* that meets the following condition: every point $p \in B$ has an open neighborhood U whose inverse image $f^{-1}(U)$ is a union of disjoint open sets of A, everyone of which is mapped homeomorphically onto U by f.

6 For the sake of mathematical expediency, Glymour's proof concerns the so-called *chronological* past of the points in question, which comprises, for any given point p, every point to which one can go from p along a past-directed timelike curve. Of course, the source of empirical data available at a point p is contained in p's *causal* past, i.e the set of points to which one can go from p along past-directed timelike *or* null curves.

7 The following far-fetched but very simple example shows what is at stake here. Think of Nietzsche's fantasy of eternal return. We may assume that Nietzsche's world was spatially Euclidean. What shape should we ascribe to his spacetime? It is either topologically equivalent to $\boldsymbol{S} \times \boldsymbol{R}^3$, with each one of us going through the same events again and again; or topologically equivalent to $\boldsymbol{R}^4$, with each one of us living infinitely many distinct but indistinguishable lives. Pragmatically there is no difference between these two descriptions. (Note that $\boldsymbol{R}^4$ is a covering space for $\boldsymbol{S} \times \boldsymbol{R}^3$).

8 See, for instance, Laudan 1984, Kitcher 1993. Two recent case studies by Psillos (1994, 1995) tend to undermine the main argument of Laudan's "confutation of convergent realism", viz. that respectable scientific research programs have in the past assumed the existence of such non-entities as the caloric fluid or the optical ether. However, in view of the growing compartmentalization of human knowledge, only a strong dose of animal faith can lead anyone to expect that the many disparate forms of understanding and inquiry which are alive today are actually converging.

9 Likewise, I am quite confident that five hundred years from now current measurements of the microwave background radiation will be regarded as excellent approximations, to whatever decimal we now think has been reached; but I am not so sure that the interpretation of that phenomenon as a remnant of the very hot, very dense universe of current cosmology will be still accepted and will not have been displaced by a drastically different conception.

10 See also Hawking 1975, 1976, and his article in the *Scientific American* (1977).

11 This is how this problem is described by Hawking and Israel (1979, pp. 17-18):
"...the area of the event horizon of a black hole has the property that it can only increase and not decrease with time. This led Beckenstein in 1972 to suggest that it might be connected with the thermodynamic quantity, entropy, which measures the degree of disorder of a system or one's lack of knowledge of it. He pointed out that the 'no-hair theorem' implied that a very large amount of information about a star was irretrievably lost when it collapsed to form a black

hole and he claimed that the area of the event horizon was a measure of this unobservable information which could be regarded as the entropy of the black hole. Other analogies between classical black holes and thermodynamics were found by Bardeen, Carter and Hawking (1973) but there was an apparently insurmountable obstacle to attributing a finite entropy to a black hole because that would imply that it should have a finite temperature and should be able to remain in equilibrium with thermal radiation at the same temperature. However this seemed impossible because the black hole would absorb some of the radiation but, by its very definition, it would not be able to emit anything in return. The paradox remained until Hawking (1974) discovered that applying quantum mechanics to matter fields in the background geometry of a black hole metric led to a steady rate of particle creation and emission to infinity. The emitted particles would have a thermal spectrum with a temperature proportional to the surface gravity of the black hole, which is a measure of the strength of the gravitational field at the event horizon and which is inversely proportional to the mass. This emission would enable the black hole to remain in equilibrium with thermal radiation at the same temperature."

12 Einstein, GP 5, 82.

13 In 1913, speaking to the 85th *Naturforscherversammlung* in Vienna, Einstein said that Newton's "laws of gravity and of the motion of heavenly bodies... have proved to be so exactly right that, from the standpoint of experience, there is no decisive ground for doubting their strict validity" (Einstein 1913, p. 1249). Giulio Maltese and Lucia Orlando (1995) draw attention to another, presumably stronger motivation for General Relativity, viz. the difficulties met in developing a general, consistent description of accelerated frames of reference in the context of Special Relativity.

14 For a detailed analysis of Einstein's explanation of Mercury's perihelion advance, see Earman and Janssen 1993.

15 I write down here the two sets of field equations for comparison:

$R_{ik} = -\kappa T_{ik}$ (11 November 1915)

$R_{ik} = -\kappa (T_{ik} - {}^1/_2 g_{ik} T^s_s)$ (25 November 1915).

Obviously, both sets of equations imply that $R_{ik} = 0$ wherever $T_{ik} = 0$.

16 In the polar coordinates (r,φ,θ,t) used by Schwarzschild, the solution is also undefined on the hypersurface $r = 2m$ (if the units employed are such that the gravitational constant equals 1). However, this singularity can be eliminated by using other coordinates and is thus inessential. The singularity on the symmetry axis $r = 0$ is essential.

17 I have taken the standpoint from which Einstein and his contemporaries could maintain in 1916 that Mercury's precession anomaly had been solved. Today one calculates the trajectories of a system of slowly moving particles, bound together by gravitational interaction, like the sun and planets, by means of the so-called parametrized post-Newtonian or PPN formalism. This is a method of approximative calculation which is neutral between several chronogeometrical theories of gravity, yielding the predictions of a particular theory when certain parameters are adjusted in a prescribed way (see the table in Will 1981, p. 117). General Relativity puts two of these parameters equal to 1 and the others equal to 0, but this privilege could be due to the fact that the PPN formalism stems from work done by Einstein and his collaborators on that very theory, in quest of a derivation of its equations of motion. As the term 'formalism' indicates, the PPN formalism is not meant as a conceptual framework for the *understanding* of phenomena and does not pose as a contribution to the "scientific world-view".

REFERENCES

J. M. Bardeen, B. Carter and S. W. Hawking, "The four laws of black hole mechanics", *Communications in Mathematical Physics*. **31**, 1973, 161-170.

J. D. Beckenstein, "Black holes and entropy", *Physical Review* ***D.*** **7**, 1973, 2333-2346.

J. Droste, "The field of a single centre in Einstein's theory of gravitation, and the motion of a particle in that field", *K. Nederlandse Akademie van Wetenschappen. Proceedings* **19**, 1916, 197-215.

J. Earman and M. Janssen, "Einstein's Explanation of the Motion of Mercury's Perihelion", in J. Earman, M. Janssen and J. Norton, eds., *The Attraction of Gravitation: New Studies in the History of General Relativity*, Birkhäuser, Boston, 1993, 129-172.

A. Einstein, (CP): *The Collected Papers of Albert Einstein*, Princeton University Press, Princeton, 1987.

A. Einstein, "Zur Elektrodynamik bewegter Körper", *Annalen der Physik*. (4) **17**, 1905, 891-921.

A. Einstein, "Zum gegenwärtigen Stand des Gravitationsproblems", *Physikalische Zeitschrift* **14**, 1913, 1249-1266.

A. Einstein, "Zur allgemeinen Relativitätstheorie", *K. Preußische Akademie der Wissenschaften. Sitzungsbericht*, 1915a, pp. 778-786.

A. Einstein, "Zur allgemeinen Relativitätstheorie (Nachtrag)", *K. Preußische Akademie der Wissenschaften. Sitzungsberichte*, 1915b, pp. 799-801.

A. Einstein, "Erklärung der Perihelbewegung des Merkur aus der allgemeinen Relativitätstheorie", *K. Preußische Akademie der Wissenschaften. Sitzungsbe-richte,* 1915c, pp. 831-839.

A. Einstein, "Die Feldgleichungen der Gravitation", *K. Preußische Akademie der Wissenschaften. Sitzungsbericht,* 1915d, pp. 844-847.

Galileo Galilei (EN): *Le opere,* Nuova ristampa della Edizione Nazionale, 20 vols. G. Barbera, Firenze 1964-1966.

C. Glymour, "Indistinguishable space-times and the fundamental group", in Earman et al. *Foundations of Space-Time Theories,* University of Minnesota Press, Minneapolis 1977, pp. 50-59.

S. W. Hawking, "Black hole explosions?", *Nature,* **248**, 1974, 30-31.

S. W. Hawking, "Black holes and thermodynamics", *Physical Review* **D. 13**, 1976, 191-197.

S. W. Hawking, "Particle creation by black holes", *Communications in Mathematical Physics* **43**, 1975, 199-220.

S. W. Hawking, "The quantum mechanics of black holes", *Scientific American* **236** 1, 1977, 34-40.

S. W. Hawking and W. Israel, eds. *General Relativity: An Einstein Centenary Survey,* Cambridge University Press, Cambridge, 1979.

P. Kitcher, *The Advancement of Science: Science without Legend, Objectivity without Illusions,* Oxford University Press, New York 1993.

L. Laudan, "A confutation of convergent realism", *Philosophy of Science* **48**, 1981, 19-48.

D. Malament, "Observationally indistinguishable spacetimes: Comments on Glymour's paper", in Earman et al. *Foundations of Space-Time Theories,* University of Minnesota Press, Minneapolis 1977, pp. 61-80 (Minnesota Studies in the Philosophy of Science **8**).

G. Maltese and L. Orlando, "The definition of rigidity in the special theory of relativity and the genesis of the general theory of relativity", *Studies in the History and Philosophy of Modern Physics* **26B**, 1995, 263-306.

S. Psillos, "Theories and models in scientific processes", *Studies in the History and Philosophy of Science* **25**, 1994, 159-190.

S. Psillos, "A philosophical study of the transition from the caloric theory of heat to thermodynamics: Resisting the pessimistic meta-induction", *Poznan Studies in the Philosophy of the Sciences and the Humanities* **44**, 1995, 105-133.

H. Putnam, *The Many Faces of Realism,* The Paul Carus Lectures, Open Court, LaSalle, IL, 1987.

K. Schwarzschild, "Über das Gravitationsfeld eines Massenpunktes nach der Einsteinschen Theorie", *K. Preußische Akademie der Wissenschaften. Sitzungsberichte,* 1916, pp. 189-196.

C. M. Will, *Theory and Experiment in Gravitational Physics,* Cambridge University Press, Cambridge 1981.

PETER GALISON

RANDOM PHILOSOPHY

1. INTRODUCTION

Somewhere, in any computer-based Monte Carlo simulation, is a line of code that produces random numbers. With names like RANDU, RANDOM, RANF, or RNDM, the command is innocuous, a single, practically invisible step in a program that could run to hundreds of lines. Indeed, in the 1960s, the then-standard IBM 709 came oufitted with a random generator that was used around the world to produce simulations of phenomena ranging from nuclear weapons and airplane wings to the impact of pions on protons, from weather modelling to the analysis of number theory. In this brief paper, I want to take out the philosophical magnifying glass and peer into the epistemological and metaphysical changes at work behind the code.

The idea of a Monte Carlo is not complicated: by sampling randomly from a set of points one can often approximate a volume. Throw darts at a circle inscribed in a square and count the ratio of points in the circle to the number of hits inside the square (including both those in the circle and those between the circle and the perimeter of the square). That ratio gives an approximation of the ratio of the circle area to that of the square ($\pi/4$). Similarly, one could take a function $f(x)$ and rough out its integral $\int_a^b f(x)dx$ by "throwing darts" at the rectangular region defined from $x = a$ to $x = b$ and by an interval of values of $y = f(x)$ from some point below the minimum of f(x) in [a,b] and the maximum of $f(x)$ in [a,b]. The ratio of hits below the curve $f(x)$ to the number of hits not below the curve then approximates the ratio of the area $\int_a^b f(x)dx$ to the area of the whole rectangle. In other words one uses random points to fill out a series of pairs (x_i, y_i) counting as hit points such that $y_i < f(x_i)$ and divides by the total number of pairs tried—this approximates $\int_a^b f(x)dx$.

Generalizing from such considerations, the Monte Carlo can be used to sample-estimate the definite integral of a high-dimensional definite integral, a feature valuable in all manner of problems from hydrodynamic flow to the intricacies of thermonuclear weapons design in which radiation transport, fusion, Compton effects, and a myriad of other processes are at work. Strikingly—amazingly it has seemed to some—these approximations could, with increasing numbers of sample points, converge to the "true" value of the definite integral at a rate far exceeding any previously-known method of numerical integration.

The Monte Carlo Method, invented by mathematician Stanislaw Ulam and mathematician/physicist John von Neumann towards the end of World War II, was

E. Agazzi and M. Pauri (eds.), The Reality of the Unobservable, 123-128.

initially used to calculate the physics of fission weapons, especially the physics of neutrons in the chain reaction. Such tasks gobbled random numbers at an alarming rate, and without a ready and reliable supply, the process of launching trial "throws" such as the choice of (x_i, y_i) would come to a screeching halt. Generated from electrical noise, from the alpha decay of radioactive minerals, and from the arrival times of cosmic rays, these "true randoms" were compiled and published in books. Putting aside the conceptual difficulty of taking a published hardbound volume of true random numbers off the shelf (how often could people begin their series 947826 and still consider it random?) the pedigree of these digits was impeccable. "True randoms", physically-generated random numbers, could be thought of as truly standing for the world: what could be more natural than using alpha-decay-produced randoms to model the diffusion of neutrons inside a reactor core? Even when the process to be modelled was not itself random—as in the evaluation of a finite integral—numbers were needed fast and plentifully.

To accelerate the early simulations to the point where they could be useful, von Neumann introduced the notion of pseudo-random numbers. Instead of plucking digits off a list of the true randoms, von Neumann proposed that a suitably chosen algorithm could generate a series of digits that "for practical purposes" could serve more or less as well. For example, one could take an eight-digit number (from a "true random" book, if one so wished) square it and excise the middle eight digits. This new group became the seed for the next one: square the eight digits and extract the central eight digits of the result. The computer itself could continue in this way, *ad libitum.*

Now from the instant of von Neumann's proposal, it was manifest to him that this sequence of pseudo-randoms was not and could never be "random." There are, after all, only 10^8 eight-digit numbers. So in at most 10^8 repetitions of this game, an eight-digit sequence will repeat. When it does, as sure as the sun rises in the east (more surely, actually) the algorithm will repeat the exact sequence of numbers it got the first time around. So the sequence is not random—much worse, it is an eternal, precisely repeating cycle. We are living in a state of sin, von Neumann confessed to a colleague, when we use Monte Carlos. Still, the sequences and their owners worked with a pragmatic confidence. If the sequence passed its tests all was fine. Or was it?

To certify the pseudo-random generators, physicists and their mathematical colleagues developed a series of tests. Were the digits equitably distributed—that is, were there roughly the same number of 0s, 1's, 2's, ... and 9's? Were there correlations between the ith and $(i+k)^{th}$ digit, where k was 1, 2, 3, . . . up to some reasonable value of *k*? One author somewhat cynically captured the philosophy of the idea this way: "if a pseudo-random number has passed a certain number of tests, then it will pass the next one where the next one is the answer to our problem."[1] It was, nonetheless, hoped that generators could be designed such that repetitions and correlations could be excluded at least for a long-enough sequence for the simulation to run. "Tests of randomness" would vouchsafe the "approximate randomness" of the series.

But as James's wry motto made clear, it was not at all clear *which* tests, of the infinite number that could be made, would actually be most useful as guardians of the true random faith. Just how fragile the whole apparatus was became manifest in 1963 when high energy physicist Joseph Lach became suspicious of the generators he and everyone else had been using.[2] Instead of asking after the equidistribution and k-correlations (correlations between the ith and $(i+k)^{th}$ term), Lach pursued the correlations of triplets on at least a fraction of their range.

More specifically, he called up the Fortran number generator RANNO (which produced pseudo-randoms between 0 and 1.0), and had the computer sequentially hunt down the first number less than 0.1. He then programmed the computer to plot the next two numbers, designated x_1 and y_1, as a point on an xy plot on his CRT. Then the computer hunted down the next number that was less than 0.1 and once again made a pair out of the next two numbers, designating them (x_2, y_2) and popping them onto his screen. Continuing in this way, the computer "ought" to have filled its screen with (x_i, y_i) equitably splashed over the screen like the gray static of a TV set. It didn't. Instead of "randomness" Lach saw sharp black diagonal lines—bands of no points at all. The message was as disturbing as it was clear: the black bands announced that RANNO, having passed its other tests with flying colors, had failed the triplet correlation test in the region $0 < z_i < 0.1$. Even if one cooked up a pseudo-random generator that did pass the triplet test, the question lurked: would it do so for quadruplets, quintuplets, and higher multiplets? Virtually every computer in the world was producing manifestly unrandom randoms. The source of the difficulty was, as George Marsaglia of Boeing Scientific Laboratories showed, in the very nature of the so-called multiplicative congruential generators of the form

$$r_i = [a\, r_{i-1} + b]\ (mod\ m)$$

where m was 2 raised to the word size of the computer in use. For example for a 5-bit word one held $m = 32$, and (for example) choosing arbitrarily $a = 21$, $b = 1$, $r_0 = 13$, yields

$$r_1 = [(21)(13) + 1]\ mod\ (32) = 274/32,$$

which has a remainder of 18. Now set $r_1 = 18$ where r_0 was, and we get $r_2 = 27$. In this way we obtain

$$\{r_i\} = 13, 18, 27, 24, 25, 14, \ldots$$

Using an elegant result from Minkowski's *Geometry of Numbers*, Marsaglia demonstrated that the hyperplanes would always occur, no matter what the choices of the parameters were in the algorithm. "For the past 20 years," Marsaglia sadly concluded, "such regularity might have produced bad, but recognized, results in Monte Carlo studies," as multiplicative congruential generators quietly spewed regular hyperplane bands into models of baryons, bombs, and biology.[3]

Even without the Lach-Marsaglia result, pseudo-randoms were in trouble. S.K. Zaremba, for one, denounced the belief that there are a set of properties, exhibited by a test or set of tests, that single out stochastic processes.[4] Furthermore, he argued, the deterministic aspect of the pseudo-random generator made a mockery of the very category of a calculation of variance within a Monte Carlo. For these and other reasons, he turned to replace pseudo-randoms with a new kind of generator, one known as a quasi-random generator. This shift I consider to be of greater epistemic import even than the shift from the true to the pseudo. The idea is this: instead of attempting in ever fancier ways to capture some essential feature of the random, the very ideal of random is taken down a notch, out of the exalted position as a privileged epistemic and metaphysical category. Do not look to secure the randomness of the pseudo-randoms say the quasi-ists, look for the most efficient and powerful distribution of numbers to solve the problem—whatever distribution that may demand. Indeed randomness is taken on the quasi scheme to be just a certain degree of "clumpiness"; and the question becomes (quantitatively) not how to achieve that specific amount and type of clumpiness, but rather the precise degree of clumpiness of sampling that is best at estimating (for example) an integral.

"Local discrepancy" was a term introduced by Hermann Weyl in the early 20th century to quantify the clumpiness of a distribution of points in space.[5] It is defined as

$$g(x) = v(x)/N - x_1 x_2 x_3 \ldots x_d$$

where d is the dimension of the unit cube. So for example in a three-dimensional unit cube we imagine a set of N points distributed in some manner. $v(x)$ is the fraction of those N points contained in a subvolume defined by the origin and the points of intersection of the perpendiculars dropped from x to the walls of the unit cube. $g(x)$ therefore gives the deviation from equidistribution, and it can be postive or negative. Many norms can be defined for $g(x)$ including the extreme global discrepancy defined on a set of points S as

$$D(S) = sup \{|g(x)|\} \text{ for } x \text{ in } S$$

where sup is the supremum of the set. The result — the crucial result of the new method of quasi-randoms — was that for $D(S)$ less than that of randoms (less clumpiness than that characteristic of randomness) the approximation of an integral could be better than random, and in some cases much better convergence of this error term

$$E = |\int f(x)dx - 1/N \sum_k f(x_k)|$$

occurred not with $1/\sqrt{N}$ as in a random Monte Carlo, but rather as $1/N$; and even the $1/\sqrt{N}$ was an enormous improvement over the many regular quadrature schemes in play since the time of Archimedes.

I take three philosophical lessons from this piece of buried Fortran code: an end to the epistemic and metaphysical dominance of randoms in simulations, a sharp blow to the picture of simulation as a literal re-presentation of a natural process, and finally a challenge to our deeply held notions of typicality. In order:

2. DESACRALIZATION OF THE RANDOM

The great discovery and proliferation of computer-based Monte Carlos in the 1940s was that random sampling took $E \rightarrow O$ faster than ordinary numerical quadrature. And somehow, I suspect, that virtue was laid at the door of the random. In a certain way, this no doubt seemed reasonable. Wouldn't it in some sense be best to close one's eyes (so to speak) and then pluck a ball from an urn—wasn't randomness just the notion of studied abstinence from intervention when sampling—justice blindfolded? And at the *metaphysical* level wasn't there something special about a random process, something picked out as a natural kind by nature itself? Here the move from pseudo to quasi Monte Carlo is brusquely pragmatic and utterly deflationary. Randomness becomes no more than the choice of the sample support $\{x_i\}$ yielding a particular value of the clumpiness, $D(S)$. But other schemes were also possible, and the quasi view became simply that $D(S)$ would be chosen to speed the convergence of E to zero, even if it meant taking a value of $D(S)$ that did not resemble the clumpiness of a truly random set. Forget the epistemology of blind justice, forget the metaphysics of a special natural kind corresponding to randomness. Efficiency of convergence matters, and $D(S)$ would be found to suit the problem at hand.

3. THE END OF SIMULATION AS RE-PRESENTATION

For some years after the introduction of Monte Carlos, and even in the 1990s in certain cases, authors frequently emphasized the particular value of Monte Carlos when modelling processes were themselves thought to be random. In this sense the metaphysical correspondence between model and physical system was grounded in the circumstance that both were random. The model was a true Monte Carlo because both it and the phenomenon modelled were of the same natural kind. Such a view was problematic from early on—after all Ulam knew perfectly well that the Monte Carlo could be used to track deterministic processes and even nonphysical mathematical formulae. And with the introduction of pseudo- randoms the "natural kind" correspondence ought to have faded that much more. Still, the view survived. Quasi Monte Carlos, on my reading, ought to eliminate that view for good—the choice of the support set $\{x_i\}$ is made to minimize error fast; random $\{x_i\}$ were useful for a few decades because they beat out the regular $\{x_i\}$ of quadrature, and that is all. Now it was the turn of the randoms and their pseudo cousins to be left behind.

4. THE NEW TYPICALITY

Finally, and this must here remain only a suggestive remark, it seems to me that we have the opportunity to examine the notion of typicality up close. When we choose the support set $\{x_i\}$ in order to estimate an integral, we want above all to choose typical points. For a smooth periodic function it may well be that steady Archimedean footsteps from $x = a$ *to* $x = b$ are perfectly good. The problem comes when those steady footsteps fall in resonance with the behavior of the function—the nice regular sampling of $\sin^2\theta$ from 0 to 10 π at intervals of π shows us that the integral of sine squared over that region is exactly zero. Bad job. Random selection aims to find typicality by avoiding all and any possible resonance between function and sample. But what we have learned from Monte Carlo is that even the random has a less than maximally efficient characteristic in extracting information from the integral: it is too bunched. It seems that typicality lies somewhere between Archimedes and Chaos, between, so to speak, the ideals of systematicity and blindness. I conclude these comments with what amounts to a double twist. Some years ago, the authors of a book of random numbers started with a set of true randoms and recoiled in horror as they found certain repetitive strings. To "improve" their set, they stuck in new digits, breaking up the unrandom look of the subsequence.[6] For years this circumstance has been regaled by statisticians as the height of perversity, illustrating how foolish it was to expect random numbers not to occasionally spew out a train of 7's. But the joke is now reversed: breaking up the strings of 7's might have been not a bad idea at all, as it probably served to lower the global discrepancy and might well have improved the convergence of Monte Carlos that made use of the doctored set. Now I don't recommend preparing the $\{x_i\}$ this way, but it reveals to us the twisting path of a random philosophy.

Harvard University - U.S.A.

NOTES

1 See James 1987.
2 See Lach 1963.
3 See Marsaglia 1968.
4 See Zaremba 1969.
5 See Weyl 1916.
6 See Lopes 1982.

REFERENCES

F. James, "Monte Carlo Theory and Practice", in Ferbel, *Experimental Techniques in High Energy Physics,* Menlo Park, Addisqã-Wesley, 1987, pp.627-77, on 656.

J. Lach, *Memorandum no. 27, Random Number Generators*, Yale Computer Center, 29 October 1963.

L. L. Lopes, Doing the Impossible: A Note on Induction and the Experience of Randomness, *Journal of Experimental Psychology* **8**, 1982, pp.626-636, on p.629.

G. Marsaglia, "Random Numbers Fall Mainly in the Planes", *Proceedings of the National Academy* **61**, 1968, pp.25-28, on p. 25.

H. Weyl, "Gleichverteilung", *Mathematische Annalen* 77, 1916, 313-352.

S. K. Zaremba, *The Mathematical Basis of Monte Carlo and Quasi-Monte Carlo Methods, Studies in Applied Mathematics* **3**, Philadelphia, Society for Industrial and Applied Mathematics, 1969, pp. 1-11.

MARIO CASARTELLI

FORMAL REPRESENTATION AND THE SUBJECTIVE SIDE OF SCIENTIFIC REALISM

Key words in my talk will be "way of representation", "way of notation", "symbolical formalization", and similar ones. These concepts, deserving many distinctions in principle, are grouped here on the basis of an elementary observation: experience and common sense testify indeed that, in all scientific disciplines, the perspective on certain problems and the possibility of solving them can be drastically modified by the introduction of a new way of representation. In this notion, besides high level formalizations, I shall consider therefore also the concrete embodiment in signs, figures, graphical devices, methods for writing, short cuts to symbolization, etc. A desordered list of examples may include:

- figures in elementary geometry;
- positional notation for numbers;
- structure formulas in chemistry;
- algebraic notation for calculus;
- all kind of diagrams, from Descartes to Feynman;
- Argand-Gauss plane for complex numbers;
- Boolean algebra;
- Symbolical Dynamics;
- Turing machines;
- nonstandard notation for infinitesimals, etc.

Before going into the argument, some factual remarks are opportune. It is a truism that every problem, before being solved, must be thought. Besides making easier the solution of an existing problem, a new way of representation can open a whole range of previously *unthinkable* themes, independently of the particular argument in relation to which it has been introduced. As if in the notation there were much more than the inventors could suspect, previously invisible objects become visible. A standard example is the Argand-Gauss representation for complex numbers: as a visualization of the roots of an algebraic equation it seems almost trivial; but only through this representation, otherwise inconceivable arguments, such as the geometry of the Riemann surfaces, or the strange properties of the Zeta function, came out from pure virtuality. I don't insist on other classical examples, such as the positional notation for integers, or the habit of representing points through coordinates. Nowadays, as frequently pointed out, a great notational expansion is due to the programming languages: besides the brute computational power, the habit of looking at certain problems in terms of "steps of program" is going to change the point of view and the sensitivity about old arguments, and to introduce, in addition, a variety of new concepts. Chaitin and Kolmogorov ideas on the algorithmic complexity, for instance, and in general the rich development of

E. Agazzi and M. Pauri (eds.), The Reality of the Unobservable, 129-135.

arguments around computability, strongly depend on the ways of representation involved with programming languages. Graphical devices are going to operate in the same manner.

It is also well known that, in some important cases, the availability of the formalism fitting a problem has preceded it. Mathematics and theoretical physics offer a rich repertoire for this: non euclidean geometry, tensorial calculus and functional analysis, with respect to relativity and quantum mechanics, are almost obvious examples[1]. To cite a more recent case, dimensionality seen through the covering of sets, instead that in terms of linear dependence, appeared as a curious representation of an old concept, a mathematical toy, until recently, when its usefulness has revealed in connection with studies on turbolence and chaoticity. And boolean algebra, of course, was inventend before computers.

The first feature I want to stress in the notation is therefore its *fertility*, a sort of creative potentiality, or power of suggestion: a representation establishes in advance a guideline for thought, opening previously invisible doors to reality.

A plausible objection is that, in some examples above, I indicated the notation as a way of representation for mathematics, while in other examples mathematics itself, as a whole, was considered a way of representation. There is no contradiction. To begin with, the formalism can be, but not necessarily must be, a scientific achievement in itself. For instance, the notation for calculus, from Newton and Leibniz to Robinson, is deeply intrigued with the conceptual development of the argument, and could not be separated from it. On the contrary, certain diagrammatic techniques, as those used for electrical or electronics circuits, are hardly reminded as scientific results. They rest anonymous contributions, simply useful for practical purposes.

Moreover there subsists a hierarchy in representation, a sort of "chinese boxes" structure (not a completely ordered one, however), and what is science at a certain degree may be representation at another one. Mathematics is a way of representation and notation for other disciplines, but, *within* mathematics, the exigence of representation exhibits the phenomenon in its purest form. There, more often than elsewhere, it can be grasped *statu nascenti*: to insist on mathematical notation, therefore, means to minimize (not to eliminate) the risk of misunderstanding between way of representation as a tool for thinking and its *possible* scientific meaning. It means also to propose at the lowest hierarchical level (notation vs. mathematical concepts) the same philosophical questions usually discussed at the most demanding and difficult links of the chain (mathematics vs. physics, or physics vs. empirical experience).

Surely, this issue does not constitute a new item in historical or philosophical considerations. Under the documentary point of view, we have extensive studies for almost every aspect of scientific notation[2]. This kind of researches, however, seems closer to erudite investigations than to epistemology. On the properly philosophical side, there is the rich fields of semiotics and linguistics (but I will not deal with this domain of studies), while in the mainstream of gnoseology the panorama is not particularly crowded. However, one should recall at least the "Philosophy of Symbolical Forms" by Ernst Cassirer[3], probably the most ambitious and serious attempt to include the scientific symbolism in a general theory of knowledge. The care devoted there to situate symbolism in the evolutive growth of human culture (a tribute payed to Hegel) dims somewhere the cognitive problem, yet well present and important. Cassirer faces up to the task within the framework of kantian transcendental categories, without drawing too demanding consequences on the ontology of symbolism. Besides the insistence on the active role of symbolical intelligence, that we shall resume later, in his work there remains, as a most precious suggestion, the documented necessity for a unitary approach to myth, language and scientific representation[4].

If the cognitive problem connected to the ways of representation has been sometimes overlooked, this is probably due to the synergic action of many factors, or prejudices. A series of them can be summarized in the following scheme:

- the formalization is only a tool, with an instrumental role in the scientific discovery (in other words, it regards the way to get the result, not the result in itself); therefore
- the formalization is pertinence of history and psychology, not of epistemology; then
- the formalization may be studied within the evolutionary development of the culture, but is only collateral to the cognitive content of the "objective" knowledge; and therefore, finally,
- it is external to the debate on scientific realism.

It is precisely the pattern above that I would focalize critically, with a double purpose: to stress that the way of representation has directly to do with the argument of our conference (observability, etc.); and to examinate some correlated consequences on the constitution of the scientific subjectivity.

"Representation" (*re-presentatio*) is a term endowed with a platonic flavour: it indicates the act of "making present again" something that was not. In this sense[5] the representation (and therefore the language) is intrinsecally linked to the past, where "past", as usual in a platonic frame, is not to be intended in a strictly chronological sense. The "fleeting moment" is by definition non formal. (Someone would exceed, by saying: non real). To make an intellectual object, to present it to the mind, means to *re*-present it. And viceversa: to represent something, means to evoke an object out from an only potential existence. Actually, the perspective on such a viceversa brings the scientific praxis near to the classical debate about the reality of *Universalia*. The innocent, "technical" act of introducing a notation can touch, in principle, philosophical questions usually reserved to fundamental concepts and entities[6].

There are many aspects to clarify. For instance: does every representation deserve such considerations?

The answer is definitely *No*. Every scientist, every person, has his own private repertoire of imaginative fancies that help him in handling otherwise abstract concepts. Sometimes such a private dimension briefly appears on the scene of scientific activity[7]. But, differently from inter-subjective formalism, mental hookings of this kind are confined to a personal and singular utilization. They share the strictly private psychic *aura* that everyone attaches to objects, remaining, in general, neither communicable nor fertile. Thus, the possibility of an individual dimension of imagination does not contrast, but enforces the peculiarity of those imaginative devices we are speaking about. Such a diversity escapes investigations around "scientific imagination", as those developed (to quote a serious version of them) by G. Holton[8]. Certainly, as Holton suggests, the difficulty of reading the development of scientific ideas in the defined form of well arranged theories has to be stressed; and due relevance must be given to the great variety of psychological, cultural, sociological contexts concurring in this process. But such an imagery, once collected, disregards the directly cognitive valence of the forms. In this kind of investigations there is the risk of a psychologistic reduction of imagination to fancies, may be brilliant fancies, that only afterwords, at the moment of a genuine scientific discovery, will get their meaning (such a meaning, then, would be a corollary of the very scientific result). How such a mass of subjective, sometimes nonsensical suggestions, may drive to realistic speculations? People shadowing this possibility probably participate an implicit cultural darwinism, an attitude that dares on the efficiency of trials-and-errors processes. Now, even if questionable in general, let's suppose the validity of such an evolutionistic scheme for the selection and establishment of good ways of representation with respect to private fancies: at most, this would explain *how* representations arise, not *what* they are. The fitting of a problem (or category of problems) may follow indeed an adaptative process; but the capability of evoking new themes, of opening new intellectual perspectives, in short the creative side of representation that I have called fertility, regards the very nature of the representation in reference to the objective world, beyond intentions, expectations, exigences and ideologies of their inventors. Beyond the false universality, also, of collective psychic phenomena.

Which is, then, the nature of the *inventio* in a "good" representation? Is there a cognitive and ontological presupposition in the activity of symbolical intelligence, as a faculty of proposing new representations?

The question is situated at the crossroad between scientific subjectivity and observability. Afterwords, once a class of arguments has become the object of scientific analysis, the previous difficulty (or impossibility) in looking at them resembles blindness, or the language nuisance known as *aphasia*. But the function of a representation is not limited to force an intellectual object to appear. A new formalization changes the way of seeing[9], and this is true in many senses, first of all because it changes the feeling about what is to be a legitimate or an absurd entity. Nowadays, for instance, it is difficult for us to understand what scandalized Greek philosophers about irrational numbers. Gauss himself changed completely his view on the nature of imaginary numbers after the assimilation of *his* formalism[10]. May be that in future, once the representation of separated continua for space and time will be supplemented by other formal habits, the quantum paradoxes will be equivalently erased from the common sense (inasmuch as paradoxes, not as physical schemes). When using the algebraic notation on manifolds, one deals with many-dimensional or infinite-dimensional entities, his sensorium is being enlarged. To speak of sensorium is justified by the fact that the evoked entities, as in the realism-nominalism contest, have the formal properties of concrete objects, whose features seem to be discovered, not to be invented. Thus, were one to be devoted to the sensist principle, *Nihil est in intellectu quod non fuerit prius in sensu*, this should be at least updated by the inclusion of those objects that *have become* perceptible thanks to the representation. In terms of our conference title, representation is the missing link between observability and reality.

Moreover, the operations of symbolical intelligence escape the tight spots of an axiomatic structure, in the sense that they are not directly deducible from axioms, but at most, they result afterwords to be *consistent* with them. It is also well known that a representation transfers on algorithmically explicit rules a relevant part of the difficulties previously committed to abstract intellectual intuition. This makes "computable", and, in some cases, almost trivial what was not such[11], whereas the really new content of a problem or discovery consists in its component which is irreducible to an algorithmic procedure (obviously, such a component too could become computable, thanks to *a new* representation). There is therefore a double mouvement: the algorithmic content of a theory, on one side, is enlarged by the introduction of a good representation; but, on the other side, this very introduction remains non-algorithmic, irreducible. It is the creative component of a representation that switches to a computative attitude and praxis.

At this point, relations between representations, concepts and reality should resume a long aged discussion about the nature of signs and symbols. This escapes our present possibilities. However, the creative fertility of a good representation is again an useful key to grasp its peculiarity. An arbitrarily conventional sign (the phone number for a person) cannot be fertile, i.e. it cannot "produce reality". But also a natural sign in the classical sense (the tracks on the ground as natural sign of the animal) is a passive product of the circumstances, and it cannot actively interact with reality, as a true symbolism does[12]. These claims to realism should not be misunderstood: not even a platonic fundamentalist would pretend that the graphical mark used to indicate the transistor in a circuit diagram has an archetypal model existing *in se*. But, once this trivially conventional part of the representation has been properly weighted or erased, a circuit diagram is far from being "conventional". Its correspondence to reality (i.e. "the circuit", inasmuch it is the target of the representation[13]) is cómpletely meaningful: it allows, for instance, working projects.

Therefore, the role played by the representation may be intended as a final completion of the process of mediation between abstract entities and formalized entities that is necessary in every transmissible knowledge, particularly in scientific knowledge. The category of formal mediation is the missing point, the lever fulcrum that, starting from the Archimedean motto ("*da ubi consistam, et terram coelumque movebo*"), has assumed, not by chance, a metaphorical valence in common language.

This point has been well focalized by Cassirer just at the beginning of his *magnum opus*[14], where he criticizes the concept of the symbol as a "mirror" of reality, or, in other words, the idea of the image as a copy: idea, in his opinion, that still underlies Herz's conception (so important indeed, his real starting point). The value of the symbols, Cassirer says, does not consist in the mirroring of a determinate existing thing, but in the result they furnish as a tool of knowledge in the unity of phenomena that they produce by themselves.

It is strange that this initial point is felt by Cassirer as a withdrawal: «By such a critical conception, science has left the pretention of grasping and reproducing the reality directly. It understands that in truth every realizable objectivation is a mediation, and it has to remain a mediation.» May be that in such an embarrassment there is a further key to the underestimation of our argument in modern speculative philosophy. Since no direct correspondences are possible between pure, abstract intellectual intuition and the concreteness of phisiologically perceptible world, necessarily there is an intermediate region, that shares the features of an objective reality, but, unfortunately, also the ambiguities of a fictional and dreaming activity. Thus, this intermediate world is ill-famed with idealists because of their suspiciousness to the "copies" of the ideal world; it is ill-famed with materialists because of its immaterial constitution; it is ill-famed with pragmatists and formalists because of its pretension to an autonomous reality. Moreover, it is also ill-frequented by occultists, superficial supporters of natural olistic philosophies, etc. It is not surprising, then, that some of the best contributions on the ontology of symbolism come from somehow exotic fields, like e.g. Henry Corbin's researches on the iranic platonism and metaphysics, from Sohoravardi to Molla Sadra[15]. To remain in a more domestic frame, there is here the rediscovery of an ancient overlooked theme, that can be expressed with the concise words of Hugh of St. Victor: «If nothing there were mediating between body and spirit, neither the body could meet the spirit, nor the spirit the body.» *(Si nihil inter spiritum et corpus medium esset, neque spiritus cum corpore neque corpus cum spiritus convenire potuisset*[16]).

Going back to Cassirer's recognition of the irreducible necessity of a mediation, we observe that, surely, representation involves a sort of likeness, but not what is usually meant by "verosimilitude"; not a facsimile of appearences, but a doubling (in its proper context) of functional activity. The analogy with the principle of realism in traditional art, "Art imitates Nature in her manner of operation" [17], is striking: symbolical realism, therefore, as opposite to imitative naturalism. This also means that the fundament of realism is not to have mental copies of objects. Symbolical intelligence, in an old fashioned terminology does not imitate indeed the *natura naturata*, but participates the role of the *natura naturans*[18].

Such an attitude, which does not require the full awareness or the consent of the involved person, has to do with the inner morphology and ontology of the subjectivity as a function, not with the psichology of the subject as an individual. It recalls forgotten spiritual disciplines: to subordinate individual expressions to obiective exigences, to behave as an instrument for supraindividual (this does not mean collective) forces and inspirations, etc. Thus, may be that in scientific praxis (but *not*, generally speaking, in scientists ideology), the last remainders of intellectual habits otherwise estinguished after the decline and fall of the Renaissance natural philosophy are still surviving.

All I have said may be read at this point as an updated commentary to some principles stated by Cusanus[19]. I mean the terms where his truly pre-Kantian position about the impossibility of a direct speculation (contradiction in terms!) of the Being postulates the need of symbolical mediation: «Neither sense, nor imagination, nor even intellect can reach the way of the being [...] therefore there is no Science about the way of the being» ("*Essendi modum neque sensus, neque imaginatio, neque intellectus attingit [...] igitur de essendi modo non est scientia*" (Note here the progression from sense to intellect through imagination). However: «Everything that may be reached through any way of knowledge can be, at most, only signs of the previously existing way of the being, and therefore are not things but similitudes, species or signs of that one»: ("*omnia quae attinguntur quocunmque cognoscendi modo illum priorem essendi modum tantum significant; et hinc non sunt ipsa res, sed similitudines, species aut signa eius*"). Indeed: «Inasmuch we have an object of knowledge, it appears through signs.» (*Res igitur, ut cadit in notitia, in segnis deprehenditur*). Finally: «It is therefore convenient to seek different ways to knowledge through different signs». (*Oportet igitur ut varios cognoscendi modos in variis segnis quaeras).*

Parma University
Italy

NOTES

1 The theory of infinite dimensional spaces of square integrable functions, for example, was introduced by Hilbert, Fréchet, Schmidt and Riesz between 1907 and 1910, twenty years before the axiomatization of quantum mechanics. As to the tensorial calculus developed by Ricci and Levi-Civita, it is well known that it was introduced to Einstein by G. Pick and M. Grossman only after 1911.

2 See, e.g., Ifrah 1994.

3 See Cassirer 1923.

4 An interesting testimony of the usefulness of such a unitary approach can be found in Hübner 1985.

5 I follow here a suggestion by Giorgio Colli: see Colli 1969.

[6] Paradoxically, scientists as Kronecker and Brouwer, with their almost pathological hostility against entities introduced through formalism, testify, in this respect, a sensitivity unknown to the pragmatic boldness of others.

[7] Interesting considerations in this connection can be found in an interview given by Alain Connes, in *Sciences et Imaginaire*: see Maréchal 1994. For the sake of our argument, see also the interview to C. F. von Weizsäcker, *ibidem*.

[8] See Holton 1978.

[9] It has been observed (for example by G. Durand) that the latin term for the pupil of the eye, *pupilla*, means – inasmuch its Greek and Arab counterparts – both *little girl* and *doll*; this would prove a primitive awareness of the presence, at the core of seeing, of an *idol* (*eidolon*, from *eidos*, as in *idea*) mediating the formation of perception. A view apparently shared by those naturalists who, after Redi, called *nympha* the intermediate status of insects.

[10] "The true metaphysics of $\sqrt{-1}$ is elusive" Gauss said in 1825. But, in 1831, "the intuitive meaning of the complex numbers is completely established and nothing else is necessary to admit them in the domain of arithmetics". References can be found in Kline 1972.

[11] It is sufficient to recall the computative fallout of the representation of functions by series.

[12] The insistence on the active and creative role of symbolical intellect runs across the whole work of Cassirer.

[13] Of course, this correspondence does not concern the microphysical entities (fields, particles...) by which the circuit is made up; the diagram, indeed, does not refer to the latter.

[14] See Cassirer 1923.

[15] See, e. g., Corbin 1978, 1979. Corbin has often stressed the affinity between this "oriental" idea of an intermediate world, where symbols are real, and certain concepts (e.g., the idea of *spissitudo spiritualis*) defended by H. More and the circle of Cambridge Platonists, not far from Newton. More questionable, not to say untenable, are Corbin's references to Junghian psycology.

[16] See Hugh of Saint Victor 1974.

[17] "*Ars imitatur Naturam in sua operatione*", Thomas Aquinas, *Sum. Theol.* **I** 117, 1. The crucial relevance of this point has been particularly stressed by A. K. Coomaraswamy. See, for example, Coomaraswamy 1977.

[18] In his essay in this volume, C. P. Enz speaks of "active realism", I presume in the same sense.

[19] See Cusanus 1974.

REFERENCES

E. Cassirer, *Philosophie der symbolischen Formen*, Wissenschaftliche Buchgesellschaft, Darmstadt 1923.

G. Colli, *Filosofia dell'espressione*, Adelphi, Milano, 1969.

A. K. Coomaraswamy, "A Figure of Speech or a Figure of Thought?", in *Traditional Art and Symbolism*, R. Lipsey ed., Princeton University Press, Princeton 1977.

H. Corbin, *Philosophie iranienne et philosophie comparée*, Buchet-Castel, Paris 1978.

H. Corbin, *Corps spirituel et terre céleste*, Buchet-Castel, Paris 1979.

N. Cusanus, *Compendium*, in *Scritti Filosofici*, I-II, I.G. Santinello ed., Zanichelli, Bologna 1974.

Hugh of St. Victor, *De Unione Corporis et Spiritus*, I, V. Liccaro ed., Sansoni, Firenze 1974.

G. Holton, *The Scientific Imagination: Case Studies*, Cambridge University Press, 1978.

K. Hübner, *Die Wahrheit des Mythos*, C.H. Beck'sche Verlagsbuchhandlung, München 1985.

G. Ifrah, *Histoire Universelle des Chiffres*, Robert Laffont, Paris 1994.

M. Kline, *Mathematical Thought from Ancient to Modern Times*, 1972.

I. A. Maréchal ed.: *Sciences et Imaginaire*, Albin Michel, Paris 1994.

GERHARD HEINZMANN

CONVENTION AND OBSERVABILITY

Poincaré once again[1]

The Kantian a priori-synthetic concept has given rise to *aporiae* which logical empiricism meant to overcome by postulating its two dogmas: the dogma of the separation of observational and theoretical language and the dogma of the possible reduction of empirical meaning to experience. It is well known that these dogmas have been criticised from Quine onwards. Quine denies that a crucial experience can determine the meaning of the observational terms and he even denies that the distinction between analytic and synthetic concerns a logical difference. It is equally known that, among those who were considered as forerunners of the logical empiricists, Duhem distinguishes himself by the fact that he seems to be at the same time a forerunner of Quine's criticism. The Duhem-Quine holistic thesis concerns the calling in question of the separation of theoretical language from observational language. It is less known that Poincaré too must be ranked among these forerunners who survive Quine's criticism; for Poincaré's conventions are neither analytic nor synthetic, but they are a kind of bicephalous selection of analytical but non-logical propositions, "guided" at the same time by experience. It has but rarely been attempted[2] to ascribe to the phrase "guided by experience" another meaning than that of a metaphor and it is probably not evident to do so without having first read the second Wittgenstein.

One of the reasons of the neglect of the prospects opened by Poincaré's standpoint is perhaps the current interpretation which associates holism with conventionalism. But it should be known that holism can be accompanied — and so it was historically in Duhem's works — by an empirical realism which has absolutely nothing to do with Poincaré's constructivism. One can indeed maintain the dichotomy between experience and theory, between what is given and what is said, simply by observing that one has made a quantitative error: the adepts of such a position refuse to confront a singular hypothesis with the facts drawn from experience but deem this confrontation possible in relation with the totality of the pertinent theoretical propositions[3].

This observation will show that, to dissociate Duhem from Poincaré, it is not necessary to adopt a definite position right away in the matter of the famous debate about the experiment of the parallaxes between Grünbaum, Putnam, Giedymin and others[4]. The solution of an empirical conflict brought about by the application of a geometrical theory with the help of a remetrication is interpretable as a mere reformulation of the theory, the physical and geometrical parts being respected, or as a epistemological change which would imply that the separation of the theoretical from the observational contents of the new theories does not correspond any longer to the separation of the former theories. I shall not examine this question, namely whether conventionalism is a merely linguistic thesis concerning the intertraductibility between

E. Agazzi and M. Pauri (eds.), The Reality of the Unobservable, 137-144.

different geometrical and physical languages, or if it is also an epistemological thesis concerning the vague and conventional borderline between the theoretical language and the observational language of physics. While I stress that Poincaré often conceives convention in the linguistic sense especially in relation with the question of the metrication of geometry, I have a philosophically more fundamental aspect in view: the point is to mediatize rigour or "artifice" and objectivity, to mediatize these two requirements by determining the conventional aspects in the *pre-theoretical* genesis of the geometric theory. What is to be done to avoid the following alternative which is a traditional one: either the impressions meet the theoretical requirements, so that the mind has nothing else to do than register the data in an adequate language or the impressions do not meet the theoretical requirements so that the linguistic construction misses its aim which is to specify the contents of these data[5].

Poincaré is regarded as the father of conventionalism. If one maintains that his conventionalism includes an aspect that is susceptible of unifying fact with freedom — this is at variance with the opinion of Meyer 1992, a book which has inspired the tone of this article[6] — the term "convention" seems to be ill-chosen. Jules Vuillemin has seen this rightly and proposes the term "occasionalism" which suggests other philosophical connotations. In order not to lay myself open to additional problems of a terminological nature, I shall abide by the term "convention".

At a first approach, Poincaré's geometrical conventionalism consists of three theses:

1. Experience does not relate to space but to empirical bodies. Geometry deals with ideal bodies and so it can be neither proved or disproved by experiences. Since the propositions of geometry cannot be analytical there must consequently exist geometrical propositions which are apparent hypotheses (SH, 24) or conventions.

2. The choice between conventions, particularly between different geometries, is guided by experience (SH, 75).

3. Euclidean geometry has nothing to fear from experience, for it is the most advantageous and convenient one (SH, 95).

Generally an empirical conflict involving the conventions of the first thesis is interpreted linguistically or epistemologically in the sense we have explained just now, the second point is considered as a metaphor and the third point as a belated temptation to introduce in conventionalism *post festum* a pragmatic profile, that is to give intuition a minor preference, after presupposing a rigid separation between theoretical language and observational or intuitive language[7]. I shall refer to this latter term in due course.

On the contrary I myself suggest an interpretation which pays due tribute to Poincaré's constant care not to separate the comprehension in mathematics from the context of its logical and historical genesis. Indeed genetic considerations in relation to the second thesis allow to link the convenience of Euclidean geometry (third thesis) to the freedom of the metric choice (first thesis). To achieve this purpose, it is necessary, to begin with, to free the first thesis — be it understood in its linguistic or epistemologic version — from its underlying dualism with regard to the link between theory and data.

The first thesis suggests that geometry can be neither true or false because it contains conventional theoretical elements. By contraposition a theory would not contain theoretical elements because it is true or false. Now Poincaré stresses that one can very well speak of truth in empirical sciences after laying down conventions[8]. How do we stand? Is empiricism to be vindicated or refuted? To avoid the acknowledgement of trivial inconsistencies, I propose to speak of two kinds of conventions: with Poincaré I call the first kind *classifications*. Classifications preserve objectivity in the strong sense of reference. One perceives immediately that seemingly it is not sufficient to consider — as Michael Friedman does — Poincaré's conventionalism as a holism limited to the general statements whose theoretical part consists of geometric principles, a holism whose limits are besides determined by a hierarchy in which geometrical conventions and physical principles follow the aprioristic arithmetic and precede the empirical physical laws.

Classification is not only the mediating element between sensation and the observable that we try to find, but it allows us at the same time to restore a clarity in the otherwise vague phrase saying that the geometrical axioms are conventions: Normally one would say that this formulation is abusive because the convention concerns only the "disguised definitions", for example that of distance, but does not apply to the topological axioms. In fact, in Poincaré's works conventions do not only occur as disguised definitions but already in relation to the interpretation of the fundamental concepts of every theory. But the conventions necessary for observational language in relation to our sensations do of course not yet determine the adjunctions of theoretical language.

Poincaré shares two axioms with Helmholtz:

- Relational elements are all that can be grasped of a reality.
- The establishing of an objective link is based on certain impressions and presupposes a symbolic competence which does not reflect but articulates the link in question.

In his book *La valeur de la science* Poincaré introduces a distinction between the brute fact and the scientific fact with the help of a gradual differentiation which shows that logical empirism was not mistaken when it considered Poincaré as a forerunner. For he broaches in this book both the problematics of protocol sentences and the problematics of the transition from common language to scientific language.

Explained by means of an example, the terms "brute fact" and "scientific fact" stand, at a first level, respectively for "the [individual] impression of a [relative] darkness felt by the witness of an eclipse, and the statement: it is dark, which this impression draws from him". (VS, 157) The latter statement presupposes, as a "convenient classification", a symbolic competence which represents the price of an objective articulation of nature. But here we are told nothing more.

On the other hand, it is well known that Poincaré gives a reconstruction, which is called "psychological" of geometrical space in relation to representative or sensible space. This latter is a category of our reason allowing us to compare our sensations without measure them[9]. I shall not take into account the psychological presuppositions criticised by Torretti[10] and shall confine myself, in my brief description of this reconstruction, to the logical aspects within the scope of traditional physics. Besides, Poincaré often uses the word "psychological" to lay stress on the comprehensive dimension, with the help of genetic considerations, not as opposed to a logical genesis but to the logical *exposition* of the results, for instance in formal axiomatics where the genesis is not taken into account.

Poincaré refuses to interpret the localisation of an object in sensible space as a projection of an external object in our representation. Localising an object in sensible space means for him studying the unfolding of an action to reach this object, that is to say studying sequences of muscular and not spatial sensations. He points out that a set of impressions can be modified in two different ways: without our feeling any muscular sensations or through a motive action accompanied with muscular sensations. In the first case he speaks of an external change, in the second case of an internal change. Now this remark suggests a conventional classification of the external changes: external changes which are corrigible by an internal change are called *changes of position*, the others change of state. Through the relation "the external change x is corrigible by the same internal change as the external change y" the equivalent external changes can be brought together in the same classes which are then called *displacements*. In order that the relation may be in fact transitive and the displacements may really be equivalence classes, it has to be presupposed — as Torretti has pointed out[11] — that the changes of position do not form a physical continuum. The study of the structure of these displacements suggests now to say that the *ensemble* of displacements forms a continuous group in the mathematical sense. Given the fact that the general concept of group pre-exists in us, we have to choice one of these models. We can accept one with a view of referring to its sub-groups our sensations. So we limit the choice between groups generating to geometries of constant curvature. Between those we choice finally the group which has "an invariant sub-group, of which all the displacements are interchangeable and which is formed of all translations"[12]. In other words, the studies of sub-groups corroborate "our choice in favour of the geometry of Euclid"[13].

I shall not go into the question of the size of the group and I end here my description in order to add some comments and draw some conclusions:

1. Contrary to the opinions which Poincaré seems to express sometimes, the objects of representative space, that is to say the internal changes, are by no means a reproduction of sensations and this for the following reason: sensations are singular instances — that we undergo — but they are not individualised and the sequence of sensations — and *a fortiori* their composition — is not itself a sensation but an observation obtained through a classification, that is to say — as Poincaré puts it — "the result of a habit"[14]. Observation implies perhaps sensations but is not a sensation itself. We find ourselves confronted by Ryle's thesis. But Poincaré half opens the door only to shut it again immediately. He introduces explicitly the distinction in question not as a distinction between sensation and observation or perception, but between sensation and judgement when he writes to the editor of *Mind*

> "I do not know whether perception is a sensation or a judgement, and I am inclined to believe that the philosophers, who use this word, understand it some in the first sense, others in the second one, that is why I avoid using it". (Mind, ns. 15 (1906), 142, trad. G. H.).

We maintain that objectivity, that is to say the transition from the sensation to the observable object, must be possible or rather comprehensible thanks to the mediatization of a classification which gives, so to speak, the perspective. That is why I hesitate to speak of dualism with respect to Poincaré's theory of cognition.

Summing up this first point, I should say of a classification or of a protocol sentence that it is conventional in so far as its observational content depends on a classificatory structure and becomes therefore dependent upon the linguistic usage of a society.

2. In representative space the changes of position can never accurately carried out since, generally, the starting position will not coincide exactly with the final position. When the sequence of our sensations, brought about by an internal change, does not correspond to the expected compensation, it is either rejected or replaced by a new "artificial convention" which

considers the change as the resultant of two components: the first one satisfies strictly the compensation, the second one is a qualitative change which will be disregarded hereafter[15]. To achieve an exact coincidence, the corrections leading to changes of position must therefore be read as orders for action in view of an intention. One acts if one could realise this intention as a norm in relation to "ideal bodies [...] entirely drawn from our mind" (SH, 93). However what is realised is not the norm, but only the order to realise this norm. Experience lays here a double part: it is the occasion to introduce the norm (its *ratio cognoscendi*):

> "Experience [...] has played only a single rôle, it has served as occasion [for the construction of space]. [...] This rôle would have been useless if there existed an *a priori* form imposing itself upon our sensitivity, and which was space" (VS, 95 (70)). "Our sensations cannot give us the notion of space. That notion is built up by the mind from elements which pre-exist in it, and external experience is simply the occasion for its exercising this power, or at most a means of determining the best mode of exercising it". (FG, 1)

and experience serves as an occasion to use the norm in order to conceptualise reality:

> "These laws [concerning displacements] are not imposed by nature but are imposed by us upon nature. But if we impose them upon nature, it is because she suffers us to do so". (FG, 11/12)

3. The classifications were related to the data supplied by representative space and led us to the general group concept. Thereby the essential change-over from representative space to geometrical space is brought about thanks to a reflexion on the results of the norms adapted to our orientation in the world:

> "In our mind the latent idea of a certain number of groups pre-existed; these are the groups with which Lie's theory is concerned. Which shall we choose to form a kind of standard by which to compare nature phenomena? And when this group is chosen, which of the sub-groups shall we take to characterise a point in space? Experiment has guided us by showing us what choice adapts itself best to the properties of our body; but there its rôle ends". (SH, 107 (87/88))

Refuting at once an *a priori* form of space independent of things and an abstract geometry of real beings, Poincaré introduces a kind of "intermediate non-entities"[16] such as the solid bodies and the group of displacements. Indeed from a mathematical point of view the group concept is especially interesting because one is thus led to the lowest denominator to generate the geometries with constant curvatures. But what is the philosophical benefit of this miracle of a seemingly pre established harmony between our mathematical interests and our experience? I have in point of fact already implicitly given the answer by making a distinction in the reconstruction between sensation and observation. Since conventions as classifications are already necessary for the very constituting of the objects of observation, the latent form of the group-concept is only the fixation or better the articulation of a scheme of rules which brings together the conventions of a lower order — the results of which are the displacements and which presupposes our body as an invariable solid.

But how is the difficulty, already observed by Helmholtz, to be overcome, a difficulty lying in the fact that the use of the concept "solid body" or the fixing of a norm of the "as if" presupposes the Euclidean geometry? The displacement-group conceived by Poincaré is not a transcendental intuitive form, but rather an *a priori* condition of measurement in Dingler's sense with the difference that this condition has not only reference to Euclidean geometry but is extended to geometries with constant curvatures. Yet Euclidean geometry is not exclusively privileged *post festum* as the most convenient one according to the criteria of external simplicity, but also to a criterion of internal simplicity: in Poincaré's foundational approach of *Dernirères Pensées* the solid body is thought as the most convenient norm for the simultaneous construction of geometrical instruments and figures. This simultaneous construction is used by Poincaré to palliate the above-mentioned difficulty which is inherent to every empiricism and every rationalism and consists in the fact that the use of the measuring instrument or the respective norm already presupposes geometry:

> "[...] we should be tempted to say that geometry is the study of the properties the instruments would have if they were perfect. But for that we would have to know what a perfect instrument is and we do not know it since there is none and since we could only define the ideal instrument by means of geometry, which is a vicious circle. We shall then say that geometry is the study of a system of laws not very different from those which our instruments actually obey [...] In this sense geometry is a convention [...]. This convention defines both space and the perfect instrument." (DP, 100/101; trad. G. H.)

The scheme of the constitution of objects (displacements) corresponds with a norm and the latter is not independent of the pragmatic situation, i. e. the underlying situation of the action: the solid body is at the same time the linguistic expression of an action, that is to say the articulation of the construction of the scheme of a perfect instrument (intuitive language) and a conventional norm guiding this action: the perspective change, the language becomes theoretical.

Through Poincaré's reflexions there appears a pragmatic solution of the chief difficulty of conventionalism, the difficulty of the following not very satisfactory alternative: to choose either a meta-necessity which governs the method, or a radical conventionalism. It is necessary to go back to those actions which can indeed be freely chosen but whose adequation with the world is not conventional because its objects are not independent data. In other words, the conventional aspect is the varied expression of an invariant which does not relate to the meaning of the action but relates to the executory aspect. The latter is linked to the human situation and to the body. The universal invariant is an anthropological invariant: "A minimum of humanity is necessary" (VS, 169 (128)).

Suppose that Poincaré's occasionalism is an efficient instrument to pass from our sensations to the laws of the displacements which are isomorphic with the concept of group of motions. Then what does it mean that metrization is a disguised definition or a convention that is neither true nor false but convenient?

First, I shall venture a remark about the disguised definition: Poincaré uses this term before 1899 to convey that a language seemingly used in a descriptive manner is not really descriptive; since it constitutes the objectivity of the fact which is apparently described, a disguised definition is not contrary to an explicit definition, because it is explicit and not implicit, but it is contrary to an intuitive view of a fact. A disguised definition defines explicitly an object by properties that have the form of an axiom *describing* an intuitive truth rather than the form of a definition:

> "If an object possesses two properties A and B, and if it is the only one that possesses the property A, this property A can serve as a definition, but as it is sufficient to define the object, the property B will be a definition no longer; it will be an axiom or a theorem. If on the contrary, the object is not the only one that possesses the property A, but the only one that possesses both properties A and B, the property A is no longer sufficient to define it and the property B will be a complement of the definition, it will no longer be an axiom or a theorem. In short, for a property to be an axiom or a theorem, the object which possesses this property must have been completely defined independently of this property. Thus, to have a right to say that the so-called axioms relating to distance are not a disguised definition of this distance, distance should be defined in another way than by these axioms. But where is this

definition ?" (Poincaré 1899, 274; trad. G. H.).

The term "disguised definition" characterises an explicit definition disguised as an intuitive axiom.

Why is the disguised definition of metric not purely analytical, according to Poincaré? Because our anthropological situation has caused that we have got into the habit of interpreting the succession of our sensations in one manner and not in another manner: under the aspect of a genetic reconstruction this manner is prerequisite because the instrument of our body is the *ratio cognoscendi* of the general group-concept. Actually, experience is powerless in face of Euclidean geometry because the latter is the mover of the game which allows us to play, to see analogies:

> "How could a solid body be of service to us to measure or rather to construct space? Well! By transferring a solid body from one position to another one [...]. Thus to each possible displacement of the solid body corresponded a transformation of space in itself, which did not alter the shapes and sizes of the figures; and geometry is only [...] the study of the structure of the group formed by these transformations [...]. That being granted, here is an other group, the group of the transformations which do not alter our differential equations [...]. Do the two conceptions differ essentially one from the other? No; [...]. To define space so that a solid body retains its shape when it is displaced, is to define it so that the equations of equilibrium are only a particular case of the general equations of dynamics, which, after the principle of physical relativity, must not be modified by this change of axes". (DP, 106/107; trad. G. H.)

So it was not only the conservative character deriving from conventionalism in the sense of the Duhem-Quine thesis that prevented Poincaré from formulating in 1905 the two principles of relativity as postulates, but also his rejection of the hypothetico-deductive method. He tries, holding by occasionalism, to show the genesis of the postulates of relativity within the scope of traditional physics and thanks to a mathematical instrument which is distinguishable from the other instruments by its comprehensibility: the group-concept.

University of Nancy II
France

NOTES

1 This paper is a slightly modified translation of a conference held in Paris (J. Sebestik/A. Soulez), *Actes du Colloque International "Science et philosophie en France et en Autriche de 1880 à 1930"*, Paris et Nancy 1995. I am indepted to Mrs. Prim for her help in translating the French version of this paper.

2 Exceptions are Vuillemin 1973 and Torretti 1978.

3 Cf. Meyer 1992, 250/51.

4 Cf. e.g. Vuillemin 1979, 77sqq. or Diederich 1974, 115.

5 Cf. Meyer 1992, 194.

6 Unfortunately I read this book when I had already completed my own book on Poincaré (Cf. Heinzmann 1995). Meyer claims that Poincaré's conventionalism fails in its attempt to go beyond a dualistic realism in which theory only treats experience without being an element of its constitution. We shall see that this opinion seems too severe.

7 Cf. Meyer 1992, 238sqq.

8 Cf. Poincaré, VS, 158/159 (118): "If you put the question to me: Is such a fact true? I shall begin by asking you, if there is occasion to state precisely the conventions, by asking you, in other words, what language you have spoken; then once settled on this point, I shall interrogate my senses and shall answer yes or no".

9 Cf. Poincaré, FG, 3.

10 Cf. Torretti 1978, 340sqq., 344.
11 Cf. Torretti 1978, p. 347.
12 FG, 20sqq — An exhaustive study of this genesis is in preparation by Heinzmann and Nabonnand, Nancy 1996.
13 Poincaré, FG, 21.
14 Cf. Poincaré, SH, 81; unfortunately Poincaré speaks also of the different impressions.
15 Cf. Poincaré, FG, 11.
16 Cf. Nicod 1923, 23.

REFERENCES

W. Diederich, *Konventionalität in der Physik*, Duncker & Humblot, Berlin 1974.

G. Heinzmann, *Zwischen Objektkonstruktion und Strukturanalyse. Zur Philosophie der Mathematik bei Jules Henri Poincaré,* Vandenhoeck & Ruprecht, Göttingen 1995.

M. Meyer, *Leiblichkeit und Konvention Struktur und Aporien der Wissenschafts-begründung bei Hobbes und Poincaré*, Alber., Freiburg/München 1992.

J. Nicod, *La Géométrie dans le monde sensible*, Alcan, Paris 1923; cited edition: PUF, Paris 1962.

H. Poincaré, *Des fondements de la géométrie. A propos d'un livre de M. Russell, RMM* **7**, 1899 pp. 251-279.

H. Poincaré, (DP): *Dernières pensées,* Flammarion, Paris 1913; cited edition: 1963.

H. Poincaré, (FG): *On the Foundations of Geometry, The Monist* 1898-99, pp. 1-43.

H. Poincaré, (SH): *La science et l'hypothèse*, Flammarion, Paris 1968; translation: *Science and Hypothesis*, The Walter Scott Publ., London 1905.

H. Poincaré, (VS): *La valeur de la science,* Flammarion, Paris 1970; translation: *The Value of Science,* New York, Dover 1958.

R. Torretti, *Philosophy of Geometry from Riemann to Poincaré*, Reidel, Dordrecht 1978.

J. Vuillemin, "Poincaré's Philosophy of Space", in P. Suppes ed., *Space, Time and Geometry*, Reidel, Dordrecht/Boston 1973, pp. 159-177.

J. Vuillemin, "On Duhem's and Quine's Thesis", *Grazer Philosophische Studien* **9**, 1979, pp. 69-96.

RAMÓN QUERALTÓ

SCIENTIFIC REALISM, OBJECTIVITY, AND 'TECHNOLOGICAL REALISM'

1. SCIENTIFIC REALISM AND 'TECHNOLOGICAL REALISM'

The concept of scientific realism involves two different philosophical levels which we specially need to distinguish. On the one hand, the epistemological level, that is to say, the claim that reality can be known by scientific reason with no limit; on the other, the ontological level, according to which the world is independent both from the observer and the knower. However, if we consider the development of science throughout the 20th century, it seems clear that the two former statements do not possess an absolute meaning. Indeed, quantum mechanics has established some limits to scientific knowledge, which are derived from Heisenberg's indeterminacy relations and complementarity principles. As is well known, experimental testings of Aspect and other scientists seem to definitively confirm these limits in spite of the debate originated from them.[1] So, the understanding of the ontological independence of the physical world has become, in a sense, relative to the way of constituting scientific objectivity, that is to say, to the process of elaboration of the scientific object. Hence, scientific knowledge uses, so to speak, some special mediations which transform reality in order to inquire about it from an epistemological view-point. Likewise, it is true that this process has been a natural result due to the permanent growth of scientific knowledge.

Now, do these current traits of science mean that scientific realism is over? Would it be necessary nowadays to substitute the notion of epistemological realism in science by another conception according to the new circumstances of scientific activity? In short, is it yet possible to speak about realism? Which would be at present its epistemological meaning? I would like to propose some criteria in this respect.

Among the main causes determining this situation, it is necessary to remark here the progressive *tecnologization* of scientific knowledge. From here onwards, I shall call it «technicality» of science. This fact points out that the constitution of the scientific object needs a great quantity of technological means, without which it would be impossible to put into practice scientific activity. I have already featured this phenomenon by asserting that technology has actually become a condition of the possibility of science[2]. Up to a certain extent, this relevance of technological factor is a new element in the historical development of science; of course it does not mean that the influence of technology on science begins now. It would be a nonsense. What I really want to say is that the level of this influence has reached a decisive point, in such a way that the enlargement of scientific knowledge will no longer be possible without

E. Agazzi and M. Pauri (eds.), The Reality of the Unobservable, 145-151.

the use of technological means. Many examples could be cited. The problem of the observation of microphysical objects is undoubtedly one of them.

This situation involves some specific consequences as regards the content of scientific realism nowadays. To begin with, it is necessary to state that technicality of science is a well generalised phenomenon, which is completely rooted in scientific knowledge. It is not any partial or incidental trait of science. On the contrary, it possesses the extension of a condition of the possibility of science - in the Kantian sense. That implies that the impact of technicality is stretched along the entire path of scientific cognitive processes[3]. So, the question arises inmediately: what is the impact of technicality on the notion of scientific realism? As a first result, it seems obvious that the basic features of technological requirements will have an influence on the epistemological level. It will bring about a special convergence between both the meaning of scientific objectivity proper to every scientific branch and the basic features of technological rationality, namely, instrumentality and efficiency. In other words, scientific objects will be constructed by taking into account technological possibilities as a first requirement. In order to clarify this idea, we can state - with Agazzi[4] - that the scientific object is elaborated by means of certain «operations», which prepare physical reality to be investigated in accordance with the specific features pursued by science. The nature of these operations can be very different, it depends on the kind of scientific knowledge we are considering[5]. Consequently, there is always a necessary transformation of reality in order to obtain scientific objects properly. The epistemological value of these operations is twofold: on the one hand, they are responsible for the structure of scientific objects; on the other, they constitute the grounds assuring intersubjectivity in the development of scientific knowledge.

In this line, it is undoubtedly possible to consider technicality of science as an aspect of these operations, which is wholly unavoidable. Accordingly, the constitution of scientific objects depends on technicality, because without these technological means it will be impossible. Thus, up to a certain level, scientific objectivity will depend on technicality. The final result is obvious: scientific knowledge continues to know of reality, but of a *technologised* reality. In general, the preparation of a scientific object becomes a technological preparation as long as technology is a condition of the possibility of science. All this raises the following question: can scientific realism become a sort of *technological realism*? This latter would imply two main traits as regards the content of scientific realism. Firstly, the ontological independence of scientific objects will become an independence related to technicality to a certain extent, that is to say, relative to the field limited by technological means, while other possible fields of objectification are not taken into account. It is a new feature of scientific enterprise as far as the actual influence of technical factors in science is concerned. Therefore, properties of reality to be known by science will then be properties to be considered primarily by technological manipulation. In this way, it is necessary to introduce *technological compatibility* as a relevant characteristic for understanding the contemporary meaning of epistemological realism. As a theoretical result, we should speak about *technological* realism. This conception reveals both the fact that technology is a condition of the possibility of science and also the consequences of that fact in the elaboration of scientific objects. So, technological compatibility implies that the scientific object is especially conditioned by technological means to use, this feature being a new factor to take into account in order to understand the meaning of realism. Note that we do not assert that the introduction of technology begins now, but its influence reaches to a specific extent nowadays which determines the elaboration of scientific objects. That is why it is not hard to justify the term «technological compatibility».

As a consequence, to the content of constitutive historical criteria of scientific knowledge, namely, the search for quantitative features which can be both mathematically expressed and empirically tested, it is now imperative to add the former criterion of technological compatibility. This situation becomes especially decisive in certain fields of Physics, Chemistry, Biology, and also in Cosmology.

Nevertheless, it is indispensable at the same time to affirm that the basic epistemological realism of science is not modified by this addition. This is rather a new delimitation of the notion of realism, but not its disappearance. Indeed, science continues to investigate reality in accordance with the parameters defined by its method, namely, reality concerned with these parameters, to which a new one is being added nowadays, technological compatibility. Furthermore, this situation does not involve any exceptional circumstance if we remember the historical evolution of human knowledge. Certainly, the progressive complexity of scientific objects discovered through the historical development has compelled science to incorporate many new instrumental means, both logical and mathematical, and also material and technological. However, the essential change consists of the fact that the result of these incorporations had not been produced before the introduction of a new element implying a *necessary* criterion for the constitution of scientific objects. Hence, the consequence of technicality is the necessity of introducing technological compatibility as one of the essential traits of the operations for elaborating scientific objects. This signifies a new aspect of realism but not its refusal.

2. THE IMPLICATIONS OF 'TECHNOLOGICAL REALISM'

This theoretical description can be effectively confirmed in quantum mechanics. The epistemological debate about the impact of observability on a physical object originates from the modifications that technological means, necessary to observation processes, produce on a «natural object». On that account, many philosophers of science have argued a sort of idealisation of scientific knowledge and, consequently, the defeat of realism. But such an epistemological position forgets two very important things. On the one hand, the elaboration of a scientific object is always complex, there are no pure objects. As a rule, the need of transforming reality in order to obtain a scientific object is not an exclusive situation of present days. Consequently, the fact of technicality means rather a *controlled* enlargement of scientific method. In other terms, the limits of technological compatibility are imposed according to specific rules originated from epistemological claims. These limits are not imposed from the outside of science, but they are the result of a rational evolution of scientific knowledge. Thus, it would not be reasonable to get a certain idealistic conception about the cognitive quality of science from technicality.

Secondly, there is another relevant factor, namely, experimental testing. In fact, the idealisation of knowledge can not explain why science is finally applied to transform reality and has obtained the extraordinary success we know. It does not deal with a minor aspect, because it is a permanent result. Moreover, it represents one of the constitutive aims of science as a cognitive activity - that is to say, the pragmatic aim of scientific knowledge, together with the theoretical aim that we refer to below. This would not be possible without a certain minimum - so to speak - of epistemological realism: why then the *expected* validity of scientific applications to reality? How to explain that this validity is justified, from a theoretical point of view, *before* the corroboration of that success?

Nevertheless, it is now convenient to enlarge the meaning of technological realism. What can the concrete consequences derived from the introduction of technological compatibility in scientific processes be? A first result to note is the ineludible increase of investigations regarding instrumental aspects of science. In other words, the pragmatic aim of science becomes more and more important. Thus, the potentiality of scientific objects for dominating reality becomes the first requirement to scientific investigatiòn as well. Operational research will have priority over theoretical research, and it will be increased because of social competitiveness of the different groups of scientists. It is the direct consequence originated from the influence of technology, that is to say, the recognition of *instrumental* values as a key criterion for deciding the content of research. Nowadays it is possible to show this phenomenon by the fact that many western governments have drastically reduced research budgets in basic sciences. This means that, together with the necessary selection that every scientific objectivisation makes over reality, another type of selection appears because of technological compatibility. Or, in other words, technicality has introduced a new necessary element to take into account for the constitution of the scientific object, which is now added to the former known elements. The complete collection of these elements then constitutes the main means for establishing the result of scientific objectivisation. Finally, from this viewpoint, it is clear that the influence of technicality can not be devaluated, because it becomes a determining factor in science.

In a sense, this transformation compels one to conceive the domination of nature as the primary aim of scientific investigation. The theoretical aspect of scientific truth, namely, the knowledge of reality structures, will maintain its specific place, but it will also be conditioned by technicality. Indeed, the theoretical sense of truth is always necessary, however the «feeling» of the researcher about it can be changed. From a traditional position, truth was an epistemological value in itself, in fact, it was considered as the greatest one. But now scientific truth would be appraised as a way of dominating reality, that is to say, not as an epistemological value in itself but as a contribution to the pragmatic aim of scientific knowledge[6]. Accordingly, the theoretical level of truth hands over its side to the pragmatic level because of the influence of technicality. In other terms, scientific truth becomes more and more *technological truth*. This is a type of truth characterised by operational aspects, namely, the immediate applicability of its results to reality in order to manipulate and to transform it. It signifies the priority of efficiency over other aspects of truth[7]. Technological truth does not search *prima facie* for the correspondence to the object, for the logical consistency of propositions, etc.[8], but it does for the highest utility and efficiency of its applications to reality. So, the efficient action to modify real things is then the validity rule of technological truth. Therefore, the criterion of epistemological validity will be to construct a *tool* to fall upon reality, this tool being inmediately tested. Tools, in this sense, can be material or not, but anyway they will always be directed to the former scopes. Accordingly, technological truth especially involves the practical aim of making a certain tool which must be necessarily useful to dominate the world to some extent. In short, truth conceived as a tool to be immediately used for its specific objective is the main content of technological rationality. Note that these features are neither positive nor negative in themselves, because it all depends on the kind of objects to be considered under them. So, for instance, technological reason is completely necessary in daily life to a certain extent, but if it is taken as the main rational instrument for understanding a human being, it would then originate a reductionist anthropological view, which would be very dangerous in society. However, in practice, this process of increasing the influence of technological truth does not seems to stop in the present day. We have also analysed some philosophical consequences of this situation[9].

By this reason, technicality compels one to establish a new balance among the different aims of scientific knowledge. The range of the observability problem in Physics in the first third of the century did not only concern ontological and epistemological frameworks of scientific knowledge. In the end, it has shown a larger relevance. Nowadays we can see that one of its causes, that is to say, technicality, possesses a special significance. Therefore, the balance among the elements responsible for the development of scientific knowledge has to be found again. Of course, this is a deep change which may mean the end of an historical period and the beginning of another one. Hence, to distinguish carefully between what is disappearing and what must remain in order to maintain the historical continuity of science is unrenounceable. Some ideas will now be exposed in this respect.

Above all, we should pay attention to the epistemological value of knowledge, namely, what we have named earlier the theoretical aim. Without such a value, science can become a sort of «distinguished» technology, whose epistemological range can be relativised by technological aims. It is important to remember that epistemological progress of scientific knowledge has always presupposed the notion of truth in itself, or at least, as in the Kantian view, truth as a regulative idea which guides investigations. From a realistic outlook, we also have to add that the notion of truth is a fundamental ground to assure a full epistemological sense to scientific knowledge. Furthermore, the notion of truth is likewise necessary just to establish the validity of technological research. In speaking of technological truth, this conception refers to the kind of theoretical truth according to technological activities[10], which constitutes a basic ingredient for the epistemological content of technological knowledge.

But, at the same time, it is necessary to take into account that technological progression - and the influence of technology in scientific enterprise - must progressively increase. Indeed, technological process tries for the greatest efficiency, and, in this sense, there is no wider efficiency than encompassing the highest number of objects under the technological paradigm. It is also indisputable that technology has largely improved epistemological results of science, and so, human knowledge to an extraordinary extent. Thus, scientists, even from an ethical view, are obliged to use technological means to assure the success of their investigations.

Finally, both ideas raise an embarrasing problem. If technology has to be used more and more, and if its internal nature necessarily implies an unlimited progression, how to avoid the possible substitution of scientific truth by technological truth? It seems that the end of this situation would be a definitive technification of science. In fact, it is unavoidable. To my view there are only two possible alternatives to face the negative results of this process. On the one hand, to reduce these negative effects of technology by limiting its use according to the social and ethical validity of scientific aims. Apart from the intrinsic difficulties in putting into practice this possibility, another decisive reason exists which makes it very doubtful. Technologists and scientists, in order to carry this out, ought to accept a limitation which would be extrinsic to technology itself. But this is to tolerate the influence of something completely strange to technology in their investigations. That is clearly contrary to the nature showed by technology, becâuse it would imply something opposite to efficiency and instrumentality. In other words, that would be contrary to the nature of technology. For this reason, technology and technologists will not voluntarily accept any interference of external criteria, above all if these try to reduce the adquisition of efficacity and power over the world.

Another possibility consists of reaching the new desirable balance by following an indirect way. It deals with increasing the appreciation of truth in its theoretical level as far as scientists and technologists are concerned. We should also refer this purpose to be obtained by society as a whole. In this respect, it is necessary to restore the social value of truth, which has been darkened through the second half of this century. Nevertheless, for scientific realism, the task can be less complicated than in social field. Certainly, without the concept of truth in a theoretical sense, realism cannot resist, and without realism, the epistemological value of science is not possible. That is why supporting epistemological realism of scientific knowledge signifies in turn to defend the theoretical notion of truth, and, in so doing, to decrease the possible technological overflowing.

To conclude, current technification of scientific knowledge has actually reached a critical line at the end of the century. It no longer deals with a certain modification of reality in order to make it accesible to scientific methodology, as it has happened in former historical periods of science. The point is that this transformation now implies a new element to the right understanding of scientific objectivity, namely, technicality of science. In this sense, technicality becomes a decisive factor for the constitution of scientific objectivity, in such a way that technological compatibility is required as a condition of the possibility of the scientific object. Far from being a simple instrument without any influence in the structure of scientific objects, technicality determines objectification processes to a very important extent. Consequences for scientific realism seem to be clear: we are leading the way to technological realism.

Sevilla University
Spain

NOTES

[1] See among others: A. Aspect, P. Grangier, P. Roger, *Physical Review Letters*, **37**, 1981, p. 460; **49**, 1982, p. 91. More recently: M.O. Scully, B.G.Englert, *Quantum Optical Test of Complementarity*, Nature **351**, 6322, May 9, 1991, pp. 111-116.

[2] R. Queralto, "Does Technology 'construct' Scientific Reality?", in Carl Mitcham ed., *Philosophy of Technology in Spanish Speaking Countries*, Kluwer Academic Pub., Dordrecht 1993, p. 167 ff. Also, *La dimensión epistemológica del uso tecnológico en el proceso de conocimiento científico - Thémata* **IX**, 1992, p. 289.

[3] I have analysed some examples of this influence in my contribution to the 1994 Congress of the International Academy of Philosophy of Sciences, held at Nancy (France). See *Hypothèse, objectivité et rationalité technique*, "Philosophia Scientiae" (Travaux d'histoire et de philosophie des sciences), 1, 1996, Cahier spécial 1, p. 187 ff.

[4] See E. Agazzi, "L'objectivité scientifique", in Agazzi ed., *L'objectivité dans les différentes sciences/Die Objektivität in den verschiedenen Wissenschaften*, Éditions Universitaires, Fribourg 1988, p. 18 ff.; *Temi e problemi di filosofia della fisica*, Abete, Roma 1974 (2nd ed.), p. 350 ff.

[5] There exists a very large diversification even in experimental sciences. For instance, operations in Chemistry are not the same as in Cosmology.

[6] The crisis of truth is a characteristic trait of contemporary thought. The relativisation of truth, and also its denial,

has been practically demanded by some postmodern authors, as long as truth has been conceived of as an 'oppressive' instrument in intellectual activity (so, for instance, the 'pensiero debole'). Scientific activity, however, has nothing to do with this contemporary illness of thought. Nevertheless, the convergence between this postmodern position and the consequences of technological rationality can undoubtedly reinforce the current crisis of the concept of truth.

7 To this respect, see my book: *Mundo, tecnología, y razón en el fin de la Modernidad*, P.P.U., Barcelona 1993.

8 And other well accepted conceptions of truth depending on the kind of rational object concerned.

9 See my book *Mundo, tecnología...*, o.c., especially chaps. IV, V, VI.

10 To this respect see E. Agazzi, *Philosophie technique et philosophie pratique*, in G. Hottois ed., *Evaluer la technique*, Vrin, Paris 1988, pp. 29-49; also, *Il bene, il male e la scienza*, Rusconi, Milano 1992, p. 160 ff.

REFERENCES

J. Agassi, *Technology: Philosophical and Social Aspects*, D. Reidel, Dordrecht 1985.

E. Agazzi, "L'ideologia tecnologica", *Nuova Civiltà delle Macchine*, **2**, 1984, p. 5.

E. Agazzi, "Philosophie technique et philosophie pratique", in G. Hottois ed., *Evaluer la technique*, Vrin, Paris 1988, p. 29 ss.

A. Borgmann, *Technology and the Character of Contemporary Life*, University of Chicago Press, 1984.

P.T. Durbin, F. Rapp eds.. *Philosophy and Technology*, D. Reidel, Dordrecht 1983.

J. Ellul, *La technique ou l'enjeu du siècle*, Economica, Paris 1990 (reed. 1954).

J. Ellul, *Le système technicien*, Calman-Levy, Paris 1977.

J.K. Feibleman, *Technology and Reality*, Nijhoff, The Hague 1982.

F. Ferré, *Philosophy of Technology*, Prentice Hall, Englewood Cliffs 1988.

J. Habermas, *Technik und Wissenschaft als Ideologie*, Suhrkamp, Frankfurt 1975.

H. Jonas, "The Imperative of Responsibility", in *Search of an Ethics for the Technological Age*, Chicago University Press, 1984.

C. Mitcham ed., *Philosophy of Technology in Spanish Speaking Countries*, Kluwer Academic Publishers, Dordrecht 1993.

C. Mitcham, R. Mackey eds., *Philosophy and Technology: Readings in the Philosophical Problems of Technology*, Free Press, New York 1972 (reed. 1983).

R. Queraltó, "Razón científica y razón técnica en el fin de la Modernidad", *Anuario Filosófico* **27**, 1994, p. 683.

DUDLEY SHAPERE

TESTABILITY AND EMPIRICISM

1. THE UNOBSERVABLE IN MODERN PHYSICS AND COSMOLOGY

Much current research at the frontiers of particle physics and cosmology deals with problems and theories for which observational or experimental test appears impossible. Theories such as Superstrings, for example, hoping to provide a unified understanding of the gravitational force and the Standard Model of elementary particles and forces, deal with energies on a scale of 10^{19} GeV, far beyond any imaginable technology by which they might be tested, at least in any direct sense. For this reason, some scientists have alleged that science has reached a stage where the cosmology of the very early universe must take the place of laboratory experiment concerning extremely high-energy physics.

One can hope that such theories will turn out to be testable indirectly, through observational or experimental investigation of their low-energy consequences, even though such consequences have not yet been deduced for Superstring theory. But there is a more serious issue, for some aspects of current theories appear to be untestable in principle. This is most evident in the supposition of some widely-discussed theories today that our region of the universe, or even our universe, is only one among many. Inflationary theories, for example, postulate that our domain of the universe, which includes the presently-observable part, is only one of a multitude of such regions, in each of which the constants of nature, among other things, can differ in value from those in another, the light-travel time at the beginning of the universe having been too short to permit communication between the domains. More radically, still other theories speak of other regions of the universe, or even other universes, which are forever causally unconnected with ours. Hidden spatial dimensions, compactified to Planck-length radii, are a further example. If such theories are accepted or even considered to be legitimate objects of scientific study, regions of the total universe must be admitted which are unobservable in principle.

Nor is such theorizing confined merely to speculative ideas. For according to currently-accepted cosmology, new regions, previously unobservable, are forever entering our horizon, then and only then becoming observable and as such available to confirm or disconfirm our theories of what the universe as a whole is like. To defend the view that hypotheses about the behavior of such regions are testable because any such region will eventually - for example, in 10^{100} years or so - come within our horizon, and are therefore "observable in principle," is of course to make a mockery of the notions of testability and observability in principle. And in any case, all present evidence indicates that the universe is infinite, and if that is true then there will always be

E. Agazzi and M. Pauri (eds.), The Reality of the Unobservable, 153-164.

unobservable regions. In yet a further active and important area of research based on ideas that are by now widely accepted, the characteristics of the unobservable interiors of black holes are explored through theory, and what goes on there can provide insight into our part of the universe outside the black hole.

Prima facie at least, the most direct response of traditional empiricism to such ideas would be that, since they have to do with what is unobservable, they are untestable, neither verifiable nor falsifiable by experiment or observation, and are therefore to be dismissed as unscientific. Those sorts of theories would thus be exceeding the limits of good scientific reasoning, by entering into uncontrolled speculation, even metaphysics. I say only that this is the prima facie traditional empiricist view, for in many respects the vagaries of that clutter of doctrines allow various maneuvers (or epicycles) by which one could at least try to defend the legitimacy of such theorizing, at least in some cases.

Here I will ignore such possible maneuvers; for most philosophers today consider traditional empiricism to be dead in any case. Classical empiricism held that all our knowledge is "based on" sense-experience, observation, or experiment, in the sense that there are data given in sense-experience (etc.), wholly free of any interpretation whatever, which constitute the source and justification of all our knowledge-claims. Twentieth-century logical positivism, logical empiricism, and, at least with regard to its falsifiability claim, Popperianism, while rejecting the thesis that sense-experience or observation is the source of all our knowledge-claims, followed the classical view in maintaining that all our knowledge-claims must be testable - justifiable or rejectable - in the light of empirical test, and they embodied this claim in such doctrines as verificationist theories of meaning and justification, of a theory-independent observation-language, and of the falsifiability criterion of scientific status, respectively. Because differences between these doctrines will not be relevant to my discussion, I shall in what follows refer to all of them, including classical versions, indiscriminately as "traditional empiricism." Though vestiges of the falsifiability doctrine still remain, the other theses on this list have been rejected in most of their variants. The rejection rests on two fundamental criticisms, each the other side of the coin from the other. First, their adherents were never able to make clear what is supposed to count as a "given" in experience or observation, untainted by any interpretation whatever; indeed, many writers today no longer believe that there is a given in that sense. And second, contrary to the claim of classical empiricism and its twentieth-century descendants, not only is interpretation always present in what we call sense-experience, observation, or experiment; it is pervasive. Observation is theory-laden.

The rejection of the doctrines on which traditional empiricism based its banishment of the unobservable from science leaves us without a theoretical ground for answering the question, must science shun the unobservable, or did traditional empiricism, in truncating what counts as science by its doctrines of testability, verifiability or falsifiability, miss an important point about science, that it might have legitimate concerns with the unobservable?

Therefore in this paper I want to reconsider the role of the unobservable in science, focussing on these questions as raised by ideas prevalent in current particle physics and cosmology. Such a reconsideration of course requires a more adequate account than traditional empiricism provided of what is involved in something's counting as a scientific observation, so I will begin by reviewing and extending relevant parts of such an account which I have developed in previous writings. I will argue that the two major criticisms of traditional empiricism do not hold against this newer view, and indeed that the points of the criticisms are incorporated into the very heart of the new view. Further, the new view will show, as traditional empiricism did not, how the distinction between legitimate and merely wild speculation is made in science, and in particular why, and, in an appropriately constrained way, when appeal to the unobservable is legitimate in science. Finally, I will generalize the issue of the paper beyond that of unobservability, in order to bring out what is, in contrast to the two usually heralded criticisms of traditional empiricism, the really central failing of that doctrine. This criticism, long known but dismissed as trivial and unenlightening, gains its significance by bringing out the important but neglected point in saying that science is empirical. It, too, will be accommodated within the view I will develop.

2. THE CONCEPT OF OBSERVATION IN MODERN SCIENCE

In earlier work[1] I argued that prior suppositions are required and utilized in cases of scientific observation to shape what counts as observational, the things that can or cannot be, and are or are not, observed, the way in which an observation and its implications are described, and the way in which an observation is performed. I called the collection of prior suppositions which are utilizable and sometimes utilized in particular observational or experimental situations, *background information.* An example was the original solar neutrino experiment, a classic experiment in contemporary astrophysics designed to test the modern theory of stellar energy production. In that particular case, a host of separate theoretical, experimental, and practical items function in one or more of three ways: in the theoretical and practical conception of the possibility of the experiment, in its execution, and in the interpretation of the results. For example, the theoretical possibility of such an experiment as providing a direct means of getting information about - of observing or observing directly - what goes on in the center of the sun, as opposed to an indirect means in which what goes on there is *inferred from what we observe* of the solar surface, is based theoretically on the weakness of the interactions participated in by the neutrino, which allow the neutrinos released in nuclear reactions in the solar core to pass unaltered from their source to the terrestrial instruments that capture them, thus preserving information about those nuclear reactions in a direct way that is impossible with photons, whose information content is profoundly altered in the course of their tortuous 100,000-year journey to the sun's surface. It is also based on the knowledge, from both experimental and theoretical nuclear physics, that neutrinos can, with a finite but very low probability, be captured, and that chlorine is a good substance to capture them with, due to its reactions with neutrinos to produce radioactive argon which, conveniently for experimenters, decays in about a month. That the experiment was practical rested on considerations including knowledge of how much chlorine (in the form of cleaning fluid) would be needed to assure capture of enough neutrinos to be statistically significant, how to extract the radioactive argon atoms efficiently from the enormous tank, and where the experiment should be constructed in order to minimize background effects which could imitate neutrino captures. Similar roles are played by background information in the execution of the experiment and in the interpretation of the experimental results. Without much prior theoretical information, the experiment would have been, in the most literal sense, inconceivable; without a considerable body of background information (sometimes including portions, like weak interaction theory, which are also used in its conception), it would not be executable; and without some portions, not necessarily the same ones, it and its results would be uninterpretable.

With some minor qualifications, this entire experimental setup is referred to by scientists as an observation (or an experiment set up to make an observation) of what goes on in the center of the sun, an observation made in order to test the theory of stellar energy production in a direct way that cannot be done, physically speaking, by relying on electromagnetic information received from the solar surface. This use of the term 'observation' differs in significant ways from that with which traditional empiricism was concerned, though it is a descendant of that usage. This is the second point which I will summarize from earlier work.

The justification of this usage is partially as follows. Traditional empiricism identified two distinguishable aspects of observation, the evidential and the perceptual. It thus maintained that the problem of what counts as *evidence* is identical with the problem of what is *given in sense-perception.* This given in perception was further held to be free of all interpretation whatever, so that to explain observation was to explain what is given in experience, and simultaneously to explain what counts as evidence in testing a hypothesis or theory. The attempt to fill this out failed, of course; but my point is that, for understanding science, the attempt was misguided from the very start. For in sophisticated cases of observation in modern science, the two functions traditionally identified, the evidential and the perceptual, are separated, and the focus is on the former. There is good scientific reason for this departure. For in its concern with testing, the focus of modern science is on observation as evidence, not on observation as perception: the two concepts are separable and are concerned with different problems. Correspondingly, sense-perception being limited and unreliable in important ways, the perceptual aspect appears as an interference with the evidential. Therefore, for the scientific purposes of observation, the human eye, for instance, is replaced by detectors which are not subject to the errors to which that organ is subject. True, to be of use for scientific purposes, those data must be finally put into a form accessible to our human senses, but that is only because we are the ones who will use the evidence; the real work of evidence-collecting, the scientifically relevant aspects of observation, has already been done independently of the senses. The role left to sense-perception is minimal, though of course we must still be trained to understand the results.

Associated with this departure from the sensory and focus on the evidential aspect of observation is a generalization of the relevant concept of observation. With the discovery that the human eye is sensitive only to a narrow range of a far broader electromagnetic spectrum, and that there are other types of interaction than electromagnetic, the collection of data of observation is assigned to detectors other than the eye or other human senses, detectors made possible in the light of background knowledge because of their sensitivity to a select portion of the range of those interactions. Thus the problem of understanding the role of observation in science cannot be equated with "the problem of perception" as it was in traditional empiricism.

The scientific usage of the term 'observation' has thus changed in two ways: by departing from more traditional common-sense and philosophical usage, and by generalizing to other types of information-carrying interactions. I have emphasized that these twin changes are brought about for scientific reasons. In general, when such scientific reasons exist, I say that the concept has been *internalized* into the scientific process: in the present case, it has become possible to refine what counts as an observation so that it becomes amenable to scientific understanding and determination. In the effort to understand the scientific enterprise, it is necessary not only to understand how science determines what counts as observational in particular cases, but also why - for what scientific reasons - it has become possible to do so - the rational mechanism by which what counts as observational has been altered to assimilate it into science.

Let us contrast the present view with that of the critics of traditional empiricism. One of the two chief criticisms of the latter doctrine was that observation is "theory-laden," and that, indeed, all aspects of science are shaped by background presuppositions. Superficially, the functioning of background information in the solar neutrino experiment bears this out. But in explaining the theory-ladenness of observation, many critics of traditional empiricism missed the point of what background information is, what its sources are, and how it functions. In their conception, presuppositions could have no reasoned source, because what counts as a reason is itself among the aspects which are shaped or determined by the presuppositions. On this view, science cannot be anything more than the working-out of a particular point of view whose choice is ultimately arbitrary. This misconception is a primary source of the antiscientism prevalent among so many philosophers, historians, and sociologists of science today. A proper view of background information avoids it.

Although the view I have presented recognizes the role of something analogous to "presuppositions" in scientific problem-situations like that of the solar neutrino experiment, it differs in four major ways from the views of such critics of traditional empiricism. The first is that, just because observation is "laden" with interpretation, it does not follow that it is arbitrary which beliefs are used as background for interpretation. As the solar neutrino experiment illustrates, it is not just *any* beliefs which can serve as background; rather, only certain ideas are *qualified* to serve as background information: briefly for present purposes, they must have proved their success in past inquiry. The second difference is this: the background employed in the interpretation of observation is a piecemeal one, consisting of separate items of information, nor are they all theoretical, nor are they even necessarily consistent with one another. Each such item may perform different functions in the interpretation of a given situation, for example in the conception of the experiment, its execution, and the interpretation of the results. It is not a single overarching Weltanschauung, as it seems to be thought of by many writers. A crucial part of the analysis of what goes on in science consists in understanding the *different* roles played by *specific* items of employed background information. Thirdly, as is again illustrated in the solar neutrino experiment, the background utilized in the interpretation consists of information which is directly relevant to the situation. There are no vague and imaginative appeals, even implicit, to anything like "social context." At the most concrete level, the scientific content of the experiment and all that goes on in it can be completely understood in terms of the relevant scientific background. Even if there is determination by social factors, that would not detract from or change the role of the scientific background. The latter must be understood in any case, and its understanding is necessary and sufficient for the understanding of the scientific character of the experiment itself, of the observation and interpretation of the scientific significance of the results.

The fourth major difference between the views of many critics of traditional empiricism and the present view lies in the latter's acceptance of a given in experience. Critics of traditional empiricism tended either to deny the existence of any such given, or to consider it powerless to confirm or deny any knowledge-claims, to select or reject any one of them as correct or even as better or worse than its rivals. But there is a "given" in observation, though it is a very different sort of thing from the uninterpreted perceptual given which traditional empiricism failed so utterly to specify. It is interpreted in three respects. First, it is marked out as significant, as carrying significant information, in the light of our best available background ideas. Second, it gets its appropriate description in terms of those background ideas. And third, in terms of those background ideas, we know how to get access to it. But even with all this interpretation in the light of background information, it remains the case that the specific character or value we find it to have is independent of those background ideas, not determined by them. (Thus the theoretically-predicted number of neutrino captures per counting period differs from the observed number). This is what makes possible (in ways I cannot discuss here) the testing of a theory such as that of the sources of stellar energy, and ultimately, again in cases like that of the solar neutrino experiment, of the background information itself which had been brought to bear in the experiment, in this case our solar models, our theory of stellar evolution, and our theory of weak interaction processes. It is (again among other things) what makes it possible for science to be self-corrective. And it is this that marks the present view as a descendant of traditional empiricism despite its great departures from the latter.

3. THE SCIENTIFIC STATUS OF THE UNOBSERVABLE

Writers in the traditional empiricist tradition give the impression that, while the substantive content of science changes over the history of inquiry, the methods by which that content is determined do not. But the preceding discussion makes it clear that what counts as observable, and also what counts as unobservable, can change with changes in the content of scientific belief. What is observable (or unobservable) depends on the current state of accepted scientific belief, including not only the instrumentation involved, but also the theoretical knowledge which tells us the nature of interactions, the theoretical possibilities of detection of a particular interaction, and how the results of a particular interaction give information about its sources. With changes in that state, the scientific conception of what counts as observable or unobservable departs gradually from predecessor conceptions, and in particular from ordinary common-sense and philosophical ones. Just as the discovery of ultraviolet and infrared radiation marked the beginning of a departure from the notion that observation is to be identified with sense-experience, so the discovery of neutrinos, together with the theory of weak interactions governing their behavior and the construction of neutrino detectors, dictated that observation is not limited to the electromagnetic spectrum, but can be performed through the opportunities offered by other interactions. Whether something is observable or not is a contingent matter of fact, resting on what we have best reason to think we have learned. It is not a question of whether it is "given" in sense-experience.

How does all this apply to those ideas, common in both speculative and accepted modern physics and cosmology, that have to do with what is unobservable? At least some of the cases I surveyed at the beginning of this paper still remain unobservable even in the new sense. The gravitational physics of black holes dictates that neither electromagnetic radiation nor anything else can escape beyond the horizon of a black hole, and that black holes cannot be detected in any direct way, only indirectly via observable effects on their accretion disks or, someday perhaps, through their Hawking radiation. Cosmology and physics dictate that there are regions beyond our own horizon that cannot be observed until they enter that horizon, and, if the universe is infinite, that, at any given time, there will always be regions which are unobservable.

The new view of observation which I have outlined recognizes the legitimacy of speculation about, or acceptance of, ideas like these which, according to the science currently accepted, have to do with unobservables. But it places important restrictions on the ideas. It is the existence of background information that secures the legitimacy of speculation about the unobservable. The background information accepted at a given stage determines the limits of what it is possible to speculate about in science: an idea which is admissible as an open possibility in the light of background information is scientifically legitimate, even if it has to do with what is, according to that same background information, unobservable. And the idea of something which is unobservable, and which may even have no observational consequences, can come to be accepted if it is either (a) logically or mathematically implied by something that is observable or has observable consequences, (b) needed for consistency considerations (as in superstring group selection), even though it is not implied by the observable portions of the theory, or (c) provides answers to problems concerning the observable parts of the theory with which no other solution deals successfully. Through long experience in knowledge-seeking, we have learned to adopt these and other restrictions, to distinguish such speculation as legitimate from "wild" or "loose" speculation. But we have also learned that it is the spirit, not the letter, of the restrictions that counts. We have learned that in science we must be opportunistic in the introduction of new ideas, and one form which this lesson takes is the relaxation of constraints such as those I have mentioned. The very fact that we do not know everything - that our background information is vague, incomplete, and even subject to contradictions - implies that there will be borderline cases. The fact that background information itself is subject to revision also wisely liberalizes the restrictions: to the extent that reasons can be given for questioning a piece of background information that would otherwise rule a certain piece of speculation as being "wild," the idea in question has a claim to legitimate attention. Despite the liberality, it is by no means the case that "anything goes."

So far, the extensions of empiricism to allow untestable speculation are not major, though they go well beyond what traditional empiricism clearly allowed. The legitimately untestable is limited to what is possible in the light of background information, and can be accepted if it is connected, either logically or through problem-solving ability, to the testable areas of some accepted or promising theory. On this rather conservative picture, at each stage of the study of a particular domain of science, there would be a central core of ideas, which are constructed on the basis of background information, and which are testable, and to which the untestable ones must be connected in one of the specified ways. But other more radical possibilities emerge when we ask the following questions, more general than any we have asked so far: is it possible that changes in science could be so thoroughgoing that there need no longer be such a testable core for a particular domain, for example for theories having to do with Planck energies? Even though our scientific knowledge has so far been built on observations of nature, could the day come when the need for further observational test will become superfluous for a whole area of theoretical investigation? Could empirical inquiry, which has guided science up to a certain point in its history, lead at that point to a new stage wherein empiricism itself is transcended, outgrown, at least in a particular domain?

The acceptable background information used in shaping new ideas and directions of research could indeed, at least in principle, reach such a stage. Ideally, the background information would have become self-consistent, detailed, and constraining of further possibilities, even providing a unified perspective on the domain in question and perhaps on a wide range of phenomena; further, it would have become self-sufficient in its ability to generate and answer questions about its subject-matter and about directions of research. To the extent that these and other conditions were fulfilled, correspondingly high confidence in the theory would be justified, enough, in cases approaching the ideal, to dispense with the need for empirical test. In fact, put in these terms, many scientists today believe that the ideal is close to being realized in particle physics and cosmology. Nor can their claim be dismissed lightly, for the reasons behind it, far more than any abstract considerations, compel us to take seriously the possibility that the ideal could be achieved, even if we do not agree that it has been or is about to be. Consider the following short list from among the remarkable and growing list of constraints that have in the past quarter century drawn a seemingly ever-tightening noose around the remaining options as to what can count as a legitimate theory of those subjects. Through employment of gauge field theories, a partial but profound unification was constructed of two of the four fundamental forces of nature, the electromagnetic and weak, and a theory of the third, the strong force, was constructed in a way so parallel to the electroweak unification that the theories of all three forces could be taken to be a single unified perspective, the Standard Model. Gauge theories are renormalizable, as previous theories of the strong and weak forces were not; this discovery in itself served to rejuvenate quantum field theory. Unlike in Newtonian mechanics, where the existence of forces was an assumption wholly independent of the rest of the theory, gauge theories, by their very mathematical structure, require the existence of forces, and specific gauge theories require those compensating force fields to have the specific characteristics they do. Only in a non-Abelian gauge theory[2] - the type we have in the electroweak theory and quantum chromodynamics - can asymptotic freedom exist, a property which makes possible the detailed and precise calculations of what happens in the earliest moments of a Big Bang universe, and so brings conviction to a partial unification of the standard theory of the quantum realm with the standard theory of cosmology, providing the theoretical basis for an understanding of the history of the very early universe. The direction of further research is highly constrained, to the construction of gauge field theories unifying the forces still further through a specification of their symmetry groups and the breaking of those symmetries. Though the fourth force, gravitation, is still not fully assimilated to this program, highly suggestive progress has been made in these directions through research on supersymmetry and quantum gravity. Unlike earlier theories, the supersymmetric theory of strings holds out the promise of being free of all uncontrollable infinities, including anomalies. Though the mathematical difficulties of superstring theory have so far prevented its achieving its aim of unification of all four forces, or even of its formulation as a field theory, consistency requirements narrow the possible group structure to only two, SO(32) and E(8)xE(8).[3] Let me not be misunderstood. I am *not*, at least here, agreeing that background theories in particle physics and cosmology have, or have in some respects, actually reached a stage at which they can be used with full confidence to extrapolate to further theories without the benefit of empirical test, or even that they are on the verge of reaching such a stage (though I would argue that *some* theories available now in *some* domains, among them *some* employed in particle physics and cosmology, have reached that stage, and are, justifiably, so used). However convincing our present theories in the domains of particle physics and cosmology are, there remain many difficulties.[4] And as the solar neutrino experiment itself illustrates,[5] even if all available evidence favors a theory, and there are no specific reasons for doubting it, much study and application is still required before they can serve as bases for confident extrapolation to further ideas without any longer requiring empirical test. But what I am saying is that results attained in contemporary particle physics and cosmology give *reasonable grounds* for believing that such a stage can in principle be reached, and, further, for the claim that particle physics and cosmology may indeed reach such a stage and may have already done so in certain respects.

It is thus not obviously unreasonable that the discovery of such constraints as those described above has led many very able physicists to the view that physics is in fact approaching, or perhaps has reached, the stage where we can proceed without the need to subject our further theories to empirical test. It is all too easy to dismiss such musings out of hand, on the basis of some vague vestigial positivism or hand-waving inductive cèguments to the effect that historically all such claims have failed, so this one can be expected to fail too. Rather, if we want to resolve *that* dispute, what is needed is a detailed assessment of the present situation in particle physics and cosmology, of the dark clouds that may lie on its horizon, and a detailed comparison of it with earlier situations in which similar claims were made, and in which dark clouds also existed. That is, however, a topic for another day, and is ultimately not relevant to the issues I am discussing here.[6] My present point is merely that these contemporary ideas, and the conclusions about the future of science that are inferred from them, show that we must take the possibility of arriving at the stage in question seriously, and that the reasons why we must do so are of fundamental importance to understanding the knowledge-seeking enterprise. For behind the question I hav. been asking so far, Must science always remain empirical? Is a yet deeper one, concerning the very nature of empiricism as a portrayal of the knowledge-seeking enterprise. That further question is, *Is the empiricism of science itself an empirical hypothesis?*

Classical empiricism always faced the objection that it never made clear whether its central claim, that all our knowledge is based on sense-experience, is analytic, a matter of definition, or empirical. Either alternative appeared distasteful: if it is analytic, it can, on its own principles, tell us nothing about the matters of fact with which it seems to be concerned; but if it is empirical, it cannot be established with absolute certainty. Twentieth-century versions of traditional empiricism faced the same objection. This objection has never been taken as seriously as others; rather, it has been dismissed as something of an amusing curiosity, embarrassing, perhaps, but ultimately somehow trivial and therefore forgivable. And so, in our time, when traditional empiricism was rejected, this objection was not among the crucial ones leading to its rejection; that honor went to the theory-ladenness of observation and the non-existence or powerlessness of the given. But now we can see that this third objection cuts deeper than the two standard objections. Those are easily absorbed into a new version of empiricism such as I have given here. The third objection has to do with the necessity of a truly radical reconception of the nature of scientific method, and such a reconception has been suggested here and developed more fully elsewhere.[7] Traditional empiricism just could not have it both ways: either it was trying to dictate, not merely by definitional decree but in effect *a priori*, what kind of thing the process of inquiry must be, or else it was not a doctrine that has to be obeyed in all inquiry whatever. The point of the arguments I have made concerning the possibilities that science could arrive at a stage where open alternatives are so constrained that we no longer need to test is not to predict that we ever actually will arrive at such a stage, and much less to say that we already have. The point is to show that that possibility cannot be ruled out *a priori*, and that it is not as far-fetched as it might first appear. Philosophers and scientists alike have too often tried to legislate, on *a priori* grounds, the way nature *must* operate, and have failed. My point is that we also cannot legislate, on *a priori* grounds, how we *must* learn, what the methods of scientific inquiry *must* be. When rightly understood, empiricism itself, like the processes and conclusions it addresses, is an empirical doctrine, and can in principle bring us to a stage at which it outgrows itself. This is the only position that is truly consistent with the basic philosophy of empiricism, that all our beliefs about nature must be based on interactions with nature.

If its character as a methodology is thus contingent, why is it, then, that empiricism has been the *de facto* key to so much knowledge? In retrospect we can see that there was no alternative to empiricism (correctly understood), not because it is a logically necessary methodology, but because of the way we and our particular universe happen to be. The scientific enterprise as I have portrayed it is a sophisticated version of a far more primitive approach, dating back at least to our hominid ancestors: to succeed in our endeavors, we must study the details, whether of toolmaking or DNA replication or the abundances of the elements, build on what we have learned, or have the best reasons for thinking we have learned, and learn new ways of learning, thinking, and explaining. As we have extended our inquiries, we have begun to learn that the ways of learning, thinking, and explaining that we use in everyday experience will not work for realms beyond the everyday, the realms, for instance, of the very large and the very small; new ways have had to be developed. How it has been possible to do that is another story; for the present it is enough to note that science has been a continual process of deanthropomorphizing of our ways of learning, thinking, and explaining the universe in which we live. And the only way such deanthropomorphization could have been achieved, so far as it has been, was by interacting with nature: as we have learned from the failures of traditional philosophy as much as from biology and neurophysiology, our brains are not the kinds of things that have access to *a priori* knowledge of the way the universe is. And given the oddities physics has taught us do in fact inhabit that universe, oddities we had no way of anticipating, the only way we could learn about that universe was through interacting with it, that is, to be empiricists in the true sense. But as I have tried to show in this paper, we have also learned that, once we have learned enough, new possibilities of learning are opened to us as to how to build further. Empiricism has certainly been one stage in this process, and was present in the beginning. Whether it will be present in later stages remains to be seen.

Once again traditional empiricism had it wrong. It tended to suggest that we start over from scratch in every problem-situation, leaving our preconceptions, including everything we had previously learned, at the doorstep of our laboratory, so that those antecedent beliefs would not interfere with what nature had to say to us; or if, in other versions, it suggested that we do need background presuppositions, it either said that there is no rationale for using those "hypotheses", as they so revealingly called them, or that we use such euphemistically-named "background knowledge" merely for the sake of argument, in this one situation only.

The view sketched in this paper rests on a very different picture, one which places it in a larger context of human activity and evolution: that, as in the solar neutrino experiment and in attempts to learn about quarks, we build our ways of learning about the world, describing it, and explaining it, by building on what we have learned already. And thus, in its manner of inquiry into the knowledge-seeking enterprise, the present approach is consistent with what it says happens in that enterprise. For the present view takes advantage of what we have come to know of human evolution, of the human brain, and of the way we learn, and builds its portrayal critically, but in a spirit of recognition of what we have learned. Philosophy, like science, is not something that can be done without any context; although the knowledge-seeking enterprise is the object of our investigations, the best results of that enterprise must also form the framework within which those investigations are conducted.

Wake Forest University
U.S.A.

NOTES

[1] See Shapere 1982, 1988, 1989.

[2] Asymptotic freedom will exist in a non-Abelian gauge theory in which there are less than five fermion families.

[3] Anomaly classes will cancel one another only if the gauge group is of the right size; SO(32) and E(8)xE(8) are of the right size (496 dimensions).

[4] For a discussion of some major difficulties, and the status of those difficulties as potential Kelvinian "dark clouds on the horizon," see Shapere 1991; see also the remarks in Note 4 below.

[5] The original judgment of some scientists and officials of grant agencies was that the theory of stellar energy production was so well-established that it was obviously correct, and therefore that it was unnecessary to test it any further, despite the fact that a test was feasible. Thus, for example, a 1964 letter to William Fowler asks that advocate of the experiment to answer the following question: "... why spend a substantial sum trying to measure something which is calculated with great confidence by nuclear astrophysicists ...?" (Quoted in T. Pinch, *Confronting Nature: The Sociology of Solar-Neutrino Detection*, Reidel, Dordrecht 1986, p. 94.)

[6] I'm prepared to argue that, while it would be premature to say that current particle physics and cosmology have reached the exalted goal, it is not in principle out of reach, and indeed that there exist a host of objective reasons for believing that these areas are far nearer to such a goal than ever before (even having essentially reached it in certain areas), and, while certainly not immune to surprise and disappointment, are far less subject than late nineteenth-century scientists to dangers of the kind that overthrew their theories.

We should at least be aware of the kinds of considerations that would have to be included in such an assessment. Among them is the fact that we now have a better idea than, say, late nineteenth-century scientists had of what has to be unified; and again, that if there are dark clouds on the horizon of physics and cosmology today, they are of a very different sort from those that existed for classical physics in 1900. If there is a literal dark cloud that we do not know what most of the matter in the universe is, that is not because current physics does not provide candidates for what that non-baryonic dark matter could be, or the kinds of interactions (only the gravitational, or perhaps the gravitational and weak) in which it would participate. If there is a seeming contradiction between Big-Bang cosmology and the age of the oldest stars, or a problem of how galaxies and clusters of galaxies could have formed in the time allowed by that cosmology, we are far from being in the state of ignorance that could be recognized by deeper-thinking scientists at the end of the nineteenth century.

Still, I think that such a sober assessment would show that physics has not, certainly not yet, reached a stage at which we can be satisfied with the claim that empirical testing of frontier theories is passé. New ideas and methods must earn their status, through much thought and application, as background guiding research, and it is still not clear what the impact of such problems as those just mentioned will be.

[7] See especially the references in Shapere 1982.

REFERENCES

D. Shapere, "The Concept of Observation in Science and Philosophy", *Philosophy of Science*, December, 1982, pp. 485-526.

D. Shapere, "Doppelt Crossed", *Philosophy of Science*, March 1988, pp. 134-140.

D. Shapere, "Evolution and Continuity in Scientific Change", *Philosophy of Science*, September 1989, pp. 419-437.

D. Shapere, "The Universe of Modern Science and its Philosophical Exploration", in E. Agazzi and A. Cordero eds., *Philosophy and the Origin and Evolution of the Universe*, Kluwer, Dordrecht 1991, pp. 87-202.

JAN FAYE

OBSERVING THE UNOBSERVABLE

1. INTRODUCTION

For centuries philosophy has been troubled by the idea that experience as a whole, or parts of it, does not provide us with reliable knowledge of the objective world. The recurrent problem has always been to what extent our senses can be trusted as suppliers of information about physical things which exist independently of our mental capacities. Are they what they are perceived to be? The skeptical response was that we have no warranted grounds for relying upon the objectivity of our senses, not even in relation to everyday objects. What is available for cognitive judgement was assumed to be sense impressions which might, or might not, be causally correlated with unknown and inaccessible substances. At times, this was the British empiricists' stand as a response to the challenge of Descartes' skepticism.

More moderate, and less skeptical, empiricists see our cognitive situation in a different light, especially after the general collapse of phenomenalism. They give up on the idea that mental qualities are the only accessible entities to which we can appeal in the process of justification. Instead, they regard sense impressions as reliable mental representations of real objects, or they take everyday things themselves to be the immediate objects of perception. In either way, modern empiricists hold that perception most often yields warranted beliefs, as long as they are about things that can be seen by the naked eye. In general, we can rely on our senses as trustworthy whenever we report about what we are supposed to be able to see. Thus, the earlier empiricists thought that every belief about material objects and properties goes beyond the realm of our immediately given sense impressions, whence we can only have warranted beliefs about these sense impressions. The neo-empiricists agree with common sense that we have beliefs about material objects, but claim, at the same time, that since sense impressions pass on to us beliefs about such objects, we can use some of these beliefs to warrant other beliefs. But it is still an open question whether such justified beliefs are also to be counted as true beliefs.

Some of the neo-empiricists, like Bas van Fraassen, are indeed skeptical about the range of our senses. For although he admits that sense experience secures immediate knowledge of observable things and properties, he also holds that observation cannot go beyond what we can see and provides us with true beliefs of unobservable entities.[1] Beliefs about what we perceive with our naked eye are both true, if they are true, and warrantedly assertible. Here truth and warranted assertibility have at least the same extension. As regards beliefs concerning unobservable things and properties, truth and warranted assertibility do not necessarily fall together, not even in the actual world.

E. Agazzi and M. Pauri (eds.), The Reality of the Unobservable, 165-175.

Here truth and warranted assertibility potentially have a different extension.

Before proceeding, a few distinctions may be relevant for the discussion. A person who believes that there exist truth-securing procedures of inquiry into a certain area I shall call an *epistemic realist*, whereas a person who denies the existence of such reliable procedures will be called an *epistemic antirealist*. The skeptic is an epistemic antirealist, claiming that sense experience is not reliable as a source of true beliefs of the external world as such or of unobservable entities, since there are no methods that can take us beyond our sense impressions or immediate experience. The epistemic antirealist, like van Fraassen, is a realist about semantics and ontology; whereas an epistemic realist can be an antirealist with regards to semantics and ontology. On the one hand, the *semantic realist* and the *semantic antirealist* disagree regarding the question of whether or not linguistic meaning should be analyzed in terms of verification-transcendent truth conditions. The *ontic realist* and the *ontic antirealist*, on the other hand, quarrel about whether something exists independently of the mind. The connection between one's ontology and semantics is accounted for in terms of what one takes truth to be, whereas the connection between one's semantics and epistemology builds upon ones view on warranted beliefs.

The main question in the dispute between scientific realists and empiricists is therefore one about the relationship between ontology and epistemology; namely whether true beliefs must be confined to warranted beliefs or whether they can (sometimes) supersede them. As an empiricist, one needs not reject the ontological commitments of our scientific theories. The empiricist may or may not be an ontological realist concerning observable and unobservable things, as well as a semantic realist concerning the meaning of sentences whose assertion refers to observables and/or unobservables. But if the empiricist is an ontological realist, he or she would in general hold that we do not possess a reliable method to prove this realist assumption to be true. In other words, I take the empiricist to be an epistemic antirealist.

Because of traditions in the debate about philosophy of knowledge, the distinction between realism and antirealism is something which the modern empiricist associates with commitments regarding belief and non-belief in the existence of unobservable entities. But this association, I shall argue, is not the result of a logical implication. It arises from a skeptical assumption that it is possible that our cognitive abilities of justification fall far short of ensuring truth of our beliefs: our experience may provide us with warranted beliefs about observables, but at the same time it is of no guarantee that beliefs about unobservables are true. One reaction to this challenge would be to embrace the realist view, accepting that truth about unobservables may always transcend our most justified claims. If he is also an empiricist, the realist may argue that those procedures justifying a certain claim make certain that its truth is at least probable. This was what Carnap, Reichenbach, and Hempel all did. Or he may, like van Fraassen, argue that no matter what is believed, no procedure ensures a belief to be true, if its truth goes behind what can be justified as empirically adequate; and beliefs about unobservables are exactly such beliefs whose truth cannot be justified. The other reaction is the idealist view urging that truth and warranted assertibility can be brought together, if one reduces every belief taken to be true to those which can be regarded as justified. However, empiricists' reactions are prima facie based on a positive assumption that a distinction between truth and warranted beliefs makes sense, so that truth may be related to beliefs in unobservable entities, while cognitive warranty is only something that relates to beliefs about observable entities.

Apart from phenomenalism, according to which any kind of belief concerning the external world goes beyond beliefs about sense impressions and therefore cannot be justified, the epistemic antirealist refuses the idea that observation can take us beyond our sense experience of observable entities and that observation can in any sense prove our beliefs in unobservable entities to be true. Nevertheless, if what is cognitively warranted can be extended to include beliefs concerning so-called unobservable entities, we may bring such beliefs closer to what is true than the empiricist allows. Thus, in the rest of this paper, I shall attempt to argue that it is possible to have warranted beliefs about so-called unobservables to the same extent as we possess warranted beliefs about observables.

2. THE CAUSAL THEORY OF OBSERVATION

The Lockean concept of perception, according to which seeing consists of visual sensations caused by substances whose properties are quite different from those we see, has a modern counterpart. Science has shown, it is argued, that the real world does not look the way we perceive it. Physical objects are colorless matter made of countless numbers of atoms. When we see things as colored, it is because they have a tendency or a power to produce certain 'sense impressions' in the mind of the observer. What we perceive is the effects of the causal interaction between the external world and our sense organs. The atoms on the surface of an object respond to the incoming electromagnetic waves of all kinds of lengths by absorbing and reflecting electromagnetic waves within different ranges of intensity. Subsequently, the eyes pick up the impinging radiation from the object and various physiological processes transport information about the radiation to the brain. The sensations become the end effects of a long chain of physical processes. This story about perception, it is said, is all there is to be told. Thus, if an object can take part in these physical processes it is observable, if not, not. The properties we ascribe to things on the basis of the immediately perceptual acquaintance are not their real properties but the result of information processing in the brain.

This account of seeing may be called the causal theory of perception. It holds that seeing is nothing but causal processes from the object to the brain which are describable in scientific terms. The causal processes carry information about the object, although the information is given to us in the form of sensations that cannot be 'read off' literally as telling us how the object really is. The causal theory claims that the external world is not given in perception. Instead, the external world has the same epistemic status as electrons and quarks, it is a postulated entity.[2] What we see directly are some internal items which give us information about this postulated entity. These perceived items are internal since they are ontologically dependent on the perceiver. They may or may not be mental items, but they are the real object of perception. However, if we are not able to perceive the external world, we are able to observe it on the basis of the information the percepts provide us with. So what really constitutes the object of our observation is quite different from what is immediately seen to be the object. Consequently, there is no difference *in principle* between observing an airplane or an electron. The modality of observation has become a physical possibility. For in order to be able to observe an object we have to establish a causal connection between the object and our brain, and this can be done either by just letting light reflect from the plane, or by using instruments to amplify information about the interaction between a probe and the electron. Instruments should be considered as an extension of our senses because physically there is no principle difference between the operation of biological instruments like our senses and the working of physical instruments.

Some philosophers even take a step further, claiming that human senses do not need to be involved in the act of seeing or observation. For instance, D. Shapere summarizes his seminal analysis of observation as follows: "*x* is directly observed if (1) information is received by an appropriate receptor and (2) that information is transmitted directly, i.e. without interference, to the receptor from the entity *x* (which is the source of the information)."[3] In a similar vein, Harold Brown claims, even though he makes a distinction between epistemic seeing and observing, that "a scanner who picks out a particular pattern on a bubble chamber photograph sees something that is epistemically different from what a physicist sees."[4] (When has a scanner become a 'who'?) The idea articulated by Shapere and Brown is that seeing an object is any interaction between an object and a receptor such that some information is transferred from the object to the receptor. And a receptor is any device which can interact with an object in a way that it gathers information from the object. According to this notion of seeing, science itself has the authority to tell us what the process of seeing is. The entire act of perception is describable in terms of scientific theories that deal with the object to be observed, the nature of the interaction, as well as the construction and the function of the receptor. Science explains under what circumstances a particular object can be perceived, and how information from the object is transmitted and becomes registered in the receptor when the object is exposed to the receptor.

The notion of observation without a human observer has serious drawbacks. Think of a man taking a picture of his family. Is the process of photographing really an example of observation? One has an object and a device, namely the camera, which registers certain information about the object transmitted by electromagnetic waves. Or think of a book being printed on the basis of information stored on disks, information which an author has written down on his PC. Again, is the computer observing which of the keys on the board is being touched? And is the printing of the book an observation of the disks, or perhaps of the keys on the board? Apparently, in all these cases some information is transmitted from one object to another where the object that receives the information is acting as a receptor. These cases fit perfectly well into the notion of seeing or observation independent of human beings. Nothing distinguishes them from other cases of observation like looking through a telescope. But if we do agree to call such examples seeing or observation, we have to accept any impact leaving some information on an object about another object as a case of observation. The big crater in Arizona would therefore in this sense be the result of Earth's observation of a meteor.

Shapere acknowledges these objections, claiming that what is important here are physical interactions in general and not the use we make of some of them to gather information.[5] Nevertheless, he is also talking about the conflation of the perceptual and epistemic element in observation. But how can a physical notion of causal interaction in any sense be traded in exchange for epistemic notions of belief and knowledge? If no information is received by a cognizing being, no beliefs and no knowledge can arise in the world.

The causal story about perception cannot be the whole truth and nothing but the truth. If it were true, indeed, we could very easily explain what seeing is. Perception would be identical with any information carrying process from the object to a receptor, be it a brain or not. But such an account conflates observation with measurement. We can have measurement without an observer but certainly not perception nor observation. What is missing in the story is the acquisition of beliefs. A perceptual induced belief about the observed object has to be present for a subject to be a case of perception or observation. If it were not for such beliefs, seeing would be indistinguishable from other causal processes, it would not even be one of its kind among causal processes. Moreover, two persons do not necessarily see the same thing even if the causal chain from the object to the brain carries the same kind of information. They may still disagree about how to categorize the thing they see. In spite of the amount of information about the object which is at disposal to the brain being is the same, different persons may form different perceptual beliefs on the basis of this input. The causal picture simply does not give a satisfactory account of these beliefs.

A challenge for the causal account is that various experiments of perception have clearly demonstrated that persons with the same retinal stimuli may see different things, while experiments with lenses turning everything upside down show that persons with different retinal stimuli will perceive the same thing. One thing about perception seems to be sure. What we perceive is something we are directly aware of. By saying this I do not want to claim that perception has no physical mediation. What I want to say is that *beliefs* acquired through the senses *are normally immediately given to us*. When we perceive something we immediately have a belief of seeing that something. The belief is not something which the mind derives on the basis of other beliefs about sense-data or brain-states, it is a non-inferential belief about what we are looking at right now. Immediate belief-acquisition is an indispensable element in perception as well as observation.

Indeed, some proponents of the causal theory of perception make a distinction between perception and observation while arguing that seeing involves concepts. Thus, another argument supporting the causal theory of perception is said to be the following: when we perceive objects that exist independently of us, what we see is a function of the properties of the items under observation as well as the properties of our senses. Hence, the argument goes, we cannot assume that the items we perceive do have any feature in common with the items which initiate the causal process. Furthermore, science shows that the latter items are quite different from what common-sense beliefs take them to be. Therefore we do not observe things directly as they are. As Brown summarizes the view: "According to the causal theory, the items that we epistemically perceive are the familiar objects of our everyday world, but these objects may not exist independently of our perception of them. Our best guess at the sorts of objects that exist apart from perception is provided by the best available physical theories."[6] The defect of this argument is that if we did not see *some* things as they are, we would be deprived of every epistemic ground for saying what we really are seeing. The causal theory of perception does not have the power to explain why we see things as we do as well as the ability to explain what is in fact the object of our senses. When a physicist is watching some tracks in Wilson's cloud chamber, does he observe atomic particles passing through, or the condensation of saturated water, or white tracks in the cloud chamber, or light being reflected from the water molecules, or photons impinging on his retina, or electrical impulses in the optical nerves, or does he see physical processes in his brain? Which of these various stages of the causal process do he and we really observe; where do we have to cut the possible joints? The hard-core phenomenalist makes his cut between the sensations and our perceiving mind emphasizing that sensations themselves are the objects of perception. The scientific realist makes his cut somewhere else, including in perception everything from the electron to the brain, except sensations as being subjective and insignificant. But it is difficult to see what philosophical grounds there are for preferring his view apart from the phenomenalist's. In his eagerness for doing scientific philosophy, the realist fails to realize that perception is not particularly a causal relationship, but is an epistemic relation between a perceiving subject and a perceptually experienced object.

For even though very few people, if anyone, will deny that causal factors exist which are necessary for our perception, like electromagnetic and neurophysiological processes, it does not follow that objects do not have the properties we directly see them to have. If, for instance, you experience a pot as red, it is partly because light of certain wave lengths is reflected from the pot and impinges on your eyes, partly because electric impulses are moving from the eyes to the brain. Exposed to a different kind of light, or in case your brain were damaged, you would experience the pot to have another color. But, on the basis of the fact that a certain amount of factors external to the object apparently participate in determining which properties you experience it as having, you cannot infer that the pot does not have the color you observe. For other factors exist which also take part in determining how you experience the pot, without these facts giving us any reason to say that it does not have the properties you experience. The ceramist's choice of color is such a necessary condition. Had she chosen another color than red, you also would have seen another color. But this factor does not provide us with an argument for the pot not being red. In fact we have every reason to think that objects typically have the properties *we* immediately experience them to have under 'normal' circumstances, and therefore that we perceive them directly as they are.

It seems correct to say that things have the properties which we ascribe to them on the basis of what we experience under 'normal' circumstances. Because what we believe to experience under these circumstances can be regarded as the criterion for which properties objects really have in order for us to have warranted beliefs about these objects. Relying on such a criterion is the only way we can justify any claim that our beliefs about what we are seeing are true. If our perception were not in general reliable about the real properties of things we experience, as proponents of the causal theory would make us believe, the description of what we are seeing could never be true in the sense that there is (or would be) a correspondence between that description and the state of the world. It would only be the non-experiential, scientific descriptions of the world that might correspond with the real world states, and therefore these descriptions only that could be true. Furthermore, by taking the causal theory seriously, we could never, in principle, rely on any experiential beliefs in order to correct others, since none of them can be said to be true (or false for that matter). But how could we even use experience to support scientific statements if perception did not tell us about how the world really is? A theory of perception which, as one of its consequences, has it that people are all wrong, since they automatically take for granted that most of the beliefs they receive through their senses are true, can never be satisfactory. The concept of truth is, I submit, derived from the idea that beliefs we gain by recognizing items in our visual field are in general constitutive of what we take the world to be like, and I think that any theory of perception has to account for this fact.

The denial of the causal theory of perception, according to which the nature of reality is quite different from how we experience it, has consequences for the conception of what we must consider to be real. Eddington once suggested that based on common-sense experience we have a table possessing extension, permanence and color, but based on scientific experience we have a second table filled with emptiness among sparsely scattered atoms.[7] But a scientific table in contrast to the one our common perception is dealing with does not exist. The table is not identical to a swarm of atoms which has no color or definite shape. Instead, there are different kinds of descriptions of the table we relate to in our everyday life. It may be described as having a particular recognizable shape, as being made of wood or steel and as having a certain color as either painted or not. This description is formulated in terms of our natural language referring to our experience. Another more scientific description would be that in which, say, the wood is built up of dead cells, whose parts consist of macromolecules which again are composed of atoms. Both kinds of descriptions may be true. None of what is mentioned is therefore more real than the other. In one way of describing the table, properties are attributed to it which are not attributed in the other description. Nevertheless, it may be objected that the consideration of these two descriptions as being equal is in conflict with the principle that two physical objects having different properties cannot be at the same place at the same time.[8] However, I do not claim both the existence of an everyday table and of a scientific table as two separate objects. What I have argued is that we have two different descriptions of the same object which are logically distinct. The various predications involved in the two sets of descriptions cannot be expressed in terms of each other, and thus it is impossible to reduce properties which are ascribed to the object according to one kind of predication to those which are attributed to it according to the other.

3. THE SEMANTIC THEORY OF OBSERVATION

In contrast to the causal theory of observation, I shall propose what could be called *the semantic theory of observation*. Central to the latter is the idea that perception as well as scientific observation consist in belief-acquisition. It is debatable, indeed, whether higher animals are able to form some sort of beliefs in spite of the fact that they do not possess a language. If we grant them such beliefs, mute animals may be involved in epistemic perception whenever they receive beliefs about their environment due to the input of their senses. My contention, however, is that such non-linguistic beliefs are of no use in obtaining an account of perception which can be generalized to hold for scientific observation too. The generalization will only work if we realize that the common epistemic element in perception and scientific observation is the acquisition of linguistic beliefs. A purely epistemic theory of perception must be replaced by the semantic theory of observation.

It is part of the folklore of today's philosophy of knowledge that perception is theory-laden. Nevertheless, the expression is not a happy phrase, because it ascribes more power to any theory held at a given time than what is actually the case. What is right about theory-ladenness is that perception depends partly on our background knowledge and linguistic competence, and partly on the physical circumstances we are facing at a certain moment. We are not merely seeing something, but are seeing that something is the case. It is not so that we first have certain sense impressions that are composed into a picture, and we then make an interpretation of this picture or an inference to what we really see. Most of what we see does not go through a process of interpretation before we are aware of what we are seeing. Familiar things and events are usually identified immediately as certain things and events when they are perceived in the appropriate circumstances. As we learn to master a particular language, we, moreover, see these things and their properties in the categories of that language. The natural language provides us with the names of things and events to be found in our everyday surroundings. Therefore, one is immediately aware of a certain entity as a person, a house, a tree or a mountain whenever one sees such an entity as the bearer of one of these names. For if one's language contains terms for these entities, then the mastery of that language consists in one's ability to apply each term to the right objects without hesitation.

We see items in our visual field as specific items because we can name them. There is, in other words, a close connection between our competence as speakers about a given subject matter and our ability to see something as something specific. The connection is very close indeed, so without this linguistic competence, we would be incapable of seeing any item as something which belongs to a certain category, as we in fact are capable of doing. It would be odd to argue that a person has a belief that what he is seeing is a car, but he does not know under which circumstances he may correctly assign the term 'car' to an item in his visual field. Similarly, it would be incoherent to claim that he does know these circumstances, and have a belief that he is seeing a car, but that he does not realize that the appropriate circumstances for applying the term are fulfilled. And likewise, it would be absurd to say that he knows the circumstances that have to be fulfilled in order for him to use the term to refer correctly to an item, but that his perception of these circumstances does not provide him with any evidence as to what he is seeing. I shall therefore define *intentional seeing* as the acquisition of a belief, by means of the senses, that the conditions for the application of a certain term to an item in our visual field are satisfied. In other words,

> For all *S*, *O*: *S* perceives *O* as *N* if, and only if, *S* acquires a belief through the senses that the conditions for applying the name '*N*' to *O* are satisfied,

where *S* is any subject and *O* is any object. If we do not gain such a belief, we cannot apply the proper term to what we are seeing, and thus we will not intentionally see the item.

It is, indeed, obvious that beliefs we immediately acquire through our senses may fail to be correct. Let's say that for years I see black crows each time I take a walk in the woods, but then one day an ornithologist tells me that the birds I have been seeing were in fact ravens. Since my linguistic competence in the domain of ornithology was insufficiently developed, I used the name of crows incorrectly; and because of my deficient concept of crows, I directly saw crows, where I should have seen ravens. But most often people are warranted in the beliefs they gain by their senses, because, as competent speakers of a natural language, they are able to apply most of the names and predicates concerning their everyday environment correctly, and therefore can recognize the appropriate circumstances in which the names and predicates do refer. Such people have a warranted belief that what they are seeing is what they take it to be.

Now, a similar story can be truthfully told about observation of everyday objects described in terms of a scientific language, or about any extension of observation to things which cannot be seen with the naked eye. A person, who controls the language of physics, does know which circumstances have to be fulfilled in order for him to use terms for unobservable entities correctly. And his background knowledge tells him when he can expect those circumstances to appear. So, whether we are engaged in observation of macroscopic things or in observation of microscopic things, the experience we receive will be categorized in terms of the language available. Consequently, the fact that perception of ordinary objects is as much theory-laden as observation of so-called unobservable objects seems to question the assumption that there is a genuine epistemic distinction buried in the differentiation of observable and unobservable entities. I hold there is none.

The above remarks presuppose that what cannot be seen by the naked eye still can be observed by instruments. Philosophers have until recently focused entirely on the non-epistemic, or phenomenalist, component of perception, which we share to a certain extent with other animals, while they have downplayed perception's epistemic, or rather intentional, component. The intentional component, however, is what elevates perception to observation. A propositional element is contained both in the act of perception and the act of instrumental observation, but it is merely perception that contains the non-epistemic or phenomenalist component. Undoubtedly, nobody would say that the physicist perceives the electron when he makes an experimental record of it. Electrons do not have visual properties. In the same way as a photograph of something is not identical with what it is a picture of, an experimental record is not identical with what has produced it. The record has properties that are different from those of the electron. Yet, the trained physicist may acquire a belief that it is a record of an electron, as soon as he views the recording. The belief need not be a result of an inference or interpretation of other perceivable features - he may immediately be able to identify the experimental outcome as produced by an electron. The question we have to answer is, therefore, how it is possible for us to directly acquire beliefs of something which is imperceivable?

On the basis of his general training and his experience of the physical circumstances in which he sees an experimental outcome, maybe a photograph, the experimentalist will perhaps immediately perceive it as the effect of a certain particle. His overall knowledge of physics and of the technical design and construction of the instruments, his knowledge of the experimental circumstances under which the outcome is expected to appear, and in particular his knowledge of how different particles leave behind distinguishable effects on various instruments, may all provide him with a non-inferential belief that the record he perceives, is the effect of a particular particle. Given his background knowledge and his knowledge of the circumstances under which the particular experience appears, the physicist is capable of correctly assigning the name of a particle to the experimental outcome. Since the young student doesn't know under which circumstances he may correctly apply the various names of fundamental particles, he is not able to form any belief as to what he is seeing, and he is not able to recognize the pattern on the photograph as a picture of a track of some specific particle. The student may even possess advanced knowledge of atomic particles, but still have no knowledge of the experimental circumstances under which he can refer to any of these particles. Without this knowledge, he is not able to see the pattern on the photograph as a track of a certain particle.

The point I wish to make is that whenever the physicist is able to use a term like 'electron' to refer to the perceivable outcome of an instrument, viz., as the effect of what the term refers to, he is involved in observation of the referent of that term. As a competent speaker of the physical language, he knows under what experimental circumstances 'electron' applies correctly to the track he can see, and as soon as he realizes that he is confronted with these circumstances, he gains a belief similar to the one he would acquire, if he were perceiving the electron itself. Since the physicist is able to acquire beliefs about imperceivable objects through his senses as if these objects were perceivable, it is reasonable to say that under circumstances in which he acquires such a belief, he is observing what cannot be perceived by the naked eye. Thus, we always perceive items in a context, and phenomena receive their meaning only through such a context. The white delineations on a photographic plate become understandable to the physicist as a picture of the track of an electron when he knows the conditions which have to be satisfied in order for him to refer to the conspicuous delineations as tracks produced by the electron, and he is aware that the context of the phenomenon corresponds with these conditions. This awareness is, I submit, a case of observation in which the belief gained by the very skilled physicist may be non-inferentially acquired. But the belief may also be inferentially obtained either because the observing person has too little experience, or because the item of observation has not been observed before, or because the reliability of the instrument is unknown. Whether or not the student perceives a white line on a photograph as a track of an electron depends on whether or not he knows about the ionization which occurs along the path of a charged particle passing through oversaturated vapor in a container. If he doesn't know this, and doesn't know how a bubble chamber works, he will not know that the term 'electron' is applicable in the given context. Thus, he will not acquire a proper belief about what he is seeing and, hence, he will not intentionally observe an electron.

Parallel to the definition of perception I shall say that we are engaged in observation if, and only if, on the basis of the act of perception we come to have a belief that the conditions for assigning a name or a predicate to either a perceptual or a non-perceptual item are fulfilled. Thus,

> For all *S*, *O*: *S* observes *O* as *N* if, and only if, *S* understands by the senses alone, or by the senses and certain instrumental procedures, that the conditions for applying the name '*N*' to *O* are satisfied.

As a competent speaker of a certain explicitly defined language, the scientist knows the circumstances in which the names and predicates of that language apply correctly to something in his visual field. But he would never have become a competent speaker if he could not recognize those circumstances whenever they were present. However, recognizing the appropriate circumstances that determine the correct use of a certain declarative sentence - a sentence containing some of these names and predicates - is the same thing as acquiring a *warranted* belief that the sentence is assertible.

A little more can be said about the circumstances under which the application of observational terms becomes established as the correct one. The sense of scientific terms is specified by the theoretical structure in which the terms are situated. In my opinion, however, various observational and experimental practices fix the reference of observational terms. These practices identify the bearer of the names and determine whether the assignment of a certain predicate to the bearer is satisfied. Thus, a term functions as an observational term, whether it stands for something perceptible or not, if its application can be associated with certain canonical experiments and observational situations. The meaning of such a term is not identical with any instrumental procedure as the operationalist is urging. But the existence of canonical procedures makes certain that we can establish the referent of the terms standing for theoretical entities, and by doing so we are involved in a genuine act of observation of these entities. Recognizing these procedures is what justifies a belief that we observe what is imperceptible and give us the opportunity to make correct statements about them.

We are now in a position to tell which entities are observable and which are not. Any object *O* is an observable entity if, and only if, based upon our sense perception and/or upon instrument readings, *O* can give rise to a belief that the circumstances in which it is correct to apply the name of *O* are realized. Likewise, any event *E* that is capable of producing in the same way a belief that *E* is identical to an event which is the referent of the name of *E*, is an observable entity. And, similarly, any property *P* which can induce the belief that the circumstances in which it is correct to ascribe *P* to an object are satisfied, is an observable attribute. Hence it follows that genuine unobservable entities or properties would be those which are not capable of provoking any such belief because of the lack of the appropriate visual and operational circumstances of the application of their names. The observable/unobservable distinction is not bound to the physiological equipment available to human beings for receiving information about the external world.

The upshot is that observation may or may not involve instruments. As a rule, observation secures warranted beliefs about perceptible as well as imperceptible entities. The question is therefore whether it is not possible that such warranted beliefs, though justified, may not be true. The history of science is strewn with examples of abandoned entities that apparently were considered well-established before they were given up. Think of phlogiston and caloric, just to mention two well-known examples. Such cases seem to imply that what is true and what is warrantedly believed at a given time may not always have the same extension, especially when the belief concerns imperceptible entities. Either one can take the realist position which argues that the truth is not epistemically constrainable in any way (hence truth and warranted assertibility are two different things that may not even have the same extension), or one may feel attracted towards an antirealist position which maintains that truth has to be epistemically constrained even if it is not synonymous with warranted assertibility. No matter what one's answer is, the semantic realist and antirealist can both agree that we can observe imperceptible entities.

Copenhagen University
Denmark

NOTES

1 Cf. van Fraassen 1980.
2 Cf. Brown 1987.
3 Cf. Shapere 1982.
4 See Brown 1987, p.88.
5 See Shapere 1982, p. 510.
6 See Brown 1987, p. 185.
7 Cf. Eddington 1928.
8 See, for instance, Hooker 1974.

REFERENCES

H.I. Brown, *Observation and Objectivity*, Oxford University Press, New York and Oxford 1987, p. 183.
A. Eddington, *The Nature of the Physical World*, New York 1928, p.ix f.
B.C. van Fraassen, *The Scientific Image*, Oxford University Press, Oxford 1980, pp. 15-19.
C.A. Hooker, "Systematic Realism", *Synthese* **26**, Dordrecht 1974, p. 463.
D. Shapere, "The Concept of Observation in Science and Philosophy", *Philosophy of Science* **49**, 1982, 485-525, p.492.

GIOVANNI BONIOLO

WHAT DOES IT MEAN TO OBSERVE PHYSICAL REALITY?

1. A PARTICULAR REALISTIC COMMITMENT

Facing the problem of realism is an enterprise which seems desperate if we think about the hundreds of attempts made in the course of history of philosophy. The duality idealism/realism seems difficult to grasp too, unless we limit ourselves to the specific work of a specific author or to a specific debate. I will not go into this long history. Instead, I will propose a realistic approach - also in order to go beyond the unsatisfactory dissolution of reality suggested by many authors. For example, Quine [1976] argues that, given contemporary physics - particularly Special Relativity, Field Theory, and High Energy Physics - we are forced to replace the usual notion of material physical objects with the four real numbers indicating its coordinates. Quine himself agrees that this is a very discouraging conclusion. But he believes that it is unavoidable, since he assumes that the physical object is what the last physical theorisation talks about. On this question, one is reminded of a short paper by Poincaré [1906], emblematically titled "La fin de la matière". Here, the French physicist-philosopher discusses, ironically enough, W. Kaufmann's comments on his own 1902 experiments on the ratio between charge and mass of electrons. According to Kaufmann, the direct consequence is the resolution of inertia into the electromagnetic field and thus the end of material objects. Probably, certain philosophers would agree with this conclusion. But as Poincaré urged caution and suggested waiting for new experimental and theoretical results to be produced, so the same suggestion should be addressed to such philosophers.

In my opinion, the principal cause of the dissolution of physical objects I mentioned, and other difficulties and misunderstandings, are to be imputed to a lack of clarity on the difference between the notions of *reality*, *objectivity*, and *observability*. Moreover, such confusion leads to further difficulties in trying to explain how to talk about the existence or non-existence of entities such as Pegasus: a problem which has been on the scene of philosophy since Plato's discussions on non-being and Aristotle's goat-stag. Surely, this problem may be reduced to the question of the truth of a specific proposition; but is this linguistic approach, proposed by Russell [1905] against Meinong [1904] and reconsidered by Quine [1948], truly satisfactory? Is it truly satisfactory to reduce the problem of existence to a mere analysis of propositions?

Let us come back to our problem. What does it mean to be realist? *Prima facie*, it might seem that it means to believe, or to assume, that an external world exists. Nevertheless, Berkeley's position [1713] confutes this claim (§§ 34-41). Berkeley, the

E. Agazzi and M. Pauri (eds.), The Reality of the Unobservable, 177-190.

idealistic philosopher *par excellence*, did not deny reality at all, even if he claimed that *esse est percipi*. Rather, he denied the materiality of the world and therefore its independence of the knowing subject. It follows that one might be an idealist and a realist at the same time: it is sufficient to suitably define the terms "idealism" and "realism".

According to Berkeley, the opposite of "idealism" should not be "realism", but "materialism". Already from this example, one should be aware that terms such as "idealism" and "realism" have so many different meanings that the only way out seems to leave the whole matter in the historians' hands. Doing so, we could probably get to know how the problem was tackled, but we could not try to solve it.

After these considerations, I would like to propose my realistic approach starting from three statements:

1. there is a primacy of theorisations (by "theorisation" I mean any form of conceptualisation, at whatever level);
2. there is a difference between a theory of what is known and a theory of knowledge;
3. both metaphysical idealism (according to which there is no material world independent of the knowing subject), and metaphysical realism (according to which there is a material world independent of the knowing subject), affirm that there is a something; they diverge on what this something is.

The first point states my complete acceptance of Kant's "Copernican revòlution", according to which we are able to know, also perceptively, only what we have a theorisation about. This is the strongest assumption I will make.

The second statement indicates that introducing a theory of knowledge means engaging in philosophy of knowledge, that is, epistemology; on the other hand introducing a theory of what is known means engaging in metaphysics, or, at least, in a certain kind of metaphysics which contains an intuition of the world with a precise and explicit *ontology*.

At this point, it might seem easy to assign labels such as "idealism" and "realism". In actual fact, this is what happens in the case of the theories of what is known, where metaphysical realism and metaphysical idealism may be considered as the two opposite poles. But, whenever we deal with the theories of knowledge, it seems difficult enough to assign labels. Who is the idealist and who is the realist on the epistemological side? It is sufficient to look through a handbook of epistemology to become aware of how the versions of epistemological idealism and epistemological realism are intertwined with empiricist, phenomenist, sensist, constructivist, etc., positions. The result is that we have a plethora of realisms and idealisms: empirical realism, critical realism, realism of the entities, realism of the theories, etc., transcendental idealism, phenomenist idealism, etc.

Let us go on to the third statement. Both the metaphysical idealist and the metaphysical realist have to admit that there is something, even if they disagree as to the extension of this something. Both of them admit that mental events exist.[1] According to the metaphysical idealist, these are the only things that there are, whereas according to the metaphysical realist there is much more. However, each kind of metaphysical realist states a different extension of this "much more". Some assume that in this "much more" there are also mathematical entities. Others deny. Some believe that there are only chairs and tables; others include theoretical entities. And so on and so forth. Ontology is a country populated at times by few entities, at times by many entities: it depends on the particular theory of what is known.

I shall limit myself to assuming exactly what the metaphysical idealist and the metaphysical realist agree on. In this way, my epistemology is unavoidably based on a metaphysics, but this is so indeterminate that it does not require any precise and explicit metaphysical commitment. This means that I will not specify the extension of what really is and therefore I will put the connected question into brackets, suspending judgement. It follows that I will neglect the problem of whether reality around us is completely produced by mental acts, or if it has a material foundation. On the other hand, we could be victims of a Cartesian devil. Nevertheless we could not know it owing to the limit of human reason or, to use an approach loved by the philosophers of language, owing to the fact that we are prisoners of a linguistic glass in which, and by means of which, we live. Certainly, it is possible to find sophisms by means of which we may try to escape from the glass. Descartes, for example, found them in the *cogito*. In more recent times, others, for example H.Putnam, have tried to show the contradictoriness of this devil. The problem of all such attempts is that they use language to show that beyond language there is something else. They are theorisations on how to escape from the glass. Therefore, they are but a greater glass which contains the way out of the smaller one. But how to escape from this greater glass?

There is a passage in which Poincaré [1902] cuts the question of the existence of ether short by affirming that this is a problem belonging to metaphysicians and that he, as a physicist, is interested only in calculus. It could seem a very reductive claim, but this is not true. We have to recall that the problems which cut across different disciplines are tackled by means of different methods and techniques. *The question of the existence of an entity is tackled differently by physics, epistemology, and metaphysics.* That is, we could answer it as physicists, as philosophers of knowledge, and as metaphysicians, i.e., from three different points of view which allow for three different conclusions.

As mentioned above, I will answer as a philosopher of knowledge and thus I shall leave the question of the metaphysical existence of the entities to metaphysicists. *I am putting it into brackets; I am suspending judgement on it.* However, my not looking into the metaphysical question does not mean, of course, that I am relegating it to the insignificant or negligible, but that, just now, I am shelving it in order to tackle another side of the problem.

2. PROBLEMATIC APPROACHES

What sofists and sceptics have challenged since the beginning of the epistemological reflection is that we know the essence of things. Despite such criticism, the idea that theorisations mirrored being in itself pervaded the entire history of philosophy, and resolved the theory of knowledge into the theory of what is known, or rather, into its realistic version. The essentialists, both empiricists and rationalists, considered that their aim was to unveil the world in itself and to realise a true theorisation, isomorphous to such a world.

Yet, the empiricists were wrong, since they believed that the originary source of knowledge was bare empirical perception. They had to face the problem of transforming subjective perception into objective knowledge, which - moreover - had to be regarded as a product of passive receptivity. On the other hand, and at least for certain authors, the privilege given to the perceptive moment involved the problem of justifying the generalisations, which had to preserve the truth of what was known to be true only for a few cases. In such a way, these authors had to face another problem: the justification of a chimerical inductive process which they necessarily had to invoke.

On the one hand, the criticism of this path, by means of the objection that what we know by means of the senses is only appearance and therefore different from what really is, entailed the elimination of the resolution of the theory of knowledge into the theory of what is known. On the other hand, it also entailed a gap between what really is - but what we cannot know - the *noumenon*, and what can be known - but is only appearance - the *phenomenon*. As a consequence, a differentiation between theorisations and world in itself had to be made, and thus the truthful status of theorisations was weakened, since true knowledge was only knowledge of the world in itself.

However, such a weakening was intolerable to some authors. Analogously, the abyss between the subjectivity of perception and the objectivity of theorisation was also intolerable. Such an epistemological dualism between the noumenic world and the phenomenal world did not have to entail the loss of true knowledge. Therefore, truth had to be recovered and the way to do this was not to start from the empirical world, at this point depreciated to pure appearance, but to try from the other side, that is, from the side of the understanding of the knowing subject.

It was I. Kant who tried to fill the gap between sensibility and understanding and to save the possibility of apodictic knowledge. We know very well the cost of his enterprise: on the one hand, the necessity of a canonical and binding categorial apparatus whose truth had to warrant the truth of knowledge; on the other, the artifices and obscurities of the transcendental deduction which had to eliminate the gap between the passivity and receptivity of sensibility and the activity and objectivity of understanding. However, the truth and definiteness of the Kantian categorial apparatus exploded into a thousand categories which are true only because they are imposed, and *a priori* only because they come logically before other concepts.

Moreover, in order not to resolve the application of the categories into a radical constructivism, there had to be a connection between the phenomenal world and the noumenic world. According to Kant, this connection exists only insofar as there is an undeniable stimulus by means of which the noumenon activates the passivity of receptivity. Yet, as an example, in his *Die Welt als Wille und Vorstellung* - following F. Jacobi and J. Schultz - A. Schopenhauer stigmatised that this stimulus was nothing but a causal connection surreptitiously introduced. Nevertheless, how can I state that there is a causal connection between the noumenic world and the phenomenal world if I cannot know anything about the first one? It might be assumed *a priori*, but in that case we would have the consequence that knowledge in its entirety would be a construction. It is an *aporia* which seems to be unsolvable and that may be overcome only by reinterpreting the Kantian notion of category, as certain authors have proposed.

This is the very problem of epistemological dualism:[2] the relationship between what can be known, but that cannot warrant the truthfulness of knowledge, and the unknowable, which has to be postulated, so has to avoid falling into radical constructivism. Whoever embraces a position in which there is some kind of epistemological dualism has to explain how theorisations and world (independent of them) are connected and how the validity of the former is justified. The latter problem has become even more important in the post-Kantian tradition, in which the certainty allowed by the apodicticy of the Kantian categorial set-up had been denied. Consequently, there has been a loss of reliability in the theorisations. They can no longer be qualified as true theorisations whether starting from the bottom - since their empirical verification is an impossible dream which crushes the endless number of necessary confirmations, or starting from the top - since we do not have a true principle they can be deduced from. Thus, theorisations, having lost their centrality, have been considered by some authors as instruments, as conventions, as attempts made to save the phenomena, etc. The explosion of the Kantian categorial apparatus has transformed epistemological dualism into an insane running towards the destruction of the cognitive value of the theorisations.

However, the post-Kantian tradition has another degenerate son, that is, constructivistic rationalism. But it is difficult to accept the resolution of the world into exactly what the theorisations construct. I find difficult to accept that everything around us is intellectual *invention*. I find difficult to accept that the Medicean satellites did not exist before Galilei "invented" them, or that the intermediate bosons did not exist before Weinberg, Glashow e Salam "invented" them.

Therefore, we have either an epistemological (idealistic or realistic) monism which resolves knowledge into what is known, or the difficulties of epistemological dualism, *à la Kant*. A way out has to be found. Probably, this is possible only by tackling the problem in a different way. What I will propose is precisely an attempt to escape from such a dilemma.

To conclude these preliminary remarks, I would like to stress that my proposal is a germ of a theory of knowledge in which the notions of reality, objectivity, and observability should be valid, both at the ordinary level, and at the most advanced physical level. I believe that what needs to be proposed is a unitary intellectual framework, in which the question of the reality of the macroscopic ordinary entities is linked with the question of the reality of the microscopic and macroscopic physical entities.

3. THE PHENOMENOLOGICAL DESCRIPTION

In the *Prolegomena zu einer jeden künftigen Metaphysik*, Kant reconsidered a classical distinction: the difference between the analytical (or regressive) method, typical of the *Prolegomena*, and the synthetic (or progressive) method, typical of the *Critique* (cf. Kant [1783], § 5). The first one starts from the conditioned to arrive at the conditions, the second one starts from the conditions to arrive at the conditioned. In this third section, I will ascend from what is conditioned - the perception - to its conditions, which I will synthetically arrive at in the fourth section. However, the analytical "ascension" has to be considered a simple description: it is not said that what appears for us to be first - i.e., perception - really is epistemologically first. Nevertheless, it is worth following such a path to *phenomenologically show* what will be synthetically explained, in the fourth section, by starting from what really is first, that is, the conditions for the possibility of knowledge: the theorisations. This is the reason why the fourth section will be titled "The Transcendental Explanation".

3.1. *From Subjective Perceptions to Observed Entities*

According to a phenomenological point of view, to perceive something is to perceive an indistinct bundle of a multiplicity of features. However, according to the first Kantian statement I presupposed, these features are due to the applied theorisations and to the perceiving subject's physio-psychological state. Such an indistinct bundle, namely, the *perceived entity*, is that on which the subject has not yet exercised his capability of objectivation, i.e., his capability of separating what is objective from what belongs to his own perception. It might also be called *phenomenon*, even if not in the Kantian sense of *Erscheinung* - i.e., such as appearance, but in the Brentanian sense of *Phänomenon* - i.e., such as something which presents itself to the mind exactly how it presents itself. Nevertheless, such a perceived entity is still a sort of bolus which has to be digested. In fact, it is presented to the mind by the senses and it has to be objectivated by emending the subjectivity of perception. Up to now, we have the *originary brute fact* embedded with subjectivity, but this - being a fact (from the Latin *facere*, i.e., to make) - is something already made, in the sense that it has been made meaningful by means of theorisations.

If perception were theoretically neutral, human understanding would be an *intellectus ectypus*, but this, as Kant [1781-1787] showed, is not true. On the other hand, to talk about the theoreticity of perception does not mean to talk about the complete production of the perceived entity: human understanding is not even an *intellectus archetypus*.

The perceiving subject does not construct reality at all, but he selects, makes meaningful, that is, he constitutes[3] *(by means of the applied theorisations) certain aspects of reality which he subjectivises (by means of his own physio-psychological state).* Thus, the perceived entity is quite different from what methaphysically is. This is the necessary but obscure and indeterminate background of perception and, therefore, of knowledge. Only our theorisations and our physio-psychological states may determine aspects (*only aspects*) of such a metaphysical reality which, in its totality, can never be grasped.

Subjectivisation enables that the perceiving subject's physio-psychological states charácterise the brute fact in a peculiar way. Given the uniqueness of the subject, the perceived entity is unique and irremediably connected with him. Therefore, a problem arises: how may a subjective entity be transformed into an objective entity? Let us consider a book. It has shape, dimension, colour, etc., and it is perceived with a certain shape, certain dimensions, certain colours, from a certain perspective, etc. If one goes around it, the shape, the perspective, the dimensions, the colours, subjectively change. Nevertheless, we are aware that something remains invariant during all the perceptions. *This something of invariant is the first product of the objectivation of subjective perception.*

Now, let us consider more than one subject. Each subject looks at the book in his own way: from his own angulation, from his own perspective, with a certain shape, certain dimensions, certain colours. But there is something that all the subjects look at in the same way. There is something which is invariant: its intersubjective shape, its intersubjective dimension, its intersubjective position. Whenever the perceived entity is intersubjectively objectified, its subjective features are emended. What remains is what is shared by all perceiving subjects. Therefore, the intersubjectively perceived entity is what transcends individual perceptions and remains invariant in the space-temporal flux of the single perceptions of the same subject, or of many subjects. In this way, the subjective is excluded and becomes a sort of vague background of intersubjective knowledge: everyone knows that it exists but everyone knows that it cannot be communicated: if one tried it, the subjective would be objectified and so emended.

Since the intersubjective entity is invariant for the intersubjective perception, the possibility of self-illusion should be avoided. Nevertheless, someone could raise an objection: "In this way, the self-illusion of a single perceiving subject is avoided, but not the possibility that all the perceiving subjects are victims of the same illusion." This is a good objection, which makes us understand that not everything that it is intersubjectively perceived has the same status. In fact, there are some intersubjectively perceived entities which are not at all illusions, and I will denote these entities as *observed entities*. Therefore, for the observability of an entity, neither the invariance for the single subject, nor the invariance for many subjects is sufficient. *There should be an invariance for all subjects who want to perceive it.* Let us try to clarify by means of an example. A subject may affirm he perceives a goblin. But this perception is a self-illusion if it is not invariant for many subjects. Nevertheless, it may happen that many subjects affirm they perceive a goblin and in this case the perception would be intersubjectively invariant. But, again, *this invariance is an illusion if it is not an invariance empirically testable by whoever wants to test it.* Whoever affirms he perceives some invariance has to accept that it may be tested by any other subject. *The observability of an entity, in order to be genuine observability, has to be at everybody's disposal.*[4]

4. THE TRANSCENDENTAL EXPLANATION

4.1. *Objective Entities*

It would seem that the objective entity is only the product of the emendation of the perceived entity from its subjective features, that is, only what is intersubjectively perceived. Actually, the domain of the objective entities is different from the domain of the intersubjectively perceived entities.

First of all, what is theoretical is obviously intersubjectively invariant. A theorisation, at whatever level it may be, is always expressed (or expressible) in such a way that it may be grasped *by whoever possesses the suitable instruments*. The everyday theorisation on a chair in a given language is graspable by the everyday speakers of that language. The physical theorisation on the Higgs boson is graspable by whoever has the necessary physical-mathematical tools to understand Field Theory. The chair and the Higgs boson, thanks to theorisations, are objective entities, that is, a knot of properties stated by certain relationships constructed by theorisations. According to Cassirer [1910] (Ch. I), it is not the sensuous vivacity of perceptions which forms the features of the true objectivity, but the richness of the theoretical relationships.

Therefore, we may release ourselves from perception and cease thinking about the *objective entity* as the emendation of the perceived entity from subjectivity. Instead, we may think about it as a *knot of properties due to theoretical relationships, independently of whether these are deterministic or statistic, of high or low level, consciously or unconsciously formulated.*

In this way, there is no *diminutio* as to what I have said so far. In fact, some objective entities are exactly what we have arrived at after the emendation. Before, I defined the objective entity as the result of the emendation of the subjective part of a perceived entity; now, as a knot of theoretical properties. But what remains after the emendation is exactly a knot of theoretical properties. These are two ways of looking at the same concept: the first one phenomenologically, the second one transcendentally, in the Kantian sense.

However, it follows immediately that there are objective entities which cannot be perceived and, of course, they cannot be obtained by the emendation of any brute fact. This conclusion also enables us to better understand the epistemological status of "strange" entities. Let us think of Diana, the Hunt Goddess. She is an objective entity, because she is a knot of theoretical properties, even if of a mythological type. It is possible to talk about Diana, to describe Diana's body, to talk about her adventures and her loves, etc. In short, there are theorisations which allow us to think about Diana as a knot of properties, exactly as we have theorisations which permit us to think about pencils and leptons as objective entities. Aristotle's goat-stag, Pegasus, the Seven Dwarves, Moby Dick, and Captain Achab are objective entities: they are knots of properties due to theorisations.

Therefore, there are objective entities which surely refer to perceivable entities - that is, to aspects of reality - as in the case of chairs, tables, electrons, and molecules. In this case, the objective entity may be seen also as the arrival point of the corresponding perceivable entity. However, there are objective entities - such as the Higgs boson and gravitational waves - about which we are not sure that they refer to any perceivable entity. In this case, we are forced to interpret the judgement on their existence in a conjectural way. There are other objective entities which are fictional entities - such as the continuum charge distribution or the infinite conductive wire. In this case, there are no corresponding perceivable entities. There are also objective entities which are produced by fantasy or myth - such as the Seven Dwarves and Pegasus - and which are without perceivable correspondents. Finally, there are objective entities which are absolutely impossible as aspects of reality, such as the triangular square, the convex concavity, and the round square cupola of Berkeley College loved by Quine [1948]. But *all of these objective entities have the same characteristic: they are invariant - of course each class is invariant in a specific way; they are invariant knots of theoretical properties and it does not matter if some knot is composed of self-contradictory properties.*

In other words, following Meinong [1904], I am claiming that the totality of the perceivable in a meaningful way (including what will be and what has been) is less than the totality of the objective, intended as the totality of the knots of properties due to theoretical relationships, that is, to the totality of the entities constructed by theorisations. Therefore we have to take into account what Meinong calls *intentional objects*, the set of which is greater than the set of real objects.

The intentional object - that is, what I have called objective entity till now - is what to which the corresponding concept directs itself. And , of course, as Meinong [1899] emphazises, there is a difference between an intentional object (my "objective entity") and a real physical object (my "observable entity"). Meinong talks about *subsistance* for the former and *existence* for the latter. Anyway: intentional objects may be known (even if not in an empirical sense), being knots of theoretical properties constructed by theorisations, and, thus, they have to be in some sense.

To claim that intentional objects - i.e., objective entities - "are", in some sense, does not mean to ascribe them a mental or psychical reality, as Meinong teaches us. If we have to talk about their being, such a being has to be interpreted in a purely semantical way. Their subsistance is uniquely a subsistance inside a semantical domain, which allows us to talk about them ascribing them certain theoretical properties. So far I have said nothing which may ascribe them a mental or psychical reality. Moreover, even if I claim that theorisations are mental representations, I do not refer to any form of psychologism. Instead, I mean only that they, and the intentional objects they construct, are intellectual products of the human mind. To deny this is to deny the power of the human mind and the combinatory and constructive power of theorisations.

4.2. *Observable Entities*

We have seen that not every objective entity is a perceivable entity, that is, not every subsistant object exists. On the other hand, we know that perceived entities may be emended from their subjectivity and whenever this happens they become objective entities. In this way, we have arrived at an objectivity which is smaller than the one which we arrive at by starting from theorisations. This means that there is a subset of the objective entities (i.e., the observable entities, or physical objects) which may be thought both (1) phenomenologically, as the objectivation of perceivable entities, and (2) transcendentally, as referring to perceivable entities constituted by theorisations.

Therefore, from a transcendental point of view, the observable is a knot of properties due to theoretical relationships, which is graspable by whoever possesses the necessary conceptual tools. Moreover, it is what refers to aspects of reality by constituting them. Instead, from a phenomenological point of view, the observable is that part of the perceivable entity which is invariant with rèspect to the intersubjective perception of whoever wants to perceive it. If *objective entities subsist* in a semantic domain, being knots of theoretical properties, *observables exist*, being aspects of reality.

According to this approach, the problem of the existence of an entity becomes the problem of its observability. The chair exists since it is observable and not since it is objective, or perceived by one subject. The electron exists since it is observable and not since it is objective or perceived by one subject. Instead, Pegasus does not exist, since it is not observable, even if it is objective and therefore it subsists in a semantic domain.[5]

At this point, it is worth underlining again that the observable is the objectivation of the perceivable aspects of reality and not of reality in its totality. Therefore, the set of the observed entities is quite different from metaphysical reality. The former is the set of the aspects which are made meaningful to perception by applied theorisations. The latter is the metaphysically indeterminate and obscure background of perception and knowledge. Only our theorisations can determine aspects of reality, but the totality is there: mute and waiting for us, who un-veil its forms. Reality can never be grasped in its totality, only its aspects made meaningful to perception can.

H. A. Lorentz's electrodynamics constructs a particular objective entity (*the intentional object*) called the "Lorentz electron", which is the knot of certain theoretical properties. In this case, it is also an observable entity (*the physical object*) and, therefore, it is the invariant part of the corresponding perceivable entity called in the same way, which is constituted and made meaningful to perception by Lorentz's theorisation. Analogously, QED constructs the "QED objective electron" (*the intentional object*), that is, a particular knot of certain theoretical properties. There is also the corresponding observable electron (*the physical object*), and therefore this is the invariant part of the perceivable electron constituted and made meaningful to the perception by QED. Lorentz's and QED's electrons are different entities, even if they have, level by level, some common properties. Nevertheless, it is wrong to think about them as two different metaphysical realities. *Reality is unique*, although there are different observable aspects of it. *Reality is unique* and Lorentz's electron and the QED's electron refer to two aspects of such ontological reality.

The fact that, from a transcendental point of view, observables are knots of properties (constructed by theorisations) enable us to understand that we can observe only what is permitted us by theorisations. Of course, the physical observability of an entity is limited not only by theorisations but also by physical possibilities. Without a telescope, the entity denoted by "mountains of the Moon" cannot be perceived from the Earth. Without the suitable theorisations, the entity denoted by "tracks in a bubble chamber" cannot be made meaningful to the perception. Nevertheless, the two impossibilities are different. The first one is a physical impossibility, the second one is a semantic impossibility. The mountains of the Moon are not physically perceivable in any way from the Earth without a telescope. In the second case, things change. In fact, *almost the same* something may be perceived as "paths on a monitor", if one has the suitable theorisations to make meaningful such things as "paths", "monitor", and "paths on a monitor". This remark enables us to understand how the distinction between "*to perceive* something" and "*to perceive that* something is something" may be misleading. It is impossible to perceive something without this "something is something". Yet it is possible that one subject perceives something as A, and that another subject perceives *almost the same* something as B. It all depends on the applied theorisations. In fact, it is possible to perceive something as "tracks in a bubble chamber" if the suitable theorisation has been applied. But it is possible to perceive *almost the same* something as "paths in a monitor", if the suitable theorisations have been applied. *Perception is always perception that something is something.*

Therefore not only are there aspects of reality which we may perceive, but there are also aspects which we cannot perceive, owing to the limits of the applied theorisations. To modify these limits means to have access to different aspects of reality.

Theorisations, constructing new objective entities, propose them as hypothetical observable entities and, therefore, the theorisations make meaningful hypothetical aspects of indeterminate reality which previously were neglected; but they were there. *The subject does not invent anything about reality, but he dis-covers, or un-veils, aspects of metaphysical reality.* It follows that our knowledge of aspects of reality is enlarged by enlarging what is objective, that is, by enlarging theorisations.

Again we have to deal with the metaphysical indeterminateness of reality in its totality. This is a sort of metaphysical pre-requisite for the endless potentiality of knowledge which leads us to interpret scientific research as a never-ending *un-veiling* of ever new aspects of reality. We have an "unended quest" not only due to the fallibility of human being, as K.R. Popper taught us, but also due to the metaphysical indeterminateness of reality as a whole.

To work inside the objective, for example, to do theoretical physics, means constructing theorisations, and therefore objective entities, which hypothetically refer to and constitute aspects of metaphysical reality not yet un-veiled, not yet made meaningful. To work inside the observable, for example to do experimental physics, means testing if the hypothetical objective entities, proposed by the theorisations, are actual representations of certain aspects of reality not yet un-veiled, not yet made meaningful.

Thus, the test of the empirical observability of an objective entity is a test of the cognitive validity of theorisations, that is, a test of their capability to be representations of aspects of reality. In this way, the problem of the absolute truth of the theorisations loses any importance. This way of thinking hides the idea that reality may be *this " aut" that.* But, according to the points above, reality is *this "and" that.* On the other hand, this is a direct consequence of the metaphysical indeterminateness of reality.

By means of applied theorisation we know aspects of reality, but this knowledge is tentative knowledge. Due to the fallibility of the human subject, our theorisations are only conjectures constructed to make meaningful hypothetical aspects of reality. And, obviously, knowing subjects belonging to different cultures and different ages construct different theorisations. That is, theorisations are never absolute but always related to a particular culture and to a particular age. This remark involves that the aspects of reality we observe *hic et nunc* may be partially or completely different from the aspects of reality observed by knowing subjects belonging to other cultures and other ages. But this is a problem that will not be faced here.

4.3. *Unobservables*

Theorisations, being constructed to represent aspects of reality, should enable us to actually observe the entities they refer to. Therefore, the theorisations which construct unobservable entities should be criticised from the point of view of the theory of knowledge. Usually, these unobservables serve the purpose of supporting some internal consistence, but their introduction creates the suspicion that the theorisation might be replaced by another one which, besides representing the same aspects represented by the first, does not involve the introduction of unobservables. Of course, the presence of unobservables does not compromise the observability of the other objective entities of the theorisation. It leads only to the idea that the theorisation might be replaced. Moreover, it should be noted that, sometimes, theorisations with structurally unnecessary unobservables are kept, even if they are pruned of such unobservables. This is the case of Classical Mechanics, pruned of the absolute references, and of Classical Electromagnetism, pruned of the ether.

I have just mentioned a theoretical unobservability, as in the case of the ether proposed by Lorentz [1909]. But there is also a technological unobservability, that is, an unobservability due to technological limitations, as in the case of the Higgs boson. Here we are facing a methodological problem because often what is technologically unobservable today may be technologically observable tomorrow. That is, we do not have any precise criterion to demarcate such cases from those whose technological unobservability will persist forever. Therefore, we are forced to submit each individual case to the judgement of a particular physicist, or a particular community of physicists.

It should be noted that by "unobservable" I have meant what can not be observed at all, both owing to theoretical constraints, and owing to technological limits. Usually, "unobservable" is interpreted also as "non-directly observable", but I prefer not to follow such a convention. Obviously, the observability of a Z^0 particle is different from the observability of a chair, or the mountains of the Moon. Although I agree that there is a difference between the two kinds of observability, I would continue to affirm the observability of the Z^0, even if it is an *observability by inference*. In this case, the entity is observed not only by means of an experimental apparatus (as in every other case), but also by means of a causal theorisation enabling us to infer the non-direct observable cause from the direct observation of its effects. As we may infer fire by observing the smoke and by means of a causal theorisation, which is testable independently of this specific case, analogously, we may infer the Z^0 by observing its tracks and by means of a causal theorisation, which is testable independently of this specific case. In both cases, the fire and the Z^0 are observable by inference. One may object that we can approach the smoke and observe the fire directly, whereas we cannot do it in the case of Z^0. But if we accepted that only what is directly observable is an aspect of reality, we would practically know all of reality. Is this a good position to support?

By the way, it should be remembered that viruses were observable by inference until the thirties. Fortunately enough, the biologists, from Pasteur onwards, have considered them as real entities. This case also teaches us the fact that *what is observable by inference today may become directly observable tomorrow*. At this point, we may distinguish among:

- *direct observability, in a physiological sense*: observability limited to what can be observed without any instrument beyond our natural "observative instruments" (eyes, ...);
- *direct observability, in a non-physiological sense*: observability connected with the use of instruments (microscopes, telescopes, ...) which does not enable a support of theoretical causal inferences;
- *observability by inference*: observability connected with the use of causal theories which allow us to infer the observability of the causes from the observability of their effects; as happens in the case of the elementary particles observed by inference from their effects in a bubble chamber;
- *technological unobservability*: in a given time some entities may be non-observable because we have not yet reached the technological stage which is needed to construct the suitable experimental apparatus, or we do not want to construct it; in this case, we may also speak about their hypothetical observability (maybe, hypothetical "observability by inference");
- *unobservability in principle*, or theoretical unobservability: there may be entities constructed by theorisations which are such to be never empirically testable; as in the case of the Lorentz's 1909 ether; in this case the entities are not even hypothetically observable.

But, what is real? Of course, an "unobservable entity in principle" is not real, whereas a "technologically unobservable entity" may be real. Surely, the "directly observable entities" are real, but what may we say about the observables by inference? Some, for example van Fraassen [1980], suspend judgement on their reality. Others, for example Cartwright [1983], argue that they are real by using the argument of the "inference to the best cause". I believe that this is a good reason. I do not want to enter the question more deeply. Instead, I want to sterss that *reality which we are speaking of is not metaphysical reality but only what I have called aspects of metaphysical reality which we have constituted in a meaningful way by means of our theorisations*.

5. MORAL CONCLUSIONS

According to my proposal, reality, or rather, metaphysical reality, is always the same for each theorisation. It is unique and indeterminate. It is the perceptive starting point of different physical objects and of physical knowledge. Therefore, we have to distinguish between *metaphysical reality, the domain of experience* and *nature*. The second one is the totality of the brute facts perceived by the single subject. The third one, *à la* Kant, is the totality of the observable entities, that is, of physical objects. Therefore, *it is not metaphysical reality which changes, but the domain of experience and nature*. The domain of experience changes if the perceiving subject's physio-psychological states change. Nature changes if theorisations change. It is nature and not metaphysical reality that is void or full of elements, entirely of electromagnetic origin, or completely reducible to a few elementary particles, simple or complex, causal or acausal, deterministic or indeterministic.

In this way, the world around us is not resolved into theorisations because they do not give us reality, but objectivity, and, in some cases, aspects of reality. Therefore, from this point of view, it is easy to stigmatise whoever talks about the indeterminism or determinism of metaphysical reality, or about its causality or acausality, its being chaotic or ordered. Reality, in its totality, is neither indeterministic nor deterministic, neither causal nor acausal, neither chaotic nor ordered. It is the theorisation which constructs one of the two poles of each duality. Thus, we perceive aspects of reality which are indeterministic or deterministic aspects, causal or acausal aspects, chaotic or ordered aspects, because theorisations have made them meaningful in such a way. To infer that reality in its totality is deterministic or indeterministic, causal or acausal, chaotic or ordered only because a handful of its aspects are made meaningful to perception in a certain way, is a fallacy: the fallacy that requires that the whole is equal to its part only on the basis that the part is constituted in a certain way.

The knowing subject constructs the objective entity which, in some cases, refers to aspects of reality, but in this way he does not construct reality, he constitutes such aspects of reality. Therefore, neither reality is resolved into theorisations, nor is theorisations resolved into reality, and we may avoid the Scylla of radical constructivism and the Charybdis of naive realism. Concluding, reality is always similar to itself; it is there. Reality is not dissolved into physical theorizations, but it is our misleading interpretation of the status of the latter which induces us to conclude that reality is dissolved in them.

Padua University
Italy

NOTES

[1] It shoud be noted that a metaphysical realist may argue for the reduction of mental acts to brain processes. However, this possibility does not have any importance for our discussion.

[2] Cf. Agazzi 1969.

[3] There is a deep difference between "constitution" and "construction". According to Kant [1787], while a (mathematical) concept is *constructed*, an object, in order to be a known object, is *constituted* by the concepts.

[4] Of course, there are entities, for example a celestial body before a gravitational collapse, which can be perceived only once. However, this does not mean that they are not observable. In fact, they may be observed by more than one subject and moreover they can be recorded, and the record may be observed and checked by whoever wants to observe and check it.

[5] There are those who argue that Pegasus is observable because we may observe that it is not. Obviously, here the definition of observability is different from mine. I have argued for a notion according to which "observable" is what is observed in an invariant way by whoever wants to observe it. Nevertheless, I could admit that Pegasus was hypothetically observable before we realised its non-observability.

REFERENCES

E. Agazzi, *Temi e problemi di filosofia della fisica*, Manfredi ed., Milano 1969.

N. Cartwright, *How the Laws of Physics Lie*, Clarendon Press, Oxford 1983.

E. Cassirer, *Substanzbegriff und Funktionsbegriff*, Bruno Cassirer, Berlin 1910.

B.C. van Fraassen, *The Scientific Image*, Oxford University Press, Oxford 1980.

I. Kant, *Prolegomena zu einer jeden künftigen Metaphysik, die als Wissenschaft wird auftreten können*, 1783. Engl. transl. by P.G. Lucas, Manchester University Press, Manchester 1953.

I. Kant, *Kritik der reinen Vernunft*, 1787. Engl. transl. by N. Kemp Smith, MacMillan, London 1989.

H.A. Lorentz, *The Theory of Electrons*, Leipzig 1909.

A. Meinong, *Über Gegenstände höherer Ordnung und deren Verhältnis zur inneren Wahrnehmung*, 1899. Now in Meinong, Vol. II, 1913-1914.

A. Meinong, *Über Gegenstandstheorie*, 1904. Now in Meinong, Vol. II, 1913-1914.

A. Meinong, *Gesammelte Abhandlungen*, Voll. I-II, Leipzig 1913-1914.

H.J. Poincaré, *La science et l'hypothèse*, Flammarion, Paris 1902-1907.

H.J. Poincaré, *La fin de la matière - Atheneum*, 1906, pp. 201-212. Now in Poincaré, ch. XIV, 1902-1907.

W.v.O. Quine, "Wither Physical Objects", in *Boston Stud. in the Phil. of Sci.* **39**, Reidel, Dordrecht 1976, pp. 497-504.

W.v.O. Quine, "On What There Is", *Review of Metaphysics*, 1948. Now in *From a Logical Point of View*, Harvard University Press, Harvard, 1980, ch. I.

ALBERTO CORDERO

REALISM, AND THE CASE OF RIVAL THEORIES WITHOUT OBSERVABLE DIFFERENCES[1]

1. INTRODUCTION

Rationality and truth: two themes realists easily turn into mad fire. In 'real science' rational theory choice proceeds largely in an eliminativist fashion, helped by basic logic and a battery of learned criteria for choosing between competing theories. Evaluations of this kind can be, however, notoriously treacherous: A rational verdict may turn out to favor a false belief. Truth ascription, expected to be less contextual than rational evaluation, is a more ambitious affair still. The existence of an initial tension between the issues of rational scientific change and the search for truth is thus hard to deny.

In this paper, I want to explore the noted tension and discuss its impact on realist interpretations of scientific theories. For this purpose I will focus on a recent case study from the foundations of quantum physics. The case concerns three approaches which take the wave function as a physical field: Bohm's quantum theory of motion, the physicalist version of the 'many decohering worlds' approach, and a family of theories that incorporate wave function collapse. Although these approaches predict different phenomena, they are contingently prevented from predicting different *observable* phenomena by the way decoherence between different wave function fronts works in the actual world. Further analysis reveals that the three theories are not just observationally equivalent, but also equivalent on more general prevailing scientific criteria for theory choice. Elsewhere I have explored a possible resolution of this case in favor of the collapse approach[2]. My purpose there was to expose the limited dependence of scientific theory choice on observable structures, and more specifically to comment on what this limitation teaches us about the defeasibility of prevailing criteria for preferring one theory to another. This time I want to concentrate on the noted tension between reason and truth in the realist project. Specifically, I will focus on the issues of underdetermination of theories by empirical criteria of theory choice, and incorrect rational selection. To make my exploration viable, within the space available here, I will concern myself primarily with the project of giving the wave function a realist interpretation. I think the chosen case, though specific, throws light on both the scope and limits of theory underdetermination and truth ascription in mature science, and on the impact of these last two features on the issue of scientific realism about entities and theories.

E. Agazzi and M. Pauri (eds.), The Reality of the Unobservable, 191-206.

2. THE WAVE FUNCTION AS A REAL OBJECT

In the standard version of quantum theory [SQT], the wave function (represented by the symbol 'Ψ') sticks out half-way between a representation of a physical field and a representation of a probability measure for possible experimental outcomes (via Born's rule). Its ontic status has been a matter of dispute since the early days of quantum theory, with some critics rejecting Ψ as a possible physical field, and other critics maintaining that it simply measures of the incompleteness of SQT as a theory of *physical reality*. To this date, all attempts at proving Ψ nonphysical have failed, however, frustrated by theoretical analyses, experiments on individual quantum systems, or both[3]. But not all is well with a realist interpretation of Ψ, let alone the entire SQT. Foremost among the encountered difficulties is the peculiar way in which measurement processes figure in the theory. According to the so-called 'Projection Postulate', when a magnitude is measured on a system, the latter's wave function collapses onto one of several possible 'eigenfunctions' linked by SQT to the magnitude being measured[4] - a strange role for measurement to play in a theory that passes for fundamental. SQT contains two different dynamical laws. One is linear, deterministic, and in harmony with the spirit of classical physics; this part of the dynamics is expressed by Schroedinger's equation (let us represent this part by the symbol 'QT_0'). The second law is nonlinear, stochastic, measurement-relative, and totally at odds with the traditional spirit of physics; this 'embarrassing' part of the dynamics is expressed by the Projection Postulate.

Exactly where, how, and when, does the Projection Postulate become operative? Largely because this question is left unanswered by SQT, many thinkers favor instrumentalist or nonphysical interpretations of Ψ, despite the otherwise enormous success of the theory. Realism about the wave function is further complicated by the fact that Ψ is a *wave* only in a very nonstandard sense of the term - the most comfortable representation of Ψ being as a vector in a Hilbert space of infinite dimensions. In addition, Ψ displays significant nonlocal behavior. Still, there is a growing acceptance that Ψ should be understood as a physical field. One old reason comes from quantum interference phenomena; favorable evidences from these have got better in recent years, thanks to vast technological improvements in the manipulation and study of single systems[5]. Another motivation is the devaluation of classical waves as physical models. Simply put, it has become hard to deny that electromagnetic waves are quantum fields; classical waves are now regarded as epiphenomenal disturbances in phase space, where representations of Ψ display such traditional properties as effective sizes and shapes, all changing according to a linear law of evolution (as long as no measurement-like situations are involved).

There remains, however, one major obstacle to taking Ψ physicalistically: the measurement problem. In the next subsection I review three responses to this old difficulty; all of them present Ψ as a physical field. Two of the proposals completely drop the notion of wave function collapse; the third approach naturalizes it.

2.1. *Three Theories*

The first theory is a nonlocal hidden variables account of quantum phenomena that assumes a mixed ontology of particles and the wave function. Introduced by Bohm in the early 1950's, it has developed since into a detailed quantum theory of motion ('BQTM' in what follows). In the earliest version, observable models agreed with the standard quantum algorithm by construction[6]. More recent presentations partly address this shortcoming by deriving some of the originally imposed coincidences with SQT from *probable* initial conditions of the universe, allowing the observable models of the theories to deviate from SQT, at least for *improbable* conditions. Improvements like these arguably give BQTM some differential conceptual fertility, particularly in cosmology[7]; no differences with applications of SQT to this world are expected, however. BQTM explicitly assumes an ontology of particles, interprets Ψ substantivally (along the lines of the 'pilot wave' concept proposed by de Broglie in the early days of quantum theory), and rejects the Projection Postulate. The theory includes adequate provisos for EPR nonlocality and quantum indeterminateness (superpositions of classically incompatible states), which is why BQTM, unlike local hidden variables proposals, is not challenged by the Bell experiments. Bohm's particles move about largely as in classical physics, though with diminished independence, due to the action of Ψ, which functions as a guiding nonlocal field that literally pushes its corresponding particle in configuration space. Physical properties are peculiar in the world of BQTM, in that only position and trajectory stand as intrinsic magnitudes of particles. The latter may be attributed other properties as well (say, linear momentum and spin), but on condition that their values be relativized to the total physical context at the time - i.e., they function as 'contextual' properties[8].

The second theory on my list is an improved version of the old 'Many Worlds' approach by Everett and other thinkers[9]. It comes partly from physical analyses of the phenomenon of decoherence (i.e., the blurring of the relative phases between different wave function fronts that occurs when systems interact with complex environments[10]), and partly from objectivist philosophical investigations of the old many worlds approach[11]. I will call the resulting proposal the 'Many Decohering Worlds' theory [MDW]. MDW presents QT_0 as the complete theory of the quantum domain, interpreting the wave function of the universe as a physical field which constitutes physical reality. Following Everett, MDW addresses the measurement problem by trying to explain away phenomena of wave function collapse and quantum probabilities. We encounter the latter in ordinary experience, defenders of MDW say, because QT_0 (along with some modest physicalist conception of the mind) predicts that wave function collapse will be 'experienced' by almost all observers in the superposition of states that quantum entanglement generates at the cosmic level[12]. According to MDW theorists, ordinary experience of a classical macroscopic world results from the fact that *deviant* superpositions (i.e., ones of the Schroedinger's cat variety) get massively reduced by decoherence effects before anyone has time to observe them. Clearly, however, decoherence cannot, by itself, reduce Ψ. All it can yield is states that mimic mixed states. This need not be a fatal flaw, but it is a warning sign that the theory needs improvement. Specifically, a coherent descriptive metaphysics of MDW is required, including (a) an explicit physicalist account of how we come to experience a 'classical' (or 'quasi-classical') world, (b) a story about how quantum probabilities are to be understood, and (c) an explication of identity through time (especially personal identity)[13]. Though a great deal of philosophical work remains to be done here, some promising advances are already apparent in the literature[14].

The third response to the measurement problem is a family of proposals that side with MDW in limiting the quantum ontology to just Ψ, yet take the dynamics of QT_0 to be incomplete. Let us represent this family by the symbol 'QT_1'. QT_1 supplements QT_0 with a second law of evolution - via a generalization of the Projection Postulate. The most developed proposal of this kind to date originates in a seminal paper by Ghirardi, Rimini and Weber[15]. Work on this approach has continued steadily over the last decade, with substantial input from many additional supporters[16]. Ghirardi *et al.* pick position as the preferred basis for wave function collapse, but not all QT_1 theories do this[17]. In all these proposals, however, Ψ stands for a physical field that undergoes events of spontaneous collapse in accordance with some nonlinear stochastic law. Such events occur everywhere in nature. They are exceedingly rare in isolated systems with few degrees of freedom; but, their frequency increases exponentially with particle number and mass - a feature central to the behavior of ordinary macroscopic objects. In Ghirardi *et al.*'s proposal, the spontaneous rate for a single structureless particle is fixed at a mere 10^{-16} collapses per second. Accordingly, single atoms and other simple systems evolve almost exactly as QT_0 dictates. By contrast, ordinary macroscopic objects behave almost exactly as classical physics dictates, because even the smallest macroscopic objects contain more than 10^{20} bound quantum particles. In such objects, therefore, the frequency of spontaneous wave function collapses is extremely high, and this effectively prevents them from developing significant superpositions of the 'Schroedinger's cat' variety. Since QT_1 theories explicitly modify the standard dynamics, their predictions generally differ from those of QT_0. For example, Ghirardi *et al.*'s model predicts statistical spreads which are often different from those obtained from Schroedinger's equation, as in the following[18]:

$$(\Delta q_{GRW})^2 > (\Delta q_{Sch})^2$$
$$(\Delta p_{GRW})^2 > (\Delta p_{Sch})^2$$

In the late 1980's deviant predictions such as these seemed to point the way toward crucial tests to decide whether wave function collapse is an objective process[19]. As we will see shortly, however, the envisaged experiments turned out to be very problematic. Also, the approach advocated by Ghirardi *et al.* presents some conceptual difficulties[20]. Arguably, the most serious problem is that the world of Ghirardi *et al.*'s specific model is *essentially of the Many Worlds variety* - just with one dominant branch in terms of amplitude[21]. If so, excessive ontological proximity to MDW would defy the quest for parsimony and modesty behind the QT_1 program. This shortcoming, however, seems an artifact of the preferred basis chosen by Ghirardi *et al.*; more natural options for the preferred basis appear to be both viable and promising[22].

As the above presentation suggests, none of the reviewed approaches is yet without blemish, though (as noted) promising work is underway on all of them. Moreover, the exercise of developing the above proposals already serves to secure a claim that seemed dubious only a decade ago: methodologically decent physical models of Ψ are possible, after all, or so I will try to argue.

3. EMPIRICAL EQUIVALENCE

For a while there was a growing expectation that experimental tests, by seizing on quantum interference effects, were soon going to decide between QT_1 and QT_0. But, then, realistic calculations showed that such effects wash away too quickly to be any predictive value. The immediate culprit is phase blurring provoked by the simplest interaction of quantum systems with their local environment. For reasonable models of measurement situations complete blurring is calculated to develop in less than 10^{-12} seconds[23]. Decoherence theorists have this figure in mind when they say that the standard dynamics yields history spaces which are appropriately 'quasi-classical'[24]. The latter obtains whenever the probability for a final event is approximately the same, whether or not collapses of Ψ occur at intermediate times (for this to happen the off-diagonal elements of the decoherence functional must approximately vanish[25]).

TABLE 1

PROPERTIES OF VARIOUS SCATTERING PROCESSES [taken from Tegmark (1993)]

(Coherence time $\tau = 1/\Lambda$; ϕ = flux of particles; Diffusion parameter $\Delta = \Lambda/\lambda_{eff}^2\ [cm^{-2}s^{-1}]$; damping goes as $exp\ [-\Delta\ |y-x|^2\ \tau/2]$).

Cause of decoherence	λ_{eff}	$\phi\ [cm^{-2}s^{-1}]$	$\tau_{electron}$	$\Delta_{free\ e}$	*10 mm dust*
300K air at 1 atm pressure	*0.1 A*	10^{24}	$10^{-13}s$	0^{31}	10^{37}
300K air in lab vacuum	*0.1 A*	10^{11}	*1s*	10^{18}	10^{23}
Sunlight on earth	*900nm*	10^{17}	*6 months*	10^{1}	10^{20}
300K photons	*0.02mm*	10^{19}	*1 day*	10^{0}	10^{19}
Background radioactivity	$10^{-14}m$	10^{-4}	10^{11} *yrs*	10^{-4}	10^{15}
GRW effect	*100nm*	*n/a*	10^{9} *yrs*	10^{-7}	10^{9}
Cosmic microwave background	*2mm*	10^{13}	10^{4} *yrs*	10^{-10}	10^{6}

Approximate classicality of this type, however, need not spoil the prospects of testing QT_1 against QT_0 experimentally. For that, the referred approximate diagonalization of the density matrix would have to be accompanied by a total screening of the effects induced by the Ghirardi *et al.*'s mechanism. Unfortunately, there is proof that this second condition obtains in interactions with typical measurement environments. The proof comes from some detailed scattering analyses. Recent work by Tegmark, for example, shows that, when even tiny specks of dust are involved, models with and without collapse converge on a common fuzzy observable surface within extremely short intervals of time, as indicated in Table 1.

Notice how the effect of air molecules dominates at room temperature, even in a laboratory vacuum of 10^6 particles/cm^3. The message is clear: QT_1 and QT_0 do not predict different *observable* phenomena[26]. And so, the foundational project of testing these theories seems to succumb to the coarse-graining imposed on the theories' models by the way we and our instruments function in the world[27]. We thus end with a case of *contingent* observational equivalence[28].

3.1. *The Case for Empirical Equivalence*

Still, observational equivalence does not automatically amount to full empirical equivalence. Perceptual observation, by itself, matters little in science. It is an important part of experience; but, as the critics of positivism convinced us long ago, observation lacks guaranteed foundational power. The interesting question is, thus, whether the situation at hand is also one of *rational-empirical* equivalence. What makes two theories empirically equivalent in this sense? Authors disagree about the specifics here, but standard scientific criteria for theory choice focus on such *epistemic virtues* as conservatism, simplicity, modesty, generality, refutability, and precision [29]. I have done a rough assessment of the pros and cons of our three theories in a recent paper - hence I will be brief here, referring interested readers to Cordero (1995).

BQTM conserves as much of the ontology of particles as well-established quantum physics permits. On the negative side, however, it is a highly nonlocal proposal (due to the *wholistic* way in which the total wave function reacts to changes); BQTM thus clashes head-on with special relativity. In addition, Bohm's conception of Ψ as a field of force runs counter to the quantum exchange paradigm of modern physics. Also, BQTM predicts a disturbing proliferation of idle, undetectable, wave function branches.

The main virtues of MDW are its ontological and dynamical simplicity, traits which are enhanced by the natural compatibility of this theory with relativity physics. But, the proliferation of 'worlds' (effective self-containment of branches) contemplated by the approach offends parsimony. MDW demands a remarkable leap of faith; this negative feature would remain even if MDW theorists succeeded in smoothing the theory's problematic metaphysics.

Proposals of the QT_1 variety basically agree with the world revealed by the senses. The most conceptually coherent and foundationally fertile versions of QT_1 are, arguably, those that focus on energy as the preferred basis for wave function reduction. Unfortunately, however, the intervals that need to be postulated for the process of wave function collapse are generally problematic in two ways. First, they are short enough to render the approach incompatible with the spirit of special relativity[30]. Second, the whole process takes long enough to allow decoherence phenomena to compromise prospective crucial tests against QT_0 theories.

As noted, there is some conceptual tension between SQT, BQTM and QT_1 theories on the one hand, and relativity theory on the other. How important, however, is this tension? At the phenomenological level, neither SQT nor QT_1 proposals allow for superluminal signalling; in this sense, both theories enjoy a *peaceful coexistence* with relativity[31]. BQTM is more problematic in this regard[32], but the theory does not allow for faster than light signalling under standard conditions. The scientific community is divided about the significance of all this. Whatever the eventual verdict, however, there is no evidence that natural laws display more than just stochastic Lorentz invariance. This, I suggest, both renders moot present charges of strict Lorentz invariance violation, and strengthens the 'leap of faith' argument against MDW.

Although the three reviewed theories have yet to forge their way through their respective conceptual difficulties, the possibility of BQTM, MDW, and QT_1 developing into coherent fundamental theories of material systems can no longer be disregarded. Nor can the prospective situation of full empirical equivalence be taken lightly. For the sake of analytic exploration, therefore, I am going to consider the possible scenario in which the three approaches, having succeeded in solving their present conceptual difficulties, remain empirical equivalent. My main goal is to use this case to consider three larger ***antirealist*** lines of reflection: (1) on how a minute blurring at the level of observable models can allow for massive degeneration at the level of deep ontology; (2) on the antirealist import of this kind of underdetermination; and (3) on ontological commitment and 'descriptive depth' in mature science. I will conclude the paper with some comments on the nature of realist strategies in ongoing science.

4. IMPACT ON REALIST INTERPRETATIONS OF SQT

Suppose, then, that our three rival approaches, each having secured an adequate level of scientific coherence, end with equally meritorious score cards as empirical theories, forcing us to declare a situation of contingent empirical equivalence. What would this development do to the initial realist hopes behind the approaches in question? Since, by hypothesis, the observable models of BQTM and MDW coincide over the range of prevailing physical conditions, it will be best to focus on the general confrontation between QT_1 (theories of wave function with objective collapse processes) and QT_0 (theories of wave function without objective collapse processes). Of course, the empirical equivalence assumed here, being based on scientific findings, is potentially revisable by possible changes in auxiliary assumptions about experimental contexts, in observability, etc. (Although the prospect of bringing QT_1 and QT_0 to a crucial test seems presently bleak, such an event cannot be ruled out a priori[33]).

How devastating can the envisaged situation of empirical equivalence be to the realist underpinnings of SQT and the reviewed programs? I think there is a good deal of disorientation about this in the literature. In an otherwise delightful recent book, for example, Albert, commenting on the status of BQTM and the Many Worlds hypothesis, goes as far as to claim that the two approaches are not so much empirically equivalent as really *incommensurable* proposals[34]. The claim is specifically centered on Albert's own version of the Many Worlds hypothesis ('Many Minds'), but the point is general enough to apply to QT_0 and QT_1 (once the imposed blurring on observable substructures is considered). Albert's attack seems the most exiting an antirealist could try - it goes right to the jugular of the objectivism and realism behind the quantum approaches outlined in the previous section.

By definition, incommensurable proposals refer to no common body of data. When serious relativists point to this feature, therefore, their claim is that the ontological fancies of the relevant theories so permeate their respective models that even simple observation reports are hopelessly theory-dependent. Khunians have this in mind when they maintain, for instance, that Galileo and the Aristotelians were really talking about different worlds. If sustainable, then, Albert's charge would mean that the scientific imagination is simply too fertile and obtrusive to allow for any significant distinction between illusion and reality concerning the quantum domain.

Are the reviewed rival theories *incommensurable*? The suggestion strikes me as odd. The authors of the proposals in question agree that the theories share common data, and indeed function as if they had common data. All we clearly have, therefore, is a situation of substrate neutrality grounded in contingent empirical equivalence. We are left with the basic tenets of SQT floating like a raft on the current world of experience, unhindered by ongoing controversy about the three proposed deep ontologies. But, incommensurability? That charge cannot get off the ground without all sorts of notorious assumptions about meaning and human cognitive capacities. Whatever may have been the case in the 1960's, strong relativist or overdemanding logicist views on meaning and cognition are no longer easy to maintain. In particular, radical forms of semantic holism - so dear to incommensurabilists - have been systematically discredited by the critics of post-positivist relativism and sociologism[35]. Simply no grounds have been articulated for claiming that change from one of fundamental/ultimate interpretation to another of the quantum domain (or any mature scientific domain for that matter) automatically stages the substitution of one self-contained set of concepts for another, without anything that could be described as increase in knowledge or at least significant invariance. Of course, it is always possible to *stipulate* that conceptual links be as strong and thorough as relativists wish, and so deny progress. But, then, relativism would succeed only by convention.

If, instead of arbitrarily stipulating philosophical problems one looks at ongoing scientific concepts in their natural contexts, a very different picture comes into view. Something that particularly strikes many students of modern science is how multiple, partial and incomplete are actual conceptualizations and descriptions of empirical situations[36]. In physics and other mathematized empirical disciplines, models routinely display discernible epistemic *levels* - just of the type we found in our three rival theories. Not only competing theories naturally share partial models down to certain explanatory level, but their respective authors generally *mean* them to do so. Descriptive practice reveals itself as being, on the whole, sensitive to underdetermination boundaries. *Interpretations* are no different in this regard - the interpretation of a theory is itself a theory, and as such displays a distinct family of models within which the behavior of appropriate 'possible systems' can be represented. A second relevant point is this. When empirical concepts are initially proposed, criteria of applicability are framed for presently foreseeable circumstances. Not all circumstances can be anticipated, however, and so the epistemic context of an empirical theory generally includes implicit *ceteris paribus* clauses. This affects the status of everything in a theory, from "natural kind terms" to foundational principles. Even for terms for which precise criteria of application exist, using different criteria in limiting conditions often yields divergent verdicts,the result being vagueness. Think, for instance of the way criteria for the term 'iron' focus on such demonstrably contextual traits as specific gravity, X-ray spectrograph, solubility in liquid substances at different temperatures and pressures, etc.[37].

So, what do theoretical claims *mean*? The topic is spectacularly complex, but agreement about some pertinent issues exists. Consider the following points. There is no evidence that when a person thinks of something, of necessity his entire theoretical network is (let alone must) be brought to bear. Nor is there any reasonable case for claiming that, at any given time, a person functions with just one conceptual network[38]. Confronted with a concrete situation, the mind typically *zooms* on some partial network highlighted by local context. In ordinary situations, the focus of attention is confined to some widely shared subnetwork of concepts - as when, for example, a Darwinian and a creationist chat over a drink about the speed of horses or about the crystalline properties of rocks. The fact that model structures are naturally fuzzy further helps this form of sharing. In scientific situations, the focus of attention is often specific, but not qualitatively different.

These remarks apply nicely to our case study. We have a family of theories whose models actually diverge from each other only from a certain level of descriptive depth down. Above this level, the models of the three theories converge over a robust (stable) intersection. Modesty would thus seem to press us to suspend judgement about claims that lie below the model-theoretic intersection of our three rival theories. Does this bring us back to SQT? Well, not exactly. Developing the theories in question has changed SQT a little. I am going to argue that this difference, though small, nevertheless helps one type of realist project within quantum physics. Let us represent the ensuing version of SQT with the symbol 'SQT*'. (Yes, I am comfortable with the anti-essentialist view that empirical theories, like biological species, change by degrees).

4.1. *Progress without Ontological Resolution*

SQT*, I suggest, is a clearer, more coherent, and more credible theory than SQT. Progress is discernible in at least three respects. First, improvements in the metaphysics of the reviewed rival ontologies have sharpened up common central models and structures. Hilbert space models, stochasticity, and nonlocality structures, for instance, used to epitomize opacity and artificiality. The situation is different now, thanks in no small part to the clarity and precision brought on the relevant concepts by the development of the theories under consideration. Especially during the last decade, BQTM, MDW, QT_1 have themselves gained in precision and made structural advances through mutual confrontation, confrontation with the rest of science, and philosophical analysis. In the process, the ideas of linear superposition and spontaneous collapse have been further articulated, expanded, clarified, quantified and deepened, each of the competing theories becoming stronger every time it overcame a challenge. Additionally, as we saw, BQTM, MDW and QT_1 have put (or are close to putting) forward proposals which are sufficiently detailed and developed to count as *existence proofs* for their respective models. This connects with a second line of progress. Within SQT* the conception of the wave function as a physical field is now in better shape than ever before. Less than two decades ago, even the *possibility* of addressing the measurement problem in terms of a physicalist model of Ψ seemed doubtful. Today, by contrast, no less than three such solutions to the measurement problem are available. All three account for the emergence of sharp property values in general and measurement outcomes in particular, and do so entirely in physicalist terms. Finally, look at what has happened to the issue of descriptive depth; I have already alluded to this, but more needs to be said. The robust intersection shared by the models of the reviewed theories is partly an artifact of their construction and only partly a discovered fact. Nevertheless, the constitutive structures here clearly reach well beyond the superficial level associated with the bare quantum algorithm. In particular, the intersection of QT_1 and MDW includes whole families of Hilbert space models, especially in the region of simple single quantum systems. BTQM, though more conceptually distanced from MDW and QT_1 by its mixed ontology, still shares models with them under a nontrivial common uniformity at the observable level. Construing SQT* as the intersection of agreement between the reviewed rival theories unveils a proposal which displays more robust resources for ontological commitment than SQT ever had. The objects and structures common to the deep ontologies fulfilling SQT* clearly include Ψ, besides all the standard quantum claims about quarks, atoms, molecules, condensed matter, bulk properties, the superposition principle in systems with few degrees of freedom, wave function nonlocality, stochastic Lorentz invariance, etc. So, the confrontation between the QT_1 and QT_0, though frustrated at the ontological level by contingent equivalence, does seem to have led to some theoretical progress at an intermediate (but hardly superficial) level.

5. CONFIDENT BELIEF, HISTORY AND THE CHARACTERIZATION OF REALISM

Field-theoretic conceptions of Ψ may be now in better epistemological shape than ever before. But, if so, can this be good enough for holding a proper realist position about Ψ or any significant part of SQT*? Too many different positions, most of them very problematic, go by the name 'realism', and so we must tread carefully. One common idea behind realist philosophies, however, seems to get at something important: the notion that our most accurate and reliable empirical theories manage to describe some real states and structures of nature.

In a recent book, Kitcher's articulation of this basic idea shows the advantages as well as the perils of realism as a philosophy of science[39]. Kitcher presents scientific advance as something that begins with (i) a superficial picture of some natural process, and (ii) a program for addressing empirical problems in terms of this picture. Progress, Kitcher convincingly stresses, comes as improvements on the initial picture by correction and deepening that refines distinctions and adds information, converting assumptions into conclusions, extending the scope of application. The result, he maintains, is a scientific disclosure of successive levels of dependence among natural processes. Like most other realists, however, Kitcher blends this basic picture of scientific progress with contentious metaphysical claims about reality, truth and progress. In particular, he tries to extend the notion of scientific deepening to include the discovery of natural kinds. Compelling critiques of this type of metaphysical move have been in existence for some time, but Kitcher does not seem to take them seriously[40]. Censors notwithstanding, I agree that the basic idea behind realist proposals does some justice to current standard views on, say, genes, molecules, atoms, quarks, the standard linear quantum dynamics for systems with few degrees of freedom, stochastic Lorentz invariance, and much more. Nevertheless, one cannot deny that the critics of realism have a point.

Central to the basic intuition behind realist interpretations of science is the issue of cognitive confidence. There are complementary synchronic and diachronic sides to this. The synchronic side has to do with presently having successful theories, especially in terms of (a) explanatory power, particularly regarding 'unespected' facts that those proposals were not expressly designed to explain; (b) novel predictive power; and (c) fertility in opening up new investigations. The diachronic side has to do with having as background a cumulative succession of reliable past theories (as opposed to a consistent history of total failures). In our case study, and in the larger stage of contemporary natural science, theories do seem to generally take over the observations that had led to their predecessors and preserve a good deal of the old theory, succeeding each other, plausibly as better and better approximations to a *correct* description of their intended domain. Such current concepts as 'electron', 'quark', and 'Schroedinger's dynamics', for example, all stand for historical successions of structures and ideas - veritable takeovers that show significant levels of overlap.

However, philosophical realists tend to have in mind more than the down-to-earth type of confidence just outlined. In particular, many of them think essential to resort to *philosophical theories of truth*, and this complicates matters a great deal. Some popular formulations of philosophical realism seem particularly misguided in this regard.

5.1. *Metaphysical Realism*

Consider, for example, the position known as 'metaphysical realism'. Defenders of this point of view stipulate that a posit is objective if it exists, and is the way it is, *independently of any knowledge, perception, conception or consciousness there may be of it.* This may initially look like a good explanatory position to have about the manifest success of our best scientific theories. Unfortunately, the ambitious metaphysical links envisaged here are marred by a host of old questions. Metaphysical realists maintain that the objectivity accorded by scientific practice to theoretical claims (as in, say, evolutionary biology, or SQT*) rationally *necessitates* a higher (metaphysical) level of objectivity for the entities and regularities the relevant theories invoke or range over. On this version of realism, therefore, scientific objectivity is to be elucidated not just by appeal to the existence of scientific facts, objects, properties, events and the like, but by these existing or obtaining independently of any cognitive access we may have to them. But, how can one hold this? The anti-contextualism built into the conception of metaphysical realism is a major source of trouble. Every sensible claim we seem able to make manifestly rests on some background knowledge, presupposes *ceteris paribus* clauses of one kind or another, and so forth. The further historical contextuality of epistemic confidence, alluded to above, only adds to the tension here. The 'super-realist' demand that we supplement scientific narratives with talk about what exists independently of any knowledge *at all levels* just seems arbitrary and misguided.

One excess invites another, and so there are also 'super-antirealists', who react by demanding that scientific credibility be limited to the observable sphere or some other foundationalist-like restriction[41]. I think such metaphysical actions and reactions, though professionally alluring, call attention away from the important point that philosophers and practitioners of successful scientific disciplines try to get at when they support realism.

5.2. *The Realist Attitude as an Expression of Cognitive Achievement*

Let us not lose sight of the original issue about realism. The central contrast is between observational claims and unobservable ones. Our relevant concern, therefore, should be the credibility of the most careful scientific statements about such postulated entities as Y and quarks, and about such regularities as the principle of quantum superposition and quantum nonlocality. Accordingly, in the remainder of this paper I want to concentrate on the following question: Are our best *theoretical* claims generally as credible and deserving to be called *true* (whatever this means) as our most credible statements about dogs, mountains and other components of ordinary experience? I am construing realism in any area of thought as the view that entities associated with that area are *real*, but I do not want to arbitrarily require standards higher than those used when we ordinarily say that dogs and mountains are real. I want even less to enter into the separate (and for present purposes unnecessary) task of spelling out a metaphysical theory of truth. Better deal with one small but seemingly feasible task than attempt the gigantic chore imposed by the whole philosophical issue of realism. My approach here is akin to Fine's 'natural ontological attitude', though I am more unashamedly contextualist and much less sociologistic[42].

Against subjectivists and idealists, the basic realist picture holds that the objects posited exist and are independent of individual wishes and dispositions to assert and believe things about them (objectivist thesis). Against instrumentalists, basic realism holds that discourse introduces distinctive posits and that claims involving them fail to come off if there are no such entities (descriptivist thesis). Among the paths seemingly available for claiming objectivist independence and descriptive power, our previous explorations have pointed toward the path provided by such scientifically learned traits as internal consistency, coherence with the rest of current knowledge, successful critical testing, and (last but not least) coherence with a robust history of past successes.

Judgements of objectivity so construed are, however, defeasible. In mature science, the epistemic status of many central assertions depends, among other things, on *current beliefs about what is thinkable and what is not*. But, the presumed universality of modal beliefs has repeatedly proved to be highly contextual. Let me mention just two instances. Recall how, in the *Dialogues Concerning Natural Religion,* the usually cautious Hume (speaking through Philo) finally succumbs to the argument from design. Hume just *could not imagine* how all the marvelous adapting of means to ends he saw in nature could possibly come from anything but God. More than a century later, in 1920, we find another generally alert thinker, N.R. Campbell, declaring himself *incapable of figuring out* how something like the relation of absolute simultaneity might be challenged[43]. The obvious lesson is, of course, that context and prior knowledge matter (or should matter) a great deal in philosophy of science. It is only on the basis of notions like "the evidence that is presently available", "the experiences so far encountered", and "the concepts we have actually acquired" - ultimately, the present body of knowledge - that the objectivity of beliefs is understood in scientific practice.

5.3. *Present Success is not enough*

Now, to the extent that judgements about the objectivity of scientific claims are contextual, realism faces this old challenge: What has contextualized empirical objectivity got to do with truth? As Leplin emphasizes, success relative to an eliminativist methodology of theory choice does not confirm except by reducing the space of possibilities identified by present theoretical commitments[44]. In this sense, the role of evidence is simply to reduce theoretical options. Our current methodology may thus be enough to sustain the rationality of change. But, as critics of realism rightly ask, is the admission that a theory is rationally preferable because the evidence disqualifies its competition the same as full endorsement? Rationally defensible theories were not enough to save Hume and Campbell from holding on to views that later came to be scientifically regarded as false and misguided pictures of the world. Realists thus need to find a coherent way of making such claims as, for example, (a) that evolutionary biology is a better conceptual framework than the one Hume had, and (b) that this framework was not entertained in his days because good reasons then existed for not doing so.

To speak of truth, a realist needs to claim more than objectivity for the best theoretical claims. Can he do this? Again, we should avoid the temptation of arbitrarily raising the standards against theories at this point. The working paradigm, let us recall, is provided by our low-level practice of truth attribution - if you like, our everyday *realism* about dogs, mountains, ordinary observable regularities, etc. So, what goes on in the practice of truth ascription at the ordinary level? One relevant point concerns levels of description. We do not understand dogs exactly the way Aristotle, Hume, or even our great-grandmothers did. Nevertheless, our concepts and theirs manifestly share a good deal of common structure. At intermediate levels of description, significant measures of coincidence are still apparent. Conceptual differences tend to become wild only when we turn to foundational ontologies. Think, for example, of the shift from traditional teleological biology to contemporary evolutionary biology. Though radical at the deepest ontological level, this transition has left many intermediate structures remarkably close to the ones held under the prior teleological biology. Instances in point include structural similarity at the level of groupings, nominal essences for species, physiological and anatomical description, etc. Clearly, the situation resembles what we found earlier in the shift from SQT to BQTM, MDW and QT_1, and also from SQT to SQT*. And this type of correlation between truth ascription and descriptive level seems both general and favorable to the realist outlook. Successful theories are, however, defeasible at all levels, and so we need to be cautious. When, in our case study, we encountered the ghost of past failures, I advocated making a double move. There, the rival theories displayed significant structural similarity (amounting to coincidence over a hole range of approximate partial models). I have compounded this first move with a zooming conception of scientific meaning.

I suggest that the type of structural similarity we found in our case study occurs everywhere in mathematized disciplines. It is emblematically displayed, for example, by the kinematic trajectories derived from a family $\{x\}$ of approximate models of a system of particles, when the trajectories in question are covered by a topological uniformity u, as in the following structure:

$$u_{\varepsilon}\left\{< x,x' >:(p)\ \left[p_{\varepsilon}\, \varepsilon\ P_{x} \cap P_{x'}\ \&\ \left(S_{x}(p) - S_{x'}(p)\right) < \varepsilon\right]\right\}.$$

In the above expression, $< x,x' >$ stands for pairs of models, P_{Ξ} stands for the set of particles subsumed under model Ξ, and ε is the degree of fuzziness marked by the uniformity. Expressing the structural and conceptual similarities between theories governed by different principles (like, say, thermodynamics and statistical mechanics) is a more complicated exercise - but I think a viable one, nonetheless, thanks to valuable insights and techniques developed by the so-called 'Structuralist School'[45]. My point is that controlled fuzzy models, as in the above family of trajectories subsumed under u_{ε}, seem good structures to focus on if we want to articulate the feeling of progress that some theory successions leave on most of us. Such models clarify and advance talk about stability of partial theoretical structures without trivializing the issues involved. They help us see how and why the ghosts of past failures and underdetermination, though real, are not nearly as destructive as they seemed at first.

Still, can this limited structural stabilization ground a proper realist position? I have accepted that physical realism about, say, electrons, Ψ or the best symmetry principles, turns into a burdensome position if one tries to adopt an overambitious metaphysical stance. But, I have insisted, there is no reason for going that way. Notice that the rationale for metaphysical realism, if taken seriously, would compromise much more than the best scientific assertions about electrons and symmetry principles - it would also compromise *ordinary claims about cats, trees and what have you.* The point is that *observational talk* is no less metaphysically problematic than *theoretical talk*, unless one arbitrarily frees the former from conceptual difficulties (via some discredited foundational move, for example).

5.4. *Concluding Cheers for Deflationary Realism*

The issue, then, boils down to this. If we agree to refrain from arbitrarily raising epistemic and semantic standards against theories, then a modest but nontrivial scientific realist stance appears to be viable. The realism that comes into view is an extension of our ordinary 'realism' about dogs, mountains, and heavenly bodies. Admittedly, this is realism relativized to both our current scientific knowledge and our ongoing practices of objectivity and truth ascription at the level of the common objects of experience. It is, however, a stance that seems to do justice to our most rational claims, be it about the observable or the unobservable. Furthermore, on this modest conception of realism, we may gain insight into the structure of nature even as we fail to come to a conclusion about the 'ultimate nature' of physical reality.

That looks like a cautious enough position to hold. Surely, in contemporary science, *intermediate* (less than 'ultimate') descriptions are rarely without considerable depth - just think of what we have come to make of cats, mountains, stochastic Lorentz invariance, or quantum superposition in system with few degrees of freedom.

Queens College, City University of New York
U.S.A.

NOTES

[1] Work on this article was gracefully supported in part by a Fellowship from the J.S. Guggenheim Memorial Foundation.

[2] Cordero 1995.

[3] See, for example, Cushing & McMullin 1989, and Squires 1994.

[4] The particular eigenfunction chosen results from a nonlinear and stochastic selection. It is nonlinear, because the prior probability for a particular outcome is proportional to the intensity of the corresponding eigenfunction in the initial wave function.

[5] For a compelling book on the pervasive role of superposition and interference in quantum physics see Feynman 1990. For experiments involving single systems, see, for example, Cook 1990.

[6] Bohm 1952.

[7] See, for example, Valentini 1991 and Holland 1993.

[8] Albert 1992, chapter 7; Holland 1993.

[9] Everett 1957, DeWitt 1970.

[10] Zurek 1982, Gell-Mann & Hartle 1990, Omnès 1990.

[11] Lockwood 1989, Albert 1992, Saunders 1993.

[12] MDW supplements this claims with a derivation of position as a preferred quantum basis for decoherence; the physical grounds for this lie in the locality of standard Hamiltonians.

[13] Cordero 1996.

[14] Saunders 1993.

[15] Ghirardi, Rimini & Weber 1986.

[16] Relevant works include Pearle 1990, Ghirardi, Pearle & Rimini 1990, Ghirardi, Grassi, Butterfield & Fleming 1992, Ghirardi, Grassi & Benatti 1995.

[17] Gisin 1989, Cordero 1996.

[18] Ghirardi, Rimini & Weber 1986.

[19] See, for example, Shimony 1989.

[20] Shimony 1989, Albert 1992, Cordero 1996.
[21] Cordero 1996.
[22] Gisin 1989, Cordero 1996.
[23] Zurek 1993.
[24] The formulation of these involves coarse-graining of complete sets of dynamical variables.
[25] In terms of the 'decoherence functional' D_{ρ}, the condition is $D_{\rho}([P_{\alpha}],[P_{\alpha}]) \gg 0$. This diagonalization of the density matrix occurs primarily in the position basis. See, in particular, Saunders 1993.
[26] This claim holds, at least, in the two extreme domains of systems with few degrees of freedom, and systems involving complex environments.
[27] More technically put, ineliminable circumstances force a topological uniformity upon the observable substructures of models of QT_1 and QT_0, and the fuzziness thus introduced on observable models turns out to be wide enough to hide all detectable differences between the two types of theory.
[28] Theories can be observational equivalent in two ways. On the one hand, equivalence may be imposed by the theories' principles. For example, Newton's theory of absolute space was observationally equivalent to its relationist counterpart in this principled way. On the other hand, there is equivalence imposed by contingent features of the world (initial conditions, contingent human capacities, etc.). QT_1 and QT_0 are equivalent in this second type.
[29] One classic reference is Quine & Ullian 1978.
[30] If one takes Ψ as a physical field, then collapse processes alter physical structures at superluminal speeds.
[31] At least that is how optimists see the matter. See, in particular, Redhead 1987, Fleming 1992, Ghirardi, Grassi, Butterfield & Fleming 1992.
[32] See for example, Kronz 1990.
[33] The range of what counts as 'observable phenomena' is relative to the state of scientific knowledge and the technical resources available. Or, a deeper, comprehensive theory of space-time-matter may be developed in the future, and such a theory might tilt the balance in favor of one of the theories presently deemed equivalent.
[34] Albert 1992, chapter 7.
[35] See for example, Shapere's contribution to this volume.
[36] Torretti 1990.
[37] Shapere 1984, chapter 15.
[38] The contrary seems to be the case. See Torretti 1990.
[39] Kitcher 1993.
[40] Leplin 1994.
[41] Churchland & Hooker 1985.
[42] Fine 1989.
[43] Campbell 1920, p. 29.
[44] Leplin 1994.
[45] In the general case, the relevant empirical uniformity U satisfies the relation $\cup \subset \ldots \_[(M_p[T] \cup M_p[T']) \times (M_p[T] \cup M_p[T'])]$, where $M_p[t]$ stands for 'potential model of t'. See, in particular, Moulines 1982.

REFERENCES

D. Albert, *Quantum Mechanics and Experience*, Harvard University Press, Cambridge, MA, 1993.

D. Bohm, "A Suggested Interpretation of Quantum Theory in terms of Hidden Variables", parts I & II, *Physical Review* **85**, 1952, 166-93.

N.R. Campbell, *Foundations of Science*, Cambridge University Press, Cambridge 1920.

P.M. Churchland & C.A. Hooker eds., *Images of Science*, The University of Chicago Press, Chicago 1985.

R.J. Cook, *Quantum Jumps - Progress in Optics XXVII*, Elsevier Science Publishers Amsterdam 1990, pp. 361-416.

A. Cordero, "On Science and Scientific Hypotheses", in G. Heinzmann ed., *Henri Poincaré*.

A. Cordero, *World Collapses*, Kluwer Academic Publishers, Dordrecht 1996.

T. Cushing & E. McMullin, eds. *Philosophical Consequences of Quantum Theory*, University of Notre Dame Press, Notre Dame 1989.

B. DeWitt, "Quantum Mechanics and Reality", *Physics Today* **23**, 1970, 30-35.

H. Everett-III, "Relative State Formulation of Quantum Mechanics", *Reviews of Modern Physics* **29**, 1957, 454-62.

R. Feynman, *QED: The strange theory of light and matter*, Penguin, London 1990.

A. Fine, "Interpreting Science", in A. Fine & J. Leplin eds., *Philosophy of Science Association 1988, Volume 2*, pp. 3-11, Philosophy of Science Association, East Lansing, Michigan 1989.

G.N. Fleming, "The Objectivity and Invariance of Quantum Predictions", in D. Hull, M. Forbes & K. Okruhlik eds., *Philosophy of Science Association, Vol. 1*, 1992, pp. 104-13.

M. Gell-Mann & J.B. Hartle, "Quantum Mechanics in the Light of Quantum Cosmology", in W.H. Zurek ed., *Complexity, Entropy and the Physics of Information*, Addison-Wesley, Reading 1990.

G.C. Ghirardi, A. Rimini & T. Weber, "Unified Dynamics for Microscopic and Macroscopic Systems", *Physical Review* ***D* 34**, 1984, 440-491.

G.C. Ghirardi, P. Pearle & A. Rimini, "Hilbert Space and Continuous Spontaneous Localization of Identical Particles", *Physical Review* **A42**, 1990, 78-89.

G.C. Ghirardi, R. Grassi, J. Butterfield & G. Fleming, "Parameter Dependence and Outcome Dependence in Dynamical Models for Statevector Reduction", *Foundations of Physics* **23**, 1992, 341-364.

G.C. Ghirardi, R. Grassi & F. Benatti, "Describing the Macroscopic World: Closing the Circle Within the Dynamical Reduction Program", *Foundations of Physics* **25**, 1995, 5-38.

N. Gisin, "Stochastic Quantum Dynamics and Relativity", *Helvetica Physica Acta* **62**, 1989, 363-371.

P. Holland, *The Quantum Theory of Motion*, Cambridge University Press, Cambridge 1993.

P. Kitcher, *The Advancement of Science: Science without Legend, Objectivity without Illusion*, Oxford University Press, New York 1993.

F.M. Kronz, "Hidden Locality, Conspiracy and Superluminal Signals", *Philosophy of Science* **57**, 1990, 420-44.

J. Leplin, "Philip Kitcher's The Advancement of Science", *Philosophy of Science* **61**, 1994, 666-71.

M. Lockwood, *Mind, Brain and the Quantum*, Blackwell, Oxford 1989.

C.U. Moulines, *Exploraciones Metacientíficas*, Alianza Editorial, Madrid 1982.

R. Omnès, "Consistent Interpretations of Quantum Mechanics", *Reviews of Modern Physics* **64**, 1992, 339 - 82.

W.V.O. Quine & J.S. Ullian, *The Web of Belief, 2nd Ed.*, Random House, New York 1978, pp. 64-82.

P. Pearle, "Toward a Relativistic Theory of Statevector Reduction", in A.I. Miller, ed., *Sixty-two Years of Uncertainty*, Plenuum Press, New York 1990, pp. 193-214.

M.I.G. Redhead, *Incompleteness, Nonlocality, and Realism*, Oxford University Press, Oxford 1987.

S. Saunders, "Decoherence, Relative States, and Evolutionary Adaptation", *Foundations of Physics*, 1993, 1553-85.

D. Shapere, *Reason and the Search for Knowledge*, D. Reidel, Boston 1984.

A. Shimony, "Conceptual Foundations of Quantum Mechanics", in P. Davies ed., *The New Physics*, Cambridge University Press, Cambridge 1989, pp. 373-95.

E. Squires, *The Mystery of the Quantum World*, 2nd edition, Institute of Physics, Bristol 1994.

M. Tegmark, "Apparent wave Function Collapse Caused by Scattering", *Foundations of Physics Letters* **6**, 1993, 571-90.

R. Torretti, *Creative Understanding*, The University of Chicago Press, Chicago 1990.

A. Valentini, "Signal-Locality, Uncertainty, and the Subquantum H-Theorem", *Physics Letters* **156**, 1991, 5 - 11, and **158A**, 1991, 1-8.

H.D. Zurek, "Environment-Induced Superselection Rules", *Physical Review* **D 26**, 1982, 1862-1880.

H.D. Zurek, "Preferred States, Predictability, Classicality, and the Environment-Induced Decoherence", *Progress in Theoretical Physics* **89**, 1993, 281-312.

MAURO DORATO

MEASURABILITY, COMPUTABILITY AND THE EXISTENCE OF THEORETICAL ENTITIES

1. INTRODUCTION

In the history of the philosophy of science of our century, few questions have undergone a more profound change than the problem of the relationship between theory and observation. One can perhaps measure the distance that has been travelled if one regards, as the starting point of the debate, the attempt at defining the so-called theoretical entities in terms of directly observable entities, and, as the final point, the current suggestion that experiments and theory represent two different and partially autonomous *cultures* within the larger discipline of physics (Galison 1987: 255).

I should state at the outset that my approach to the question of the existence of what I shall still call "theoretical entities" endorses the shift in emphasis from observability and reference to experiment and causal intervention, preached by the now dispersed Stanford group — Hacking (1983), Galison (1987) and Cartwright (1983, 1989). Rather than reiterate their arguments against the exaggerated importance that the notions of direct observability and reference have received until recently, I will here take for granted that the distinction between direct and indirect observability is vague, contextual, and not very relevant to the problem of realism about entities. Likewise, I will also assume without discussion that the distinction between theoretical and observable entities is one that historically has often been replaced, as Hempel once suggested, by the pragmatic distinction between entities that have been recently introduced, and more entrenched, older entities.

However, while I am also inclined to maintain that it is up to the community of experimental scientists to decide, in each particular historical circumstance, when a set of experiments can be taken to confirm the existence of some postulated "theoretical" entity, I will try to argue that the experimentalist's attitudes of realism about entities are not the mere by-product of changing historical fashions or styles of thought, but can be traced back to some "sufficient" conditions that can be clearly specified and tested against historical cases.

With respect to this purpose, I will urge that besides Hacking's manipulability criterion, it is important to consider another, independent condition for the existence of some postulated entity x, namely measurability, to be discussed together with its close cognate, computability. The quotation marks around "sufficient" are supposed to express the idea that the conditions in questions are abstracted from the experimentalist's implicit practice in modern laboratories, so that they should not be regarded as infallible,

E. Agazzi and M. Pauri (eds.), The Reality of the Unobservable, 207-217.

algorithmic criteria that have been invariably applied in each and every controversial historical case. My aim in what follows is essentially to make explicit standards and methods that physicists have implicitly (and often collectively) adopted in the course of the history of physics, *vis à vis* the question of the existence of what is unobservable with the naked eye. The validity of the general conditions proposed here can be decided only by checking them against concrete, individual cases.

Consequently, I will first illustrate why Hacking's criterion must be supplemented by a second condition of measurability (see ch. 2) and then try to explore the relation between the highly indirect causal intervention realized by measurement of distant, non-manipulable objects (like stars) and the peculiar kind of "virtual" manipulability instantiated by computer simulations in current laboratory practice.

This point is of crucial importance: given the current importance of computers both in very complex theoretical calculations and in the analysis of experimental data, it would seem that some new kind of synthesis between the theoretical and the experimental "cultures" can be attained by paying due attention to the notion of *computability*. Such a notion applies both to the constructive character of the mathematical model used in the theory and to the finite set of instructions that are used by the experimentalist to measure some of the magnitudes of an entity postulated by the theory. Considering the new interest for a computable analysis and physics (Pour-El, Richards 1989, Svozil 1993), and the recent discussion of an interesting case of a measurable but non-computable physical magnitude in quantum gravity (Geroch and Hartle 1986), it could be suggested that the notion of the computability of a theory should replace the old notion of observability in providing a sufficient and perhaps also a necessary condition for its applicability (see section 3).

Finally, in the last section, I will consider the age-old metaphysical problem of the existence of quantities as mind-independent properties of things and events. By claiming that quantities exist just in case we have a set of linear order relationships and some empirically established correlations between the quantities, I will further justify my claim that measuring or computing at least two of the magnitudes of some entity x is, together with its manipulability, a sufficient condition to ascribe existence to x (see section 4).

2. MANIPULABILITY AND MEASURABILITY

In the formulation of the problem of scientific realism, it is customary to separate the question of the existence of theoretical entities from the issue of the truth of scientific theories. Let me call this two questions R_1 and R_2 respectively. Following Boyd and Putnam (1984), one can formulate these two questions as follows:

R_1 "Terms in a mature science typically refer."

R_2 "The laws of a theory belonging to a mature science are typically approximately true."

Various philosophers have stressed that the two questions are independent of each other, and Hacking and Cartwright grant R_1 but deny R_2. In order to draw the distinction, they need to rely on a postulated independence of phenomenological from fundamental laws (Cartwright), or low-level from high-level causal laws (Hacking), something which can also be used to ground the autonomy of the experimental from the theoretical culture (Galison). The main idea in this approach is that fundamental laws expressed within theoretical models may come and go, and that, at any given time, certain models may even contradict one another, while the common core of the low-level causal connections, used at the boundary between engineering and physics, remains stable through major theory change. Truth of the low-level causal laws need not extend to fundamental laws.

In order to assess the tenability of this distinction, we can look in more details into the argument proposed by Hacking. By criticizing the "spectatorial theory of knowledge", Hacking offers a causal and pragmatic criterion for the existence of postulated entities by discussing the example of the electron, a particle that has been regarded by various positivistic philosophers as a typical instance of a useful fictional instrument to save the phenomena. Says he: "*We are completely convinced of the reality of electrons when we regularly set out to build — and often have enough success in building — new kinds of device that use various well-understood causal properties of electrons to interfere in other more hypothetical parts of nature*" (Hacking 1983: 265)[1]. Notice that the existence criterion invoked by this quotation is essentially the same that we implicitly adopt for the existence of macroscopic objects: we manipulate them, we know how to use them to achieve our goals, and we are affected by them.

One could object to Hacking that we cannot apply the causal version of the argument of the "continuous spectrum" from macroscopic to microscopic objects once suggested by Grover Maxwell for direct and indirect observability (1962). Although in ordinary language we often use the expression "I observe a plane by observing its trail", or "I observe a ship by observing its wake", this talk is justified because it presupposes the possibility of observing ships and planes independently and directly (see Van Fraassen 1980). Likewise, so the objection goes, we believe in the existence of middle-seize objects because our causal interaction with them is direct, while the use of electrons as tools must only be indirect. As a consequence, the manipulability criterion for the existence of an entity should only be trusted when the entity in question can be manipulated directly.

To this argument, I reply that the distinction between direct and indirect causal interaction is a matter of degree and, as such, is too vague to be used effectively as a general argument against Hacking's kind of causal realism. Exactly as in the case of observation, if the distinction between *direct* and *indirect* manipulability were a criterion of existence, we would have to admit that also the separation between existence and non-existence is a question of degree, which is patently absurd. Therefore, either all entities which can be causally manipulated (directly or indirectly) exist, or they all don't. Although it must be admitted that a general increase in the directness of experimental interaction is usually taken as an argument in favor of the reliability of an experimental result (Galison 1987: 261), judgments of degrees of directness as existential evidence must be contextualized to a particular historical and experimental situation and cannot be used as general criteria of existence. This could be seen as a limitation of Hacking's criterion.

However, what is more important to our purposes is the following remark. To be able to reliably achieve some goal by using a postulated entity T as a *tool* implies, undisputably, knowledge of T's causal powers. The latter knowledge, in its turn, is knowledge of the properties of T, since "what makes a property a property it is, what determines its identity, is its potential to contribute to the causal powers of the things that have it." (Shoemaker 1980: 114). *It would be highly unnatural to presuppose the existence of properties without an existent* (not necessarily a *continuant or an enduring substance*) *that carries or instantiate those properties*, despite the admittedly bizarre fact that, according to the theory, an electron is a pointlike, structureless particle which before measurement need not be localized at all!

The nature of this something (the existent) carrying the properties or the causal powers in question has never been really specified in terms of the traditional metaphysical categories of objects, substances or events, and here I can just point out to some difficulties that deserve further study. In her contribution to this volume, B. Falkenburg has invoked the notion of "substance" meant as "an empirical collection of properties." However, one can interpret such collection of properties instantiated by an entity T at some time-place as a *collection of events*, according to the well-known theory by J. Kim (1993). For Kim, an event is defined as a triple $< S, P, t >$, where S is a substance, P a token property instantiated by S and t is a time, or rather, to make the definition consistent with special relativity, a time-place. In this definition, the notion of events presupposes the notion of a thing or substance, and Falkenburg's characterization would turn out to be incoherent: if we need the notion of substance to characterize that of a property (event), even a microscopic substance cannot be regarded merely as a "collection of properties" (events). Furthermore, in ordinary language things have *spatial* but not temporal parts (they are wholly present), while temporally extended events have *temporal* parts and are not wholly present. Given the evanescent duration of some elementary particles, one might want to suggest that the notion of event is to be preferred to that a substance, since the category of a substance is linked to the concept of permanence in time or duration.

Unfortunately, the real problem in making sense of the ontological status of the microworld is that *both* the use of a thing-talk *and* of an event-talk lead to various difficult puzzles linked to the problem of localization in quantum field theory. Some recent results point toward the fact that the localizability of microscopic particles can be obtained only at the cost of violating a locality requirement of the (serious) kind "act-outcome correlation" (Malament 1996). Consequently, insofar as relativity conflicts with what Shimony called parameter-dependence (see Cushing and McMullin 1989), all talk about "particles" should be replaced with talk of properties of quantized fields. Then the following questions arise: how are we to make sense of the experimentalists' talk of "trapping" single atoms or even photons in small regions of space? Is this conflict yet another proof of the autonomy of the experimental and theoretical cultures? These important metaphysical questions have received little attention, especially within the very same tradition that considers the issue of realism as having to do with R_1 rather than with R_2. If we are to grant existence to micro-entities, as I think we should, we should also try to figure out how this existence relates to the familiar "logic of the solid bodies" (Bergson's expression) to which we adapted during evolution.

The question of the status of properties of entities in the microworld can be further developed in a direction that takes us back to R_1 above. We should notice, in fact, that manipulability of electrons is as of today hardly the only reason to believe in their existence. By now, we have measured many of their quantitative properties, like mass, charge and spin, and this fact certainly reinforces our belief in their existence. While I accept Hacking's criterion of manipulability as a sufficient condition of existence, I think that it should be supplemented by some other conditions, covering the properties of those objects that are just too big or too far away from us to be used as *tools*[2]. Certainly, we cannot manipulate a black hole or a distant galaxy in the same way in which we can use neutrinos to investigate the core of the sun (Shapere 1982). Hence, Hacking concludes his analysis with a certain dose of skepticism about non-manipulable entities: "I suspect there might be another representation of the universe, equally consistent with phenomena, in which black holes are precluded" (Hacking 1983: 275).

I would like to urge that, despite a concession to the fact that reality is bigger than our capacity for manipulation — so that the latter for him is not necessary for existence — Hacking seems unduly cautious about the importance of the criterion of measurability. The reason for his prudence probably lies in the fact that one can obviously measure a magnitude of an entity x without *directly* causally interacting with it, or without manipulating it[3]. Hacking's omission of the extra-condition of measurement is justifiable in microphysics, since in this branch of science *measurement just is a particular form of causal interaction*. However, the manipulability criterion does not apply to entities postulated by astronomy and cosmology, and I would not want to deny the existence of galaxies external to our own if it is accepted by contemporary science.

The modification required to cope with this difficulty must take into account the possibility of using the causal powers of more familiar but non-manipulable entities y — of which we have measured some properties — to inquire into the properties of some less familiar entity x. This procedure is well-known to astronomers, and will be illustrated by the following example. Consider Hubble's use of Cepheids stars to calculate the distance of remote galaxies[4]. In the case of these variable stars, the almost-linear relation between intrinsic brightness and periods can be used to calculate their distance and the distance of galaxies containing them. Since Leavitt established in 1908 that the longer the periods of the cepheid stars, the greater their absolute luminosity, from knowledge of the apparent luminosity one could calculate the distance of the star after calibration. The period-luminosity method of calculating distances, which was developed at the beginning of our century, was far more accurate than previous parallactic methods (Berendzen, Hart and Seeley, 1976: 23) and was of paramount importance in solving the controversy about the *existence* of spiral galaxies external to our own (Smith 1982).

The point of this example is two-fold. First, our knowledge of the properties of the Cepheid variable stars (their causal powers) allowed astronomers to measure the distance of galaxies, and that *measure*, when found in accord with other data about the internal motions of spirals, was thought sufficient to regard the existence of the natural kind "island universes" as established. This measure did not warrant the existence of something, but rather determined what kind of intrinsic and relational properties (size and relative distance) that something possessed. However, once such properties of galaxies were established, their existence as objects separated from our Milky Way was taken for granted, and could be "used" to inquire into other more unknown domains. *This kind of bootstrap procedure of successive calibrations characterizes the progress in the precision with which we measure the capacities of nature.*

The second point concerns black holes and is another instance of the bootstrap procedure characterizing measurement in science. There is nothing in principle that prevents us to use black holes as the cepheid stars of the example. If we knew the mass of an object that is collapsing toward a point in space where we have good independent reason to expect the kind of intense gravitational field attributed to black holes, knowing the distance of the object we could calculate the intensity of the gravitational pull exerted by the black hole. If such a measurement of the force gave some precise value, that value could later be used to calculate the unknown mass of another star attracted by the black hole.

Notice that in this example we must have independent reasons to believe in the existence of black holes; measuring just one property of that point in space — say, the gravitational pull —

is not enough to regard the existence of black holes as established: astrophysicists might be measuring the gravitational field created by *something else*. I take it that this is a nice illustration of the fact that scientists may be wrong in identifying what kind of object they are really measuring, *although they are measuring the magnitude of something existing*. According to some philosophers, such an imagined situation really happened in concrete historical cases, concerning "measurement" (or "manipulation") of entities like the caloric, the flogiston, or the *N* rays, that were subsequently found to be non-existing[5].

The important point is that when physicists measure *only one property* of what we take to be an instance *x* of the natural kind *X*, they may be wrong about its other properties, so that we may be attributing the entity other features that, unbeknownst to us, are really incompatible with the possession of the magnitude that has already been measured. Cases like these show either that the experiments have to be repeated, or that, more plausibly, other properties assigned by the theory have to be measured before *X* can be deemed to be existent. Otherwise, progress in our knowledge (an alternative explanation) may show that *X* possesses incompatible properties, and is therefore an impossible object.

In a word, all I need to claim for my purpose is that measuring quantitative properties of *x* implies that *x* exists, although we may be wrong in identifying what *x* is in cases in which only one magnitude has been measured. Furthermore, if we can successfully inquire into the properties of *x* by using a measured property of *y*, then *y*'s existence should be taken for granted. Later, however, *x* can become in its turn an instrument to inquire into the properties of yet unknown entities *z*, so that also *x* should be regarded as existent. What holds for the Cepheid stars may hold as well as for black holes.

3. COMPUTABILITY AND MEASURABILITY

Among modern historians of experimental physics, Galison has insisted more than others on the exponential growth of instrumentation in twentieth-century laboratories, and on the major role played by computers in the changing mode of gathering experimental evidence. Despite Galison (1987), such an innovation has not received sufficient philosophical attention. As an indication of a possible way to remedy this situation, this section will focus in particular on some philosophical problems opened by the current, pervasive use of computer calculations and simulations as an interface between theory and experiment. With respect to this question, an interesting problem that should be considered lies in the attempt at elucidating the relationship between the computability of a physical theory and the notion of the measurability of a magnitude of an entity discussed in the previous section. Once this link is clarified, there is a reasonable hope that an increased understanding of the role of computers in the application of theories may provide us with a new bridge between the theoretical and the experimental cultures. The best candidate to take such a role is the notion of *the computability* of a physical theory, which may be regarded as a possible explication of the old notion of *observability*, in the context of a new operationistic philosophy of science in which the operations "with pen and pencils" correspond to those performed by recursive functions.

In order to provide the necessary framework to argue for these two claims in turn, I will refer to a paper by Geroch and Hartle (1986), whose main focus is a discussion of the relation between computability and physical theories. By proposing an analogy between the notion of a computable number in constructive analysis and that of a measurable number in physics, the following definition of measurable number is supposed to be the physical version of the corresponding, more rigorous definition of a computable number: "Regard number w as measurable if there exists a finite set of instructions for performing an experiment such that a technician, given an abundance of unprepared raw materials and an allowed error of ε is able by following those instructions to perform the experiment, yielding ultimately a rational number within ε of w." (1986: 542). The definition of a computable number is well known from computable analysis textbooks, and here for semplicity I will follow Geroch's and Hartle's approach. Let us say that number w is *computable* if there exists an algorithm (a Turing machine) such that for any given positive ε as input, the algorithm provides as output a rational number r, approximating the real number w in such a way that $|w - r| < \varepsilon$. In other words, the number is computable if the algorithm or the Turing machine generates rational approximations to the number[6]. Finally, the clearest characterization of the notion of *a computable theory* has been given by Kreisel: a theory is computable if every sequence of natural numbers or even real numbers which is well defined (observable) according to the theory is recursive in the data (Kreisel 1974: 11). In other words, the sequence of measurements yielding real numbers must be generated by an ideal Turing machine if the data are used as inputs.

In order to appreciate the analogies between measuring the magnitudes of some entity in physics and computing some number, we can think of "the finite set of instructions" performed by the experimentalist in the quoted definition as an algorithm implemented on a computer, and the unprepared raw material as the initial, blank memory of the computer[7]. Consider now the growing importance of the role of computers in laboratories, and assume with Campbell (1920) that physics *is* the science of measurement. The analogies illustrated above have the merit of explaining why *the existence of algorithms capable of calculating solutions to differential equations in which a law is stated offer a sufficient condition for the applicability or the empirical significance of a theory.* In some sense, they can also explain why the three-dimensional simulations of unreal, virtual experimental situations on which Galison has called attention play the role they do in experiments and in calculations. I will briefly expand on these two remarks in turn.

With respect to the problem of the empirical content or the applicability of a theory, its computability, corresponding roughly to the existence of an algorithm for implementing it (or, alternatively, for making predictions by solving a system of differential equations), guarantees that the numbers that the theory regards as measurable are computable. As an example discussed by Geroch and Hartle, note that the measurability of a number such $e^{1/2}$ depends on the prediction of exponential decay of the velocity of a projectile given by the *theory* of viscous fluid. However, if the latter theory, as it turns out, has an algorithm for its implementation, then the program that provides numbers as solutions to the differential equations can be taken to correspond to the finite set of measurement instructions quoted above. In this way, the instructions will concern the raw unprepared material that is necessary to assemble a computer and will also contain the program to be run on it. By considering the definition of a measurable number given above, *the algorithm is the finite list of instructions, and the numbers calculated by the program as output correspond to the measurement outcomes.*

I take it that this remark is also relevant as a possible explanation for the success of virtual simulations of non-existing, counterfactual situations, since those situations correspond to possible physical situations or physically possible worlds, and the simulation program corresponds, *via* the analogy illustrated above, to a *possible* set of measurements on the system in question. These explanatory ideas may be made more precise by adopting to the problem at hand some interesting speculations on the relation between physical systems and Turing machines put forward by the Oxford physicist David Deutsch in the context of a quantum theory of computation (Deutsch 1985).

Stipulate that two computer machines are computationally equivalent if and only if they compute the same function from a given input label. Suppose furthermore that "a computing machine *M* is capable of perfectly simulating a physical system *S* under a given labelling of their inputs and outputs if there exists a program $\pi(S)$ for *M* that renders *M* computationally equivalent to *S* under that labelling. In other words, $\pi(S)$ converts *M* into a 'black box' functionally indistinguishable from *S*"." (Deutsch 1985: 99). A physical version of the Church-Turing principle can then be stated by saying that "every finitely realizable physical system can be perfectly simulated by a universal model computing machine operating by finite means" (*ibid*).

The important point of this principle is that it is meant as having empirical content, since Deutsch argues for instance that classical physics, allowing a continuum of possible states, does not obey it, and is applied, as is well known, only through successive, discrete approximations. Despite this, classical physics is such that most of its theories are indeed computable, in the sense that there exists an algorithm to implement them, and every number that is measurable according to the theory is computable and *vice versa*.

According to Deutsch, Geroch and Hartle, however, there might be a physical theory, and quantum gravity is a candidate, in which measurable numbers determinable from the theory may well be non-computable. Deutsch even envisages the possibility that some physical laws may not respect the limitations of mathematical processes we call algorithms, and observes that the reasons why we can construct electronic calculators or perform mental arithmetic cannot be found in mathematics or logics, but rather in laws of nature: "*The reason is that the laws of physics happen to permit the existence of physical models for the operations of arithmetic* such as addition, subtraction and multiplication. If they did not, these familiar operation would be non-computable functions." (Deutsch 1985: 101). The empirical character of the physical version of the Church-Turing principle establishes a new, profound link between a computable law of nature and an algorithm or a recursive function, between initial data or boundary condition and inputs, and between final data and outputs. Physical systems may be regarded as computing devices, and nature itself may be a finite realizable system perfectly simulated by a computing machine.

Equipped with this analysis, we can now go back to the problem of the existence of theoretical entities. By recalling the image of a continuous spectrum going from directly manipulable middle-seize objects, to indirectly manipulable microscopical entities, to non-manipulable astronomical objects whose property we use to probe other entities, it is clear that the level of directness of causal intervention diminishes progressively. With virtual computer images, used to inquire into non-existent physical situations, we stumble in a discontinuity, as the simulation represents only a kind of virtual manipulation of a certain real physical system. However, the existential result of the procedure is established, provided that, as it is argued here, the criterion of the actual measurement of an entity x is sufficient for its existence. In this sense, the criterion of the actual computation of the value of a magnitude in a laboratory, by being sufficient for the actual measurement of a magnitude in the sense discussed above, can be considered another, more general "*sufficient*" condition for the existence of an entity.

In order to figure out whether the condition of computability is also necessary for the measurability of a number, one does not have to resort to the complicated case of quantum gravity discussed by Geroch and Hartle. These authors recall that while numbers like $e^{1/2}$ can be measured by the viscous fluid theory, some other numbers (like the fine-structure constant or the ratio of energy levels in the helium atom with finite mass and relativistic effects) are not given by the currently available theory, but can only be calculated by experimental means, and therefore cannot be regarded as computable with respect to currently available theories (1986: 544-545). For us, this is sufficient to exclude that computability be regarded as a necessary condition for the acceptability of a physical theory, independently of the controversial possibility envisaged by Deutsch that some physical system might turn out to compute functions that are Turing non computable. In other words, it may seem too restrictive to impose that every physical theory be computable, and computability, although sufficient for assigning empirical content to a theory, should not be regarded as necessary.

4. THE EXISTENCE OF QUANTITIES AND MEASURABILITY

It might be argued that unless we show that quantity is a natural category, the criterion of measurability and, *a fortiori*, that of computability, would be empty, since the attribution of a quantity to an entity would be unjustified. If the measure or the computation of the value of the magnitudes of an entity x is regarded as a sufficient condition for its existence, the magnitudes in question must be regarded as real properties of x. By a natural category, I mean in fact that objects possess quantities in a mind-independent way.

This idea has been repeatedly denied by philosophers like Carnap in his *Philosophical foundations of physics* and by Herbert Dingle in various papers. The former often insisted that the difference between the qualitative and the quantitative properties of objects "is not a difference in nature, but a difference in our conceptual system, that is, in our language." (Carnap 1966, ch. 5), while Dingle criticized the idea that objects possess a preexisting property that our operation measures. Rather, he conceives of quantity as a property ascribed to objects by our operations. Given my approach to the problem of the existence of theoretical entities in terms of the operations given by measurement and by computation, it is appropriate to briefly clarify my attitude with respect to Carnap's conventionalism and Dingle's operationism.

I think that the problem of the existence of quantities has no simple solution, and deserves to be studied more in depth. To my knowledge, this problem has been clearly posed by B. Ellis (1966, ch.2), but not answered in a definite way. His necessary and sufficient criterion for the existence of quantities amounts to the possibility of comparing any two objects in some way *via* asymmetric, transitive relations of 'smaller than', 'greater than', together with the symmetric transitive relation of equality. Although the existence of a set of *linear order relationships* is necessary to the existence of quantities, it is doubtful whether it is also sufficient. As noted by Ellis, people can be linearly ordered by taking the product of age and height, call it hage, but hage is hardly a genuine quantitative property of human beings like age and height (1966: 31).

Since I believe that quantities are kinds of properties possessed by objects in a mind-independent way, in order to eliminate such problems we must introduce some extra-condition that takes care of this difficulty.

The problem can be solved by considering that there is an ineliminable empirical element in the ascription of quantities to objects. It is up to empirical science to decide which quantities are genuine and which are spurious, and this does not involve any conventional move. It is only by invoking some empirical or statistical correlations between properties that we can legitimately attribute a quantity to objects. If we were to discover that hage, the product of height and age, is positively correlated with the occurrence of, say, cancer, so that the higher the hage, the greater the probability of getting cancer, we would not discount hage as a spurious quantity. On the hypothesis that the correlation with cancer exists only with the product of age and height, and is absent when any of the factors is considered separately, we would regard hage as a genuine quantity.

In other words, it is only by entering into a network of established empirical correlations that objects come to possess quantities. Together with the existence of a set of linear ordering relationship, this aspect points to the *relational* character of quantities. The fact that an object possesses a certain quantity to a certain degree depends not only on the existence of other objects possessing that quantity to another degree, but also on the nomological structure of empirical science as a whole[8].

University of Padua
Italy

NOTES

1 Italics in the original.

2 For instance, Galison has stressed stability and directness (1987: 261 ff.).

3 Although he says: "The best kinds of evidence for the reality of a postulated or inferred entity is that we can begin to measure it or otherwise understand its causal powers," he then writes, rather skeptically: "Perhaps there are entities which we shall only measure and never use" (1983: 274-5). I want to claim that we have good reasons to believe that such entities really do exist.

4 I first discussed this example in my Dorato 1988.

5 I here try to reply to an objection that was raised during discussion by Paul Churchland against the sufficiency of measurability and manipulability as conditions for the existence of an entity *T*. Giovanni Boniolo pointed out to me the case of the *N* rays.

6 For a more rigorous definition of a computable real number, see Pour-El and Richards 1989, ch.1.

7 "Unprepared" refers to the fact that one must be able to assemble the material from scratch: the Earth-Moon mass ratio is not measurable because the system is prepared. Likewise, instructions asking to calculate the ratios of the lenght of two calibrated metal rods would not involve unprepared raw material, otherwise, as Geroch and Hartle argue, every number would be measurable (1986: 543).

8 This observation shows that Carnap and Dingle's positions are incorrect. In some sense, it is trivial that the difference between qualitative and quantitative properties is reflected by our language, but what we refer to by using predicates or functors respectively is real if it enters into causal relations. On the other hand, to make the ascription of a quantity to an object depend on a series of operations *à la* Dingle seem to multiply quantities *preater necessitatem* by destroying their unifying power.

REFERENCES

R. Berendzen, R. Hart, D. Seeley, *Discovering the Galaxies*, Columbia University Press, New York 1984.
N. Campbell, *Phisics. The Elements*; Cambridge University Press, Cambridge 1920.
R. Carnap, *Philosophical Foundations of Physics*, M. Garnder ed., Basic Books, New York 1966.
N. Cartwight, *How the Laws of Physics lie*, Clarendon Press, Oxford 1983.
N. Cartwight, *Nature's Capacities and their Measurement*, Clarendon Press, Oxford 1989.
J. Cushing, E. McMullin, *The Philosophical Consequences of Quantum Theory: Reflections on Bell's theorem*, University of Notre Dame Press, Notre Dame, IN, 1989.
H. Dingle, "A Theory of Measurement", *British Journal for the Philosophy of Science*, **1**, 1950, 5-26.
M. Dorato, "The World of the Worms and the Quest for Reality", *Dialectica*, **42**, 1988, 171-182.
B. Ellis, *Basic Concepts of Measurement*, Cambridge University Press, Cambridge 1966.
P. Galison, *How Experiments End*. Chicago University Press, Chicago 1987.
R. Geroch, J. Hartle, "Computability and Physical Theories", *Foundations of Physics*, **16**, **6**, 1986, 533-349.
I. Hacking, *Representing and Intervening*, Cambridge University Press, Cambridge 1983.
J. Kim, *Supervenience and the Mind*, Cambridge University Press, Cambridge 1993.
G. Kreisel, "A Notion of a Mechanistic Theory", *Synthese*, **29**, 1974, 11-29.
D. Malament, "In defense of dogma: Why there cannot be a Relativistic Quantum Mechanics of (localizable) Particles" in *Perspectives on Quantum Reality*, ed. by R. Clifton ed., Kluwer, Dordrecht 1996, pp. 1-10.
G. Maxwell, "The ontological status of theoretical entities", *Minnesota Studies* in *the Philosophy of Science*, **3**, 1962, 3-27.
H. Putnam, "What is realism", in J. Leplin ed., *Scientific Realism*, University of California Press, Berkeley 1984, pp. 140-154.
M. Pour-El, I. Richards, *Computability in Analysis and Physics.*, Springer-Verlag, Berlin 1989.
D. Shapere, "The Concept of Observation in Science and Philosophy", *Philosophy of Science*, **49**, 1982, 485-526.
S. Shoemaker, "Causality and Property", in R. van Inwagen ed., *Time and Cause* Reidel, Dordrecht 1980.
R. Smith, *The Expanding Universe*, Cambridge University Press, Cambridge 1982.
K. Svozil, *Randomness & Indecidability in Physics*, World Scientific, Singapore 1993.
B. Van Fraassen, *The Scientific Image.*, Clarendon Press, Oxford 1980.

JESÚS MOSTERÍN

OBSERVATION, CONSTRUCTION AND SPECULATION IN COSMOLOGY

1. THE AIMS OF THEORY

Our most serious business is the conduct of our own life. In order to make the best decisions we need the best information on the world we live in. And nowadays we expect a large proportion of that information to be provided by science. Consequently, all reflective human beings (not just the professional scientists) have a stake in science.

We can concern ourselves with science and philosophy for a variety of reasons, like money, prestige, career advancement, technological applications, or just playful curiosity. But the deepest reason for our interest in science is our need for orientation. We find orientation in some domain when we are able to give answers to our questions concerning that domain.

A model of something is a conceptual machine for answering questions about that something. Any map of a territory is a model of that territory. A map of Parma allows us to answer questions like: "How do I get from the station to the hotel?", "Should I now turn to the right or to the left?" We find global or cosmic orientation when we are able to answer our questions about the universe, which is the broadest possible context of our life. For that sake we need a cosmic-question answering machine, i.e., we need a reliable cosmological model.

A theoretical model incorporates different layers or components: underlying mathematics, physical theories, special assumptions, specific values for certain parameters, etc.

Scope, accuracy, testability, and computability are universal goals of theories and theoretical models. Sometimes (but not always) we pursue also other aims in devising models, such as understandability (providing intuitive insight into the inner workings) and beauty. Confronted with a variety of theories or models for a prospective application, we make a choice of theory (or model) based on a trade-off between these goals. General relativity, for example, has broader scope and greater accuracy but much less computability (because of its non-linear equations) than Newtonian mechanics or special relativity. So we only use it where we cannot help using it, like in cosmology.

2. DEMARCATION BETWEEN EMPIRICAL AND SPECULATIVE SCIENCE

When we chart the surface of an only partially known planet or continent, we record some empirically observed features (like mountains, rivers, lakes, and coasts) with the

E. Agazzi and M. Pauri (eds.), The Reality of the Unobservable, 219-229.

help of some mathematical fictions or constructs that we introduce for technical reasons (like the meridians and the parallels of latitude, and the system of projection). And (at least in the past) cartographers used to complete the map through speculative outlines of the unknown land *(terra incognita)*, which only further exploration in the future would eventually confirm or prove wrong. There was an abhorrence of the blank in cartography. There still is one in cosmology.

Concerning the totality or the Universe, philosophers have stopped speculating, but physicists have taken over from them (as S. Hawking recently remarked, with only half tongue in cheek). In any case, some scientists, and specially some cosmologists, are nowadays fond of speculation, i.e. of thought (eventually mathematical thought) freed from the checks of observation.

The classical philosophers of science (like the Positivists and Popper) posed the problem of demarcation between science and non-science. Now the problem has to be posed inside science itself, as the demarcation between empirically grounded science and speculative science. Speculative science is a mathematical kind of metaphysics. It aims at becoming empirical science, but has not yet reached that status, and perhaps never will. On the other hand, speculative science is the locomotive of scientific progress. The worth of speculative theorizing inside the global scientific enterprise is not in question. But if we are to rely on science for building our world view and fixing the context for the conduct of our lives, we'd rather separate the reliable wheat from the merely speculative chaff.

3. THE MODEL NO ONE WANTED

At this moment there are in sight no viable alternatives to the general theory of relativity as the basic conceptual framework for the modeling of the universe as a whole, and there are no working alternatives to the standard big bang cosmological model (or family of models). This fact is not due to the will of the scientists who created the model, still less to the prejudices of the scientific establishment. On the contrary, it is almost exclusively due to the strong observational constraints which reality puts on the activity of model making. The standard big bang cosmological model is the model no one wanted, but which recalcitrant experience forced (almost) everyone to accept, at least for the time being.

This situation is not unique. At the end of the 19th century most physicists were happy with the idea of the ether filling space and providing an absolute framework of reference. The American physicist Albert Michelson performed several very accurate experiments with the aim of providing a definitive proof of the reality of the ether at rest in space. They should have detected movement in the ether, but not such movement was detected. In hindsight we know that the Michelson experiments struk a decisive blow against the ideas of ether and of absolute space, opening the way to Einstein's special relativity. But Michelson kept hoping he would find some explanation for the "failure" of his experiments. He never brought himself to accept relativity.

"The expansion of the universe is the greatest of all tests of Einstein's geometrodynamics, and cosmology the greatest of all applications. Making a prediction too fantastic for its author to credit, the theory forecast the expansion years before it was observed... In 1915, when Einstein developed his general relativity theory, the permanence of the universe was a fixed item of belief in Western philosophy... Thus, it disturbed Einstein greatly to discover that his geometrodynamic law predicts a nonpermanent universe; a dynamic universe; a universe that originated in a big-bang explosion, or will be destroyed eventually by contraction to infinite density, or both. Faced with this contradiction between his theory and the firm philosophical beliefs of the day, Einstein weakened; he modified his theory"[1]. Einstein went to the length of arbitrarily complicating his equations of the gravitational field with the introduction of an *ad hoc* term (the so-called cosmological constant Λ) just to avoid having a dynamic, expanding universe with a beginning in the finite past or an end in the finite future as a solution.

In 1929 Edwin Hubble empirically discovered the expansion of the universe and formulated his famous law relating distance and receding velocity of the galaxies, whereupon Einstein accepted the fact that the universe is dynamic and abandoned the cosmological constant, calling its previous introduction "the biggest blunder of my life." But Hubble himself did not like the obvious interpretation of his observational discovery, insisted in using the wording "apparent velocities" in his law and for several years tried (without success) to look for some other explanation which would preserve the static character of the universe[2].

Still later, when the expansion of the universe was already universally accepted, Bondi, Gold and Hoyle introduced different versions of the steady state theory in 1948, intended to avoid the beginning and to reconcile eternity and temporal homogeneity with expansion. The steady state model had problems because of the radio sources count and with the uniform age of the galaxies, and was struck a death blow by the discovery in 1965 of the microwave background radiation, unaccountable in the steady state model, but predicted by the big bang model. Its discoverers, Penzias and Wilson, didn't like what they discovered: "I do not think that either of us took the cosmology very seriously at first. We had been used to the idea of steady-state cosmology. Philosophically, I liked the steady-state cosmology. So I thought that we should report our result as a simple measurement."[3]

Nowadays the accurate calculation of the actual abundances of hydrogen and helium from the big bang assumptions is considered one of the main empirical supports of the standard model. It is ironic that this calculation was first carried out (in 1964 and 1967) by Fred Hoyle, the staunchest proponent of the steady state theory, who perhaps hoped to get some result contradicted by observation, and so a disproof of the big bang model. In fact he contributed to its success. (It was also Hoyle who invented the name "big bang" to refer to the adversary theory in a derogatory way during a popular BBC show. The name stuck).

The failed steady state model was dropped with much sorrow. "The steady state theory is aesthetically and philosophically pleasing to many people, to whom it is a matter of regret that observations indicate that it is not the correct model" [Islam 1983]. Even some of the scientists who have made great contributions to the standard model, like Steven Weinberg, considered that the steady state model was much more attractive, and dropped it with regret when it became incompatible with the facts[4].

4. LAYERS IN THE STANDARD BIG BANG COSMOLOGICAL MODEL

The big bang cosmological model is a multi-layered mathematical structure, which is supposed to be sufficiently similar (in relevant structural respects) to the actual universe to allow its use as a reliable guide for answering questions about the universe (how did it begin?, how will it end?, how large is it?, ...).

The following are the main different layers that can be distinguished in the standard big bang model (i.e., in the big bang model as accepted by most cosmologists and astronomers; more speculative layers will be touched upon later on):

(1) The underlying mathematical structure of the big bang model (and of general relativity) is that of a 4-dimensional differentiable manifold. The manifold is assumed to be Hausdorff, $C^{(\infty)}$, connected, paracompact, and without boundary.

(2) The manifold is provided with a locally Lorentz metric **g** of signature (-, +, +, +).

(3) Einstein's equivalence principle. It implies that the effects of gravity must be equivalent to the effects of living in a curved spacetime. This principle is common to General Relativity and to all metric theories of gravity, like the Brans-Dicke theory.

(4) Einstein's field equations, describing the generation of spacetime curvature by mass-energy distribution: $\mathbf{G} = 8\pi\mathbf{T}$ [assuming units which make the gravitational constant and the speed of light = 1].

(5) Time-orientability and stable causality condition, which imply that it is possible to introduce a cosmic time, defined at every point (or event) of spacetime in such a way that this time is strictly increasing along any time-like geodesic.

(6) Robertson-Walker metric, corresponding to a perfectly homogeneous and isotropic universe.

(7) Friedmann solutions (with zero pressure and zero cosmological constant).

(8) Hawking-Penrose singularity theorems, which imply the big bang proper.

(8) Parameters to be filled by hand: Hubble constant H_0, deceleration parameter q_0, average density ρ.

(9) Thermal history of the universe from 10^{-10} s after the big bang till now, including the passage from a radiation- to a matter-dominated age.

(10) Chemical history of the universe, including the account of primordial nucleosynthesis.

5. THEORETICAL CONSTRUCTS IN THE BIG BANG MODEL

The mathematical model for spacetime (the collection of all events), is a pair (M, **g**) where M is a connected four-dimensional Hausdorff $C^{(\infty)}$ differentiable manifold and **g** is a locally Lorentz metric (of signature - + + +) on M. Together with the existence of a Lorentz metric, the Hausdorff condition implies that M is paracompact.

Clearly, these two first layers of the model respond to computability requirements. All these properties of the mathematical structure are more of a convenience construction, postulated for the sake of allowing computations, than anything somehow extracted from observations. We simply decree that the universe (or, at least, our model of the universe) has to be such as to allow us to make computations and predictions or retrodictions with our usual mathematics.

This model corresponds to a maze of world lines of particles and light rays, and the intersections between them. One can in imagination extrapolate to the idealized limit: an infinitely dense collection of light rays and world lines of infinitesimal test particles. Continuous, differentiable (i.e. smooth) coordinate systems operate in this limit. But this idealized limit is not to be mistaken for reality. It is a fiction indulged in for the sake of computation.

For example, the evolution of physical systems is described in mathematical language by means of differential equations. It has to be possible to calculate derivatives. So, we postulate a differentiable manifold. For the definition of the differentiable manifold it is required that any open set of our topological space be mapped in a continuous and bijective way to an open set of a real n-dimensional space. Given this bijection, the notion of a derivative in a real space can be transferred to the manifold, and it becomes possible to define differentiable functions in our universe.

The order of differentiability of the metric ought to be sufficient for the field equations to be defined. The order of differentiability of the metric is not physically significant, since one can only measure the metric with some margin of error, and so one could never determine that there was an actual discontinuity in its derivatives of any order. Thus one can always represent one's measurements by a $C^{(\infty)}$ metric[5].

The manifold is assumed to be Hausdorff or separable, because in non-separable topological spaces it is not possible to prove the inverse function theorem, needed in classical analysis[6].

Many physical magnitudes are represented in terms of multiple integrals. To ensure the existence of these integrals, that is, the convergence of the sums used to define these integrals, we require our topological space to be paracompact. This assumption implies that from every atlas we can extract another atlas with the property that each point in the manifold is covered by only a finite number of charts. This finiteness property ensures the existence of the sums used to define the integrals.

Also the layers labeled fifth and sixth above are constructs adopted for easiness of computation. The general theory of relativity is compatible with universes which violate the time-orientability and stable causality condition (and hence the possibility of a cosmic time), like Gödel's rotating universe model, in which time loops (and so time travel) are possible[7]. The very notion of time used in the description of the early universe is greatly simplified, ignoring, for example, the changes in clock rates as gravitational field strengths change.

Another tribute paid to computability is the introduction of the Robertson-Walker metric, corresponding to a maximally symmetric spacetime, in which one should not be able to tell one point from another (homogeneity), or one direction from another (isotropy). General relativity allows for any variable curvature of spacetime, but computations in such models soon become intractably difficult or impossible. So we postulate that the universe is a perfect fluid whose particles are the galaxies. The perfectly uniform distribution of matter generates a maximally symmetric spacetime of constant curvature. If such an enormous simplification is made, then we can easily find computable models for the universe: the Robertson-Walker spacetimes.

6. OBSERVATION AND EMPIRICAL TESTS OF THE STANDARD MODEL

The big bang standard model does not build on untested physical theories. Its main theoretical basis is the general theory of relativity, which up to now has passed every test to which it has been submitted. From the initial detection of the deflection of light around the Sun, or the explanation of the perihelion precession of Mercury, to more recent and spectacular tests like the discovery of gravitational lenses, or the exact computation of the minute changes (due to the emission of gravitational waves) in the period of the binary pulsar discovered by Hulse and

Taylor in 1974 (for which its authors got the 1993 Nobel Prize in physics), general relativity is solidly anchored in observation.

Of course, it may be that general relativity does not provide a correct description of the universe outside of the range in which it has been tested, i.e., at extremely small lengths or in situations where spacetime warps drastically. In any case it seems to be a good principle that the prediction of a singularity by a physical theory indicates that the theory has broken down, i.e. it no longer provides a correct description of observations[8]. So, if the big bang model is correct, it surely is not correct at the big bang singularity.

The big bang standard model deals adequately with 3 large empirical facts: the expansion of the universe, the abundance of chemical elements, and the cosmic background radiation. It is still incomplete, because the values of certain parameters (Hubble's, omega) that have to be put in by hand have not yet been ascertained (and the estimates vary by a factor of three).

The empirical discovery of the uniform expansion of the universe by Hubble in 1929 was the first motivation and the only observational corroboration of the big bang model for many years. Of course the big bang model accounted for the observed expansion in a very natural way. But that was not enough. In the 40s and 50s most scientists considered cosmology too speculative a subject to be taken seriously. There was practically no contact between theory and observation (besides Hubble's law).

The age of the Earth (4.7 billion years), estimated empirically from the radioactive study of the oldest rocks, was in good agreement with the age of the universe (between 10 and 20 billion years), as retrodicted by the big bang model, taking into account that the chemical composition of the earth presupposes a previous generation of stars, which should have had time to complete their "life" cycles and scatter the heavy atoms (of which the earth is made) cooked in their interior in supernova explosions.

A major triumph of the big bang model was the computation and explanation of the correct proportions of hydrogen (75 %) and helium (25 %) in the galaxies, as observed. The theoretical calculations were first carried out by Fred Hoyle, Roger Taylor and Yakov Zeldovich in 1964, refined by James Peebles in 1966, and refined in great detail by William Fowler, Fred Hoyle and Robert Wagoner in 1967. In 1980 David Schramm calculated the expected amount of lithium produced in the big bang, and it also agreed with observations. The detection of large amounts of intergalactic helium by A. Davidsen from data collected by an ultraviolet telescope on board the space shuttle *Endeavor* in 1995 has added impressive empirical support to the big bang.

The most decisive support for the big bang model came from the serendipitous discovery of the microwave background radiation by Penzias and Wilson in 1965. This observational result was the more remarkable for having been predicted much in advance. The first predictions of the cosmic background radiation from the big bang model were made by Ralph Alpher, George Gamow, and Robert Herman in 1948. It was again and independently predicted by Robert Dicke, James Peebles, and others in 1965, shortly before the actual discovery.

7. SPECULATIVE EXTENSIONS OF THE BIG BANG MODEL

The standard big bang model (the, at least in part, empirically grounded and almost universally accepted cosmological model) consists mainly of the ten layers mentioned before. Several speculative extensions (with an increasing level of speculation and a decreasing contact with any observations) have been proposed:

(11) The explanation of the decoupling of the strong and the electroweak forces at 10^{-35} s after the big bang through some GUT. (The Weinberg, Salam, and Glashow theory of electroweak unification has received empirical confirmation in the detection of the W and Z bosons, but the GUT's remain utterly unchecked).

(12) The inflationary universe model (or scenario). It coincides with the standard big bang model in the evolution of the universe after 10^{-30} s after the big bang, but diverges dramatically in the account of what went on earlier. (More on this later on.)

(13) Quantum gravity, quantum fluctuations "before" the big bang, and quantum cosmology. Speculations based on non-existing theories.

(14) Speculations on many universes causally and informationally isolated from ours. They can be spatially, temporally, or other-dimensionally multiple (whatever all this means). These other "universes" are ineffable, beyond science, and beyond any type of checking, by definition.

8. PUTATIVE CONCEPTUAL PROBLEMS WITH THE STANDARD MODEL

Having been able to explain the observed expansion of the universe and the cosmic abundances of hydrogen and helium, and having seen its prediction of the cosmic microwave background radiation confirmed, the standard big bang model had acquired a new respectability in the scientific community. However, during the 70s some cosmologists began to have qualms about the initial conditions of the universe, some of which had to be picked up "by hand" from the observations, and could not be deduced from the theoretical nucleus of the model. These misgivings about the initial conditions concerned their being just contingent features of the world, not necessary features of the model. Others found some of the initial conditions somehow improbable. Many, finally, did not see any real problem in having some "just so" (not further explainable) initial conditions. The discussion about the status of these problems or non-problems still continues.

The horizon problem was first clearly stated in print by Charles Misner in 1969. The horizon problem is the problem of accounting for the large-scale uniformity of the universe, inferred from the observed uniformity of the cosmic background radiation.

No signal or causal influence can travel faster than light. For each point in space one can consider a sphere containing all the other points that could have had some influence or could have homogenized with the central point since the big bang. The outer edge of that sphere is called the (special-relativistic) horizon of the central point. Each point has its own horizon. Since heat exchange, or any other homogenizing process, cannot travel faster than light, a point's horizon at any moment can extend no further than the distance light can have traveled since the big bang. For example, the size of the horizon 400,000 years after the big bang (when the universe became transparent to light and the cosmic background radiation was emitted) was about 400,000 light years. Beyond 400,000 light years from any given point, there would not have been time for light or heat or any other signal to have traveled across that distance since the beginning of the universe. But the regions of the universe which produced the cosmic radiation, when the universe was 400,000 years old, were more than 50 million light years apart at that time - much too far apart to have had time since the big bang to exchange heat and homogenize. The "horizon problem" arose because the smoothness of the cosmic background radiation suggests that different regions of the universe separated by more than each other's horizon (about 150 times more) would have to have exchanged heat, so violating the horizon restriction. Of course, the difficulty disappears if we accept that the universe began with a very high degree of homogeneity in the first place, so that there was no need for posterior homogenizing, but this looks like an arbitrary initial condition to some people.

Another problem with initial conditions which has worried some (but not all) cosmologists is the so-called flatness problem. The actual universe is not very open, because galaxies have been able to condense and form. Neither is it very close, because it has not yet recollapsed. It seems to be a little open, but almost flat, in any case not very far from flat. Why should the universe today be so near the boundary between open and close, that is, so nearly flat? Why is k (the curvature of spacetime) so close to 0 and why is the measured value of Ω (the density parameter, the quotient of the actual to the critical density) today so close to 1?

The flatness problem was first stated by Robert Dicke in 1969. It was also noted by Brandon Carter and Stephen Hawking, and restated by Dicke and Peebles in 1979. "There is a broad range of attitudes about the flatness problem. Some scientists consider the initial value of Ω to be an accidental property of our universe, a value that should be accepted as a given, and they see no validity in the flatness problem. For this group of cosmologists, the flatness problem is a non-problem, an issue lying beyond the domain of science. Others agree with Dicke and Peebles that the required initial conditions seem too special to be accidental and that some deeper physical explanation is required. Among this latter group are scientists who say that for some reason the inertial gravitational and kinetic energies must have been balanced *exactly*. Ω was and is exactly 1. This view requires the existence of huge quantities of undetected mass.... Even today there is still no consensus on the meaning or depth of the flatness problem."[9]

It follows from the big bang model that as time goes on, Ω should differ more and more from 1, unless it started out exactly 1, in which case it remains 1. In an open universe, Ω begins smaller than 1 and gets smaller and smaller in time; in a closed universe, Ω begins larger than 1 and gets larger and larger. In order for the value of Ω to still remain between 0.1 and 10.0 today, after more than ten billion years, the initial value of Ω had to lie between $1 - 10^{-59}$ and $1 + 10^{-59}$. Should we search for the physical processes that could have set so fine a balance? Or should we accept it just a given initial condition, in no need of further explanation?

A very different kind of putative problem is the so-called monopole problem. This arises from the incorporation of untested theories of grand unification (GUTs) - i.e., unification of the electroweak and the strong nuclear forces - into some extensions of the big bang model. According to these theories, there should be lots of magnetic monopoles around. But where are they? We cannot detect any. Of course the magnetic monopole is not a problem of the whole family of big bang models, but only of the models which incorporate the grand unified theories.

9. THE INFLATIONARY UNIVERSE MODEL

In the 70s a radical change took place in the cosmological community, till then a reserve of astronomers and general relativists. After the standard model of particle physics had become well established with the experimental confirmation of the electroweak unification, several leading particle physicists - like Steven Weinberg - decided to enter the field of cosmology. This led to a rejuvenation of the subject and to the proposal of new extensions of the big bang model, which incorporated several tested and untested particle theories.

At the beginning of the 80s the horizon, flatness and monopole problems found an elegant solution in the new inflationary universe model (or scenario), first introduced by Alan Guth in 1981.

The inflationary expansion of the early universe is a natural consequence of some grand unified theories. Those theories predict that at the moment when the single unified force began acting as separate forces, the universe was in a peculiar state called a false vacuum, which behaves as if it had negative gravity. Negative gravity repels, so that instead of retarding the rate of expansion, it speeded up the rate of expansion. The inflationary period of exponential expansion would have ended when the universe changed from the false vacuum state back to the normal state, with attractive gravity. The story is much more complicated, involving the supercooled symmetrical state during the inflationary era, followed by a phase transition in which the strong force separated from the electroweak force, etc., but this is not the place to delve into any details[10].

According to the inflationary universe model, in the extremely short time between 10^{-35} second and 10^{-32} seconds (after the singularity) the universe went through an inflationary era, during which it expanded prodigiously by a factor of as much as 10^{50} times more than what had previously been assumed. This implies that in the pre-inflationary period the universe was much smaller than it had been assumed to be by the standard hot big bang model. In this very much smaller universe everything was within the horizon of light velocity. Any initial differences in temperature could be easily smoothed out and thermal equilibrium could be established. This equilibrium was preserved during the inflationary expansion and inherited by the later universe. In this way the horizon problem is solved.

In 1978 Guth learned about the flatness problem from a lecture by Robert Dicke. His inflationary universe model also solves the flatness problem. Regardless of the initial curvature of the universe, any observable region of the universe would be very nearly flat after the period of rapid expansion, just as any small area of the surface of a ball would appear nearly flat after the ball has been inflated to a huge diameter. The inflationary phase would have smoothed out any wrinkles in spacetime, just as the inflation of a balloon smooths out any wrinkles in its surface. The inflationary universe model firmly predicts that the universe should be extremely close to flat.

As a bonus, inflation solves the monopole problem (which in its context is a serious problem, as the inflationary universe model is based on the same grand unified theories that predict magnetic monopoles). The huge inflationary expansion has moved the magnetic monopoles so extremely far apart, that it is most unlikely that we will ever find one, even if they were produced in great numbers at the beginning.

The original inflationary universe model, as proposed by Guth in 1981, had difficulties (after inflation the universe became too inhomogeneous). These difficulties were solved in 1982 by a new version of the inflationary universe model due to Andrei Linde, and to Paul Steinhardt and Andreas Albrecht. Unfortunately, the new inflationary model also had serious problems, as pointed out by Guth. Later on Linde simplified and streamlined the inflationary scenario considerably: inflation appeared as a naturally emerging feature in many theories of elementary particles, and in any scalar field. There is no longer any need for phase transitions, supercooling, etc. All possible kinds and values of scalar fields in the early universe are considered. The ones where inflation takes place become exponentially large and dominate the total volume of the universe. This model or scenario, called chaotic inflation, really reduces the assumptions on initial conditions to a minimum. Out of an extremely large class of initial conditions (out of chaotic initial conditions) it produces the actual universe. It explains the horizon and flatness and monopole problems without resorting to any specific initial conditions. In fact, it provides a physical mechanism which makes the initial conditions irrelevant. Whatever the initial conditions, the result (after inflation) would have been the same - an elegant idea that pleases many physicists.

The inflationary universe model (specially in its chaotic version) deals successfully with the putative conceptual problems of the big bang model, and specifically with the need of a very narrow range of initial conditions. But it builds on untested theories, like the GUT's (whose predictions are either untestable in principle, or testable and seemingly proven wrong, like proton decay) or the scalar fields. And it remains empirically unchecked.

There is no direct observational evidence to confirm the inflationary universe model. In fact, the one critical prediction (the flatness of the actual universe, i.e. $k = 0$, $\Omega = 1$) of the model seems to be inconsistent with the observations made up to now, which give at the present time a measured value of Ω of about 0.1 or 0.2, whereas almost any version of the inflationary universe model predicts that the value of Ω today should be extremely close to 1 or exactly 1, even 10 billion years after the big bang. In 1986 P. Peebles reviewed the then available evidence and concluded that it favors a low-density, open universe. In 1994 P. Coles and G. Ellis reviewed the new empirical data on visible and dark matter, and reached the same conclusion: if anything, the universe seems to be open, not flat.

The standard big bang model of cosmology does not include flatness or inflation[11], still less, initial quantum fluctuations or many universes. It is true that some cosmologists consider flatness, inflation, and cold dark matter as part of the "current theoretical orthodoxy". Recently some of them went to the length of proposing a value of the Hubble constant of 30 in spite of all the observational evidence pointing to 65, just to solve the theoretical problems of "the current theoretical orthodoxy". The recent measurements of the Hubble constant by Wendy Freedman from data collected by the Hubble space telescope, if confirmed, would imply (assuming the inflationary value of $\Omega = 1$) that the universe is much younger (about 10 billion years) than the old stars of the globular clusters it contains. Obviously something must be wrong here. Speculation is necessary and nice, but, as even the speculative authors of the proposal of a value of 30 for the Hubble constant concede, "in the end, observation will, as it must, have the final say."[12]

Barcelona University
Spain

NOTES

1 See Misner, Thorne and Wheeler 1973.
2 See Berendzen, Hart and Seeley 1976.
3 See Wilson in Bertotti *et al.*, 1990.
4 See Lightman and Brawer 1990.
5 See Hawking and Ellis 1973.
6 See Fustero and Verdaguer in Meyerstein, 1989.
7 See Gödel 1949.
8 See Hawking and Ellis, *op. cit.*
9 See Meyerstein, *op. cit.*
10 See Linde 1990 or Narlikar and Padmanabhan 1991.
11 See Coles and Ellis 1994.
12 See Bartlett, Blanchard, Silk and Turner 1995.

REFERENCES

Bartlett, Blanchard, Silk & Turner, "The case for a Hubble constant of 30 km s^{-1} Mpc^{-1}", *Science*, **267**, 1995, 980-983.
R. Berendzen, R. Hart & D. Seeley, "Man Discovers the Galaxies", *Science History Publications*, New York 1976.
B. Bertotti, Balbinot, S. Bergia & Messina eds., *Modern Cosmology in Retrospect*, Cambridge University Press, 1990, [426 p.].
P. Coles & G. Ellis, "The case for an open Universe", *Nature*, 370, 1994, 609-614.
K. Gödel, "An example of a new type of cosmological solutions to Einstein's field equations of gravitation", *Reviews of Modern Physics*, **21**, 1949, 447-450.
S. Hawking, & G. Ellis, *The large scale structure of space-time*, Cambridge University Press 1973.
A. Lightman & R. Brawer, *Origins. The lives and worlds of modern cosmologists*, Harvard University Press 1990.
A. Linde, *Particle Physics and Inflationary Cosmology*, Harwood Academic Publishers, Cur, Switzerland 1990.
W. Meyerstein ed., *Foundations of Big Bang Cosmology*, World Scientific Publishing, London and Singapore 1989.
C. Misner, K. Thorne & J. Wheeler, *Gravitation*, W. H. Freeman and Co., San Francisco 1973.
J. Narlikar & T. Padmanabhan, "Inflation for astronomers", *Annual Review of Astronomy and Astrophysics*, vol. 29, 1991, 325-62.

MARK WILSON

WHERE DID THE NOTION THAT FORCES ARE UNOBSERVABLE COME FROM?

As a school boy, I was informed by my chums that alligators lived in the Brooklyn sewers, abandoned pets that subsequently grew into giant saurians. It seems to me that modern philosophy of science is burdened by a large heritage of folklore not completely dissimilar to that of my alligators. A book like Bas Van Fraassen's *The Scientific Image*[1] contains comparatively little *direct* argument for its "anti-realism" - it rests instead upon sundry appeals to historical lineage: "I, Van Fraassen, now present an up-to-date defence of the empiricism favored by Hume, Duhem and Carnap". Leaving Hume and Carnap aside, it is certainly true that opinions of an extreme anti-realist tenor can be easily extracted from the writings of many scientists - not simply Duhem - of the late nineteenth century. Passages can be culled, for example, from Karl Pearson's *The Grammar of Science*[2] that might make the most rabid modern anti-realist blush. So I asked myself, "Why were *these* guys beefing? What, so long before quantum theory, drove so many Victorians and Edwardians into antirealist arms?" A complete accounting would be very complicated, but an important part of the answer traces some peculiar puzzles that attend classical treatments of continua-that is, fluids, elastic solids, aethers and other forms of continuously flexible bodies.

Within this historical context, the anti-realist philosophizing played an important role in serving *as an excuse* for evading certain mathematical difficulties that naturally arise in the treatment of continua. The notion of *force* becomes subject to various unexpected tensions within this setting as well. Unfortunately for modern philosophy of science, the only remnants of this struggle that have managed to survive to our time are odd scraps of philosophical flotsam recounted by authorities who have forgotten the context of the original debate. Here is a good sample of what I have in mind, drawn from Max Jammer's influential *The Concept of Force*[3]:

> With the rise of Newtonian dynamics and its interpretation along the lines of Boscovitch, Kant and Spencer, the concept of force rose almost to the status of an almighty potentate of totalitarian rule over the phenomena. And yet, since the very beginning of its early rise to power, revolutionary forces were at work (Keill, Berkeley, Maupertuis, Hume, d'Alembert) which in due time led to its dethronement (Mach, Kirchhoff, Hertz). This movement in mathematical physics, from the time of Newton onward, was essentially an attempt to explain physical phenomena in terms of mass points and their spatial relations. For it became increasingly clear that the concept of force, if divested of all its extrascientific connotations, reveals itself as an empty scheme, a pure relation.

The inclusion of *Hertz* as one who favored "mass points and their spatial relations" should warn us that something has gone amiss here. But the basic attitude expressed here is not idiosyncratic to Jammer alone: the logical empiricists often tried to justify

E. Agazzi and M. Pauri (eds.), The Reality of the Unobservable, 231-239.

their focus upon logistics on the grounds that such tools might eventually provide the "proper logical analysis of *force*", as Hertz, Kirchhoff, Mach *et al.* were supposed to have sought. But this alleged need for "logical analysis" is almost purely mythological-the Victorian physicists had been concerned, albeit in a confused way, with some hard issues of physical doctrine that mere "logical analysis" could not conceivably resolve. The great mechanist Clifford Truesdell has written a scathing review[4] of some of the neo-positivist attempts to "clarify the logical foundations of mechanics". Although the gusto of Truesdell's critique is perhaps to be regretted, his analysis seems to me correct in its fundamentals.

In any event, such folklore about *force* has distorted current debate in philosophy of science to a surprising degree. Consider the central role that "logical analyses" of *mass* and *force* play in Van Fraassen's brief for "anti-realism":

> [L]et us examine the standard example of "underdetermination" to be drawn from foundational studies in classical mechanics. In the context of that theory, and arguably in all of classical physics, all measurements are reducible to a series of measurements of time and position. Hence let us designate as basic observables all quantities which are functions of time and position alone. These included velocity and acceleration... They do not include mass, force, momentum, kinetic energy...
>
> We see in [the axiomatic theories of mechanics developed in this century] many different treatments of mass... What explains this divergence and the conviction of the authors that they have axiomatized classical mechanics? Well, the theories they developed are demonstrably empirically adequate in exactly the sense I have given that term... The thesis of constructive empiricism, that what matters in science is empirical adequacy, and not questions of truth going beyond that, explains this chapter in foundational studies[5].

But make no mistake: any of Sonny Liston's opponents would have been surprised to learn that *force* is a non-observable. What has actually gone wrong is this: the potted palms, musty settees and other paraphernalia that dominate much of the stage in modern debates over anti-realism *et al.* originated as props within forgotten productions of the late Victorian melodrama, "Whatever Shall We Do with Flexible Bodies?". Such furnishings were appropriate to that original setting, but nowadays such stagecraft merely lends an illusion of venerable respectability to modern worries that bear little genuine relationship to the scientific issues that bothered the Victorians. The notion that classical mechanics happily tolerates divergent notions of *force* because of its recognized non-observationality is a largely fictional account of why all of this funny old furniture still remains upon stage.

To gain a better appreciation of these remarks, let us rehearse some of the interpretational difficulties that attach to Newton's laws of motion. Specifically, what is the term "body" in Newton's laws supposed to denote? Should a "body" be: (a) an isolated mass-point understood in the manner of Boscovitch, (b) an elementary rigid body atom; (c) a *nonisolated* point (or "particle") within a truly continuous substance, (d) any nice section (of non-zero measure) of an *extended* object? Or something else again? It is a surprising oddity of Newton's laws that, to a rough measure, they can be interpreted to accommodate any of these divergent readings reasonably well. That is, Newton's laws seem to span a large spectrum of *size scales* with tolerable success.

This apparent indifference to scale allows conventional textbooks to get away with inferential *mahem* of a certain sort. Any reasonably complete primer of classical mechanics will contain certain "landmark equations" that need to appear in any adequate survey, e.g., the standard equations of celestial mechanics, the Navier-Stokes equations, Lagrange's equations for a pendulum, etc. A text must be organized so that all of the required landmarks eventually receive a visitation. Most textbooks, I'm afraid, *cheat* to permit a quick transversal of the required Grand Tour. Consider a text that begins, as most modern primers do, with isolated point masses. "**F** = m**a**" is then properly treated as the frame upon which differential equations for these particles are constructed: specifically, terms for the sundry forces acting upon a particle are summed and equated to m**a**. So far so good, but this same procedure is often copied without comment later in the book when the Navier-Stokes equations for a continuous fluid are assembled. With nary a word of caution, "**F** =m**a**" is utilized as the frame upon which the key equations are built. But how should we understand the frictional sheering "forces" that are assembled in this mix? The orthodox answer, from a molecular point of view, is that such "forces" are not really true *forces* at all, but represent the transport of momentum in and out of the largish "particle" of finite volume from which individual point masses are free to enter and leave. That is, from a molecular point of view, the "body" to which "**F** = m**a**" applies in the Navier-Stokes case is not a point mass but instead a rather abstract unit which preserves an identity over time by retaining certain gross features, in approved "Ship of Theseus" fashion. Clearly, the terms "force" and "body" must have shifted their meanings silently somewhere in the textbook to accommodate this otherwise unjustified application of "**F** = m**a**".

An honest mass point textbook should try to prove a *new* version of "**F** = m**a**" suitable to "Ship of Theseus" particles, but this step is rarely taken because it presents formidable difficulties. Indeed, it is quite difficult to ascertain when a swarm of point-masses will display enough medium scale coherence to accept approximation as a "Ship of Theseus" particle. By shifting scale, many textbooks opportunistically exploit the vagaries of the term "body" in Newton's laws to make an inherently untidy derivational situation look cleaner than it properly can claim to be.

As a scientific notion, the concept of *force* is really quite unproblematic *as long as* one consistently adheres to a mass point viewpoint. But if one utilizes the fact that "**F** =m**a**" often "works" directly for viscous fluids as well, one has begun to move away from a "sum of forces" interpretation of *force* towards a quite different understanding in terms of "transport of momentum". There is likewise nothing wrong with the concept of *transport of momentum* either, but it cannot bear the same foundational weight as mass-point "**F** = m**a**".

Another manner in which the scale tolerance of "**F** = m**a**" tends to pull the notion of *force* away from an unproblematic mass point understanding attaches to the cavalier manner in which so-called "constraints" are typically introduced into mechanics textbooks. Consider a bead sliding along a frictionless wire - the wire acts to *constrain* the bead to follow a certain path. In beginner physics, one is supposed to calculate the force exerted against the wire by the bead. To this day, I bear a grudge against such problems, for the following reason. I did not know much serious science when I went off to college (although I had read many popular accounts), but I was well versed, courtesy of an older brother, in the philosophical literature of the period such as Israel Scheffler's *Anatomy of Inquiry*. I sensed that those toy examples - "all ravens are black" *et al.* - were not the stuff of which substantive science is made, so I signed up for freshman physics in hopes of meeting with a genuine, real life "theory T". Newton's laws - except for that hazy "action = reaction" stuff - certainly gave the appearance of a canonical "theory T" and everything went fine until we were confronted, several weeks in, with a bead and wire problem. Of course, one is supposed to find the so-called "force of constraint" exerted against the bead by using "**F** = m**a**" to calculate the acceleration required if the bead is to continue moving at a constant velocity along the wire. But I worried, "Isn't this pattern backwards? Where's the force law between particles that is supposed to generate this acceleration? Don't we need some kind of force law on the pattern of gravitation here?" And if I had been a little more sophisticated, I might have added: "The kind of 'force' obtained by this kind of reasoning will have to be *velocity dependent*, because the wire can exert the right amount of force on the bead only if it can 'see' the velocity of the passing bead. But the derivation of conservation of energy that we went through last week was predicated upon the third law assumption that forces are *not* velocity sensitive. Why have the rules of the game been changed in mid-stream? Israel Scheffler never did anything like this to me!" I became quite discouraged at my stupidity in the face of my first "theory T" and didn't look seriously at physics for a long time. I did not realize that such freshman puzzlement is common amongst students conditioned to expect greater rigor from physics.

In truth, I had simply run afoul of another example of textbook opportunism. From a mass-point viewpoint, "forces of constraint" can exist only approximately and usually only with the aid of much supplemental machinery. After all, a bead will most easily slide in the manner prescribed only if some viscous fluid *lubricates* the bead's passage, a point completely neglected in the standard textbook fluff about "internal" and "external" forces.

In truth, an "honest" passage from a mass point view to almost any of the formulas I called "landmark equations" would probably require foundational investigations comparable to those needed to shore up statistical mechanics, whose linkages to fundamental principle, even today, remain incomplete. Of course, pedagogical disaster would follow in the immediate wake of such "honesty". Unfortunately, slippery textbook uses of "Newton's laws" have the unfortunate effect of making poor statistical mechanics *look worse* in comparison to other aspects of physics than it rightfully is. That is, statistical physics receives a lot of dissatisfied grumping of the ilk: "Boy, all of these half-proved appeals to ergodic theory *et al.* don't found the subject in the rigorous manner we have come to expect in textbook classical mechanics". But, as we have just seen, the crisp "rigor" of those textbook derivations itself, often arises as an artifact of *punning* upon Newton's laws. A quite misleading expectation of what is possible in the name of "rigor" becomes generated thereby. It seems to me that *untidiness* of derivation should be expected almost *anytime* macroscopic knowledge-whether it be "Ship of Theseus" continuity, satisfaction of constraints, equilibrium with respect to observables, etc. - is harnessed to simplify the physics of some complicated underlying swarm of point masses.

In stating this, I am *not* endorsing any Nancy Cartwrightish "the laws of physics lie" position. On the contrary, I see nothing wrong with the operative laws here, except that they lead quickly to some very complicated mathematics. Such complications usually drive one to *educated guesses* as to how the mathematical consequences of such laws will sort themselves out when strong interactions or a largish quantity of particles is under consideration.

I have spent some time on these issues because one needs to recognize that the mass point interpretation of Newton's laws that is now standard orthodoxy in elementary textbooks does not organize the "landmark equations" of classical physics as effectively as one might first presume. Might it be possible to discard the mass point interpretation in favor of some variant reading of "body" that would better respect the scale indifference of "**F** = m**a**"? Our Victorian physicists were quite aware that naive "bottom up" modeling based upon point masses could not reach the required landmark equations directly, rigorously or even accurately (e.g., the wrong number of constants for an elastic solid are obtained). Rather than accepting the "action-at-a-distance" picture of *force* native to the point mass picture, perhaps "**F** = m**a**" would be better interpreted in a manner that directly exploits the characteristics of the *contact forces* that appear between *extended bodies*? Returning to our bead and wire example, one might allow its erstwhile "approximate" "forces of constraint" to stand as the proper primitive basis for the notion of *force* within physics. It was a proposal of this sort by Hertz that become transmogrified, via folkloristic transmission, into Jammer's misleading "Heinrich Hertz believed that forces were inherently unobservable and appeared as unnecessary intervening elements in physics".

But it will be more interesting for our purposes to look at the many Victorians who adopted essentially a "continuum mechanics" outlook[6], whose basic methodology is still actively pursued in modern macroscopic approaches to the science of materials. Here we interpret "**F** = m**a**" as applying to *true* continua, *not* to atomized facsimiles thereof. Oddly enough, once the course of "true continua" is selected, one can easily find oneself driven to the philosophical thesis that *successful physics must engage in inherent idealization*. In using the term "*inherent idealization*", I mean the thesis that all modelings must commence in some deliberate distortion of reality before the enterprise of physics can even begin. Let us look at an example to clarify what I have in mind. Microscopic analysis shows that steel can evince a huge spectrum of fine structures, whose formation hinges delicately upon the manufacturing process. Nonetheless, for engineering purposes, steel can be quite successfully modeled as a homogeneous continuum where all of this underlying fine structure has been ignored. Or more accurately, the continuum treatment works with certain bulk parameters that must reflect, in a complicated and averaged way, the nature of the underlying fine structure composition. Accordingly, for the sake of convenient macroscopic description, we often *smooth out* features that we *know* actually correspond to variegated inhomogeneities within the metal.

But this is merely an example of *convenient idealization*, not what I intend by *inherent idealization*. We have merely seen that a smoothed out continuum description of steel is a convenient methodology to employ in physics, not that no unfalsified description of steel can ever be framed, however mathematically intractable that honest description might prove. But the doctrine of *inherent idealization*, as I understand it, claims that an unfalsified scientific description is not a possible option. And this was a philosophical doctrine that many Victorian physicists working within the continuum approach wanted to endorse. But why? Although steel clearly contains a variegated microstructure, there is no obvious reason why the aether should not turn out to be a true continuum whose description requires no averaging of smaller scale variations (and the aether was the continuous material of greatest interest to the Victorians). Why claim that even a description of the aether can't escape inherent idealization?

The reason is essentially mathematical. Given certain background mathematical assumptions, it can easily appear as if *no* continua can be coherently described without introducing some inherently idealized structures. The reasons trace to some tricky problems that plague the "physical infinitesimals" natural to continuum physics.

Suppose we have a string that is pulled at its boundaries by a tension T (see Figure). Assume that the string can transmit that tension equally through its parts. For generality's sake,assume further that a *variable* gravitational force pulls upon the string (perhaps the string density or the field is inhomogeneous).

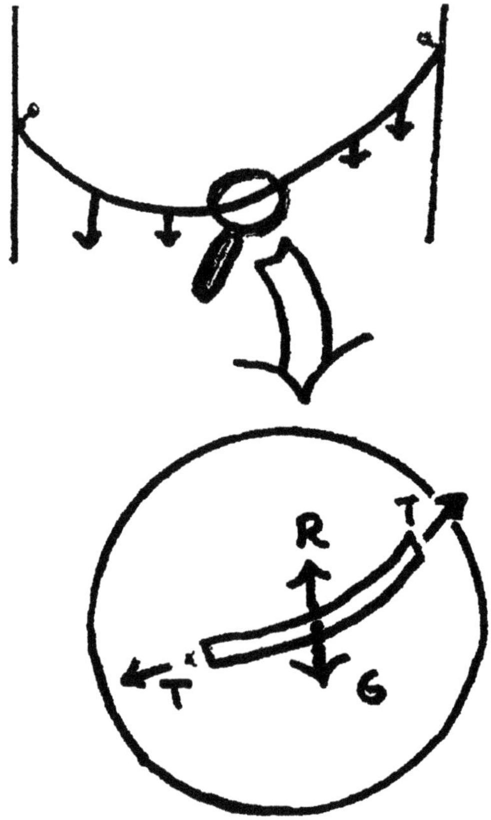

Figure 1

How will the string hang? Let's call this question the "global string problem". Following the familiar motto that "physics is simpler in the small", let's try to understand first what happens at each individual point of the string. Accordingly, let us now interpret the term "body" in Newton's laws as applying to an arbitrary point contained within the continuous string. This supplies the "local string problem" of setting up a differential equation based upon "**F** = m**a**" as a frame. The usual story runs something like this. If the string is at rest, the gravitational pull on each "point" of the string can only be balanced by the tension in the string T. But how can T do that? Well, look at the two ends of the "body" and note that T is pulling in slightly different directions. The composition of force addendum to the second law declares that the two end forces will act upon our "body" by their resultant R. To maintain the string in equilibrium, R must therefore be exactly equal and opposite to G. By filling in a few details, we can extract the standard differential equation for a string at rest.

Note the two kinds of forces: the gravitational "body" force that pulls upon the *inside* of our "body" and the two tension based forces that act upon its *boundary*. The latter are called "contact forces" because they arise from our "body's" contact with its neighbors. In a higher dimensional object such as a beam, contact forces should exist around the complete boundary of a "body".

But wait a minute-wasn't our string "body" supposed to be a *point*? How on earth does a point wind up possessing a boundary with different forces aligned along it? Frankly, if they don't think a mathematician is listening, you can still hear engineers happily chatting about "integrating around the infinitesimal boundary of the element". Clearly the "body" we are dealing with can't be a proper point, but some odd kind of infinitesimal, carrying the ghost of a departed boundary. Now it is not a trivial exercise in delta/epsilonics to straighten this out. Note that setting up our equation involved two stages: (1) a collection of *varying* contact forces was summed around the boundary of our infinitesimal; (2) their resultant was carried *inside* the infinitesimal to combine with the *unique* "body" force operating there. Note the funny business here: our infinitesimal is credited with a boundary long enough to carry a varying set of contact forces, but the calculations act as if there is only point *inside* the infinitesimal where body forces can act.

Let us try to understand our infinitesimal "body" in a more rational way. Suppose we no longer assume a solitary "inside" point in our "body"-let us stretch it out, as it were. *Where* inside the element is the resultant vector from our boundary supposed to be carried now? Within a stretched infinitesimal, shouldn't we expect the applied gravitational force to *vary* across our element and beg for some kind of principle of inter-element combination as well? In fact, a funny thing has happened on our way to the string equation: *our original "global string problem" has completely reproduced itself as the "local string problem"*. Once again we are asking "what happens inside a stretch of string when you pull at its two ends?" How can we expect to make any progress in physics if everything replicates itself as we move towards the small? And there are other puzzles that could be brought up here as well.

Of course, I've dressed matters up in terminology that rhetorically amplifies the oddness of common textbook maneuvers, but I have not been false to them (partial differential equations — the mathematics used for continua — were not well developed in Berkeley's time, but they would have provided the Bishop with many moments of hilarity if they had). Nowadays, we understand that (most of) this mess can be straightened out if we are more sophisticated in our mathematics and do not assume that Newton's laws can completely ground our procedures. Most of the relevant studies are more familiar to engineers and applied mathematicians than to physicists *per se*. Although many of the key elements required in such an approach were known in the nineteenth century, they were never pieced together in quite the right way.

So what did nineteenth century scientists do, given that they didn't see how a straightforward mathematical framework for continua might be assembled? Answer: they filled in with *anti-realist philosophy*. Consider this: suppose our infinitesimal stretch of string is really a tiny *rigid* body. Our problem of equilibrating the contact forces applied along the boundary with the body forces acting inside the element immediately becomes less severe, because a rigid body doesn't care, in terms of translation, where forces are applied against it - all forces are *translationally equipollent* within this setting. The distribution of forces across a rigid body affects only how the body will *turn*, a feature that can be treated by attending to the so-called moment of the forces.

Hence the idea suggests itself: maybe at some smallish level of analysis the behavior of continuous stuff can be *approximated* as that of either a mass point or a rigid body. Let us first set up the physics of our string by pretending that is composed of an assemblage of rigid rods, and smooth away this artificial fine structure afterwards by exchanging differential operators for finite differences. Claims that continua should be approached in this manner are sometimes called "rigidification" or "freezing" principles. Here A.E.H. Love, one of the great elasticians of the past century, expresses a mass point formulation of "freezing":

> The necessity for a simplification arises from the fact that, in general, all parts of a body have not the same motion, and the simplification we make is to consider the motion of so small a portion of a body that the differences between the motions of its parts are unimportant. How small the portion must be in order that this may be the case we cannot say beforehand, but we avoid the difficulty thus arising by regarding it as a geometrical point[7].

Love subsequently cites Karl Pearson's anti-realist philosophy tract as warrant for the inherent idealization that is central within this odd procedure.

So, as much as anything, it was the apparent refusal of continuum physics to simplify in the small without the artificial intervention of idealized entities that originally prompted many nineteenth century physicists to the doctrine of inherent idealization. Without "rigidification" idealizations, it seemed that no toehold could be found, as physics descended, like the Incredible Shrinking Man, into the labyrinth of the continuum. The acceptance of inherent idealization had the convenient advantage of providing a quick and dirty substitute for some missing mathematics that we now know how to provide. In retrospect, this ploy clearly represented a poor method of confronting the problem — greater mathematical patience was wanted, not antirealist philosophizing.

In any event, the subsequent history of physics has come down pretty firmly in favor of an elementary particle "bottom" to physics and the Victorians worries about how to halt a descent into the infinitary have accordingly been forgotten. We are left with a strange stew: *classical* physics, because of the dominating requirements of a transition to quantum mechanics, is almost invariably now taught utilizing *mass point* "foundations". Nonetheless, the *philosophical commentary* that often accompanies textbook presentations actually derives from the tricky mathematical problematic that *continua* faced in the nineteenth century. Thus a conceptually unproblematic physics becomes garnished with quotations from venerable authorities that apparently claim that some *mysterious* "idealized" ingredients lie within that physics, mysterious enough to warrant a radically antirealist reevaluation of physical method. But this is one soup in which we philosophers hardly need to swim. And likewise there is no warrant for the claim that "unobservationality" of *force* constituted a well documented problem within classical physics either.

There are actually quite a few ways in which this flexible body prehistory has set modern philosophy of science upon aberrant courses. Within the special context of inherently idealized continuum physics, a very nice version of the "underdetermination of theory" problem can be set up, as Pierre Duhem in fact proceeds to do. The "underdetermination" thesis remains widely accepted in philosophy of science today, despite the fact that it has been extracted from its original continuum mechanics context — the only context in which I have ever seen anything like a *proof* that the thesis could possibly be true. But these are matters of some complexity, so I will stop here[8].

Ohio State University
U.S.A.

NOTES

1 See van Fraassen 1980.
2 See Pearson 1892.
3 See Jammer 1962.
4 See Truesdell 1984.
5 See van Fraassen, *op cit.* 1980.
6 Actually, for the reasons I shall sketch below, the majority of late nineteenth century scientists who *seemed* to accept mass points or rigid bodies as fundamental in physics actually *believed* in the continuum point of view, but felt that its direct development was hindered by a need for "inherent idealization". Insofar as I can determine, Hertz shared this viewpoint as well.
7 See Love 1906.
8 Thanks to Penelope Maddy and Bob Batterman for discussions preparatory to this essay.

REFERENCES

M. Jammer, *The Concept of Force*, Harper & Brothers, New York 1962, pp. 241-2.
A.E.H. Love, *Theoretical Mechanics*, Cambridge University Press, Cambridge, England 1906, p.2.
K. Pearson, *The Grammar of Science*, Walter Scott, London 1892.
C. Truesdell, "Suppesian Stews", in *An Idiot's Fugitive Essays on Science,* Springer-Verlag, New York 1984.
B. van Fraassen, *The Scientific Image*, Clarendon Press, Oxford 1980.

NANCY CARTWRIGHT

QUANTUM MECHANICS WITHOUT THE OBSERVABLES

I want to talk in this paper about two widespread movements in the philosophy of science that seem to have passed us by in the philosophy of quantum mechanics - and I think much to our detriment. The first is the by-now very old attack on the theory-observation distinction. The second is the attack on the theory-dominated view of science. For my purpose, though, I want to look not broadly at the attack on theory-dominated views of science - after all, even in the hey-day of the theory-dominated view no-one denied that experiment and technology are a legitimate part of scientific activity - but rather more narrowly on the theory-dominated view of scientific knowledge. Even in a highly articulated, mathematical science like physics, knowledge is not encoded entirely - I would say not even primarily - in the abstract mathematical structure of the theory. Harry Collins[1] has taught us about the importance of implicit knowledge; Ian Hacking[2] about the importance of the knowledge encoded in our techniques for intervening in the world as opposed to the knowledge expressed in our representations of it; Peter Galison[3] has reminded us that experiments are not principally to test theories but to find out about the world, and in particular in his work on trading zones, Galison has described the negotiation between the experimentalists' and the theoreticians' knowledge that allows science to proceed even though no common understanding is articulated; Norton Wise[4] has shown how the steam engine can be seen as an actual physical embodiment of Kelvin's simultaneous understanding of both physics and political economy; and Andrew Warwick's[5] work on training Cambridge wranglers reminds us that even at the level of high theory mathematical physics is not *in* the mathematical theory, but must be molded *out* of it. I do not want to say more about the attack on the theory-dominated view right now, though I will return to it soon.

What I want to do here is to launch an attack on the Born Interpretation of quantum mechanics; and that brings to the fore the older issue of the theory-observation distinction. The Born interpretation is supposed to be just that - an interpretation. In quantum mechanics the basic descriptive device is the quantum state function, represented as a vector in the Hilbert space, or more concretely by the "ψ function".

Now we are supposed not to know what the ψ function means. Its relation to other symbols in the theory, supposedly laid out in some appropriate axiomatization, only defines it implicitly. In order to complete the interpretation, we generalize the procedures first described by Max Born. We purport to assign to each linear Hermitian operator on the Hilbert space, say $\hat{A}$, what we call an "observable" on the system, A. The generalized Born Interpretation then tells us that a certain inner product on the space

$$<\psi|\hat{A}|\psi>$$

E. Agazzi and M. Pauri (eds.), The Reality of the Unobservable, 241-249.

represents the average value

$$\mathrm{Av}_{\psi}(A)$$

that A will take in an ensemble of systems identically prepared to be in the state ψ. This is in the special case of quantum mechanics just the same regimentation of scientific knowledge that the attack on the theory-observation distinction was meant to undermine. And I think it is subject to just the same kinds of criticisms. What are these quantum observables? What are they supposed to be doing for us?

In general observables were supposed to perform two roles vis-a-vis theory.

1) They were credited with some kind of *epistemological priority*. They were supposed to be the kinds of things that we could get hold of with certainty, or at least with far greater certainty than the so-called theoretical facts which lie on the other side of the theory-observation divide. In the quantum case that is clearly no more plausible than it proved to be in general. The so-called quantum "observables" are things like the spin of an electron, and by now our methods for finding out information like that are no more nor less reliable in general than for assigning quantum states.

2) *Meaning* Axioms only define mathematical symbols implicitly. We need more to pick out their referents entirely. This claim comes from an old tradition that supposes the autonomy of mathematics in the physical world. It is the tradition of Poincaré for instance: Mathematics provides a number of alternative geometries independently of physics. To use geometry to study the natural world we shall have to find some physical interpretation of geometric symbols. The point there, though, is that the interpretation should be taking us into the things the symbols stand for, not into a set of operations that test whether the symbol applies. In the case of quantum mechanics we do that by pointing out that the ψ function stands for a quantum state, and then going on, if you wish, to tell more than you would ever want to know about the quantum state in general; but more importantly about the quantum state of particular kinds of systems in particular circumstances. The positivists confused matters. They had the wild hope that these two different jobs could be done at once. The very interpretation of what the symbols stand for would provide the procedures for securing epistemological certainty that the symbol applies. The Born interpretation is the quantum mechanical legacy of this mad positivist hope.

But if we were to give up the Born interpretation, how would we interpret quantum mechanics? The answer is that a good theory needs no interpretation. We've had 50 years of working with the quantum state function and we know very well what it means - or as well as we know anything else in physics. This is not to say that questions of interpretation never come up. After all it was perfectly reasonable for Maxwell to worry about the meaning of the displacement current or about whether the potential function had a physical interpretation or was mere mathematical baggage. But these are specific questions that arose for very specific reasons.

We are very often misled here by what I call the Hilbert-ization of theory. We think that what a theory is, is a handful of axioms, possibly with some very general rules of interpretation. That is why I began by reminding you of the attack on the theory-dominated view of scientific knowledge. Really of course what is relevant is what we might call the attack on the theory-dominated view of theory, or as I have called it, the Hilbert-ization of theory. There is a sense in which there is no quantum theory. What gets us into trouble is the thought that there is a quantum theory; for if there is then what it must be is whatever is expressed in some standard axiomation. We have a quantum theory of the laser, and of molecular bonding, and of superconductivity (indeed several of these !), etc., etc., etc. And in each of these it is clear that the quantum state is functioning as a well understood working concept and has no more - nor less - need for interpretation than any of the other concepts we work with in physics.

I have said that a good theory needs no interpretation. The Born interpretation is peculiarly bad in the kind of interpretation it provides. Recall the old demand for operational definition. For each concept there is one operation and the success of that operation is necessary and sufficient for the concept to obtain. Now we know that the demand for operational definition has been widely abandoned across the sciences because it denies to us the rich variety of different ways which we devise to find out about the same scientific concept. Operationalism belies experimental practice.

The Born interpretation just is an operational definition, although with the unfortunate features that the "measurements" we are supposed to perform to determine the quantum state do not seem to be real operations involving real interactions with real measuring devices, and it suffers all the same problems as do operational definitions in general. It is unnecessary because it has little to do with how we usually find out about the quantum state. Let me give just one example from the treatment of Josephson junctions.

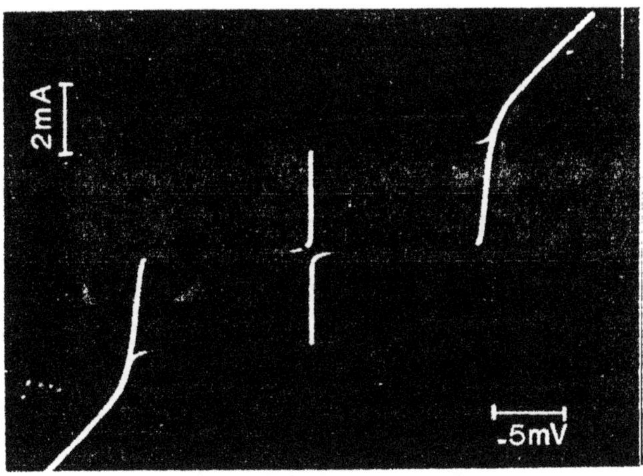

Figure 1

Typical voltage-current characteristic for a Sn-Sn_xO_y-Sn Josephson junction at T=1.52 K. Horizontal scale: 0.5 mV/div; vertical scale: 2 mA/div.

There is one central way that we learn about the quantum state function that describes the electrons flowing through a Josephson junction. Whether we want to test the original claims about the systematic phase relations between the two sides of the junction, or about macroscopic quantum tunneling, or about the limitations of the standard RSJ (resistively shunted junction) model, about Gaussian noise, shot noise, etc, we use the same techniques. We produce *I–V* - current voltage - curves like Figure 1 [6]. Figure 2 [7] is another, and the accompanying diagram makes a second point worth noting. One trick in designing an experiment is to figure out how different features of the quantum state will reveal themselves in modifications of the *I–V* curve. Another is to figure out how to ensure that what you are looking at is the *I-V* curve of the function in the first place and not that of the rest of the circuitry in which it is embedded. After all, remember that for the most part superconductivity is a very low temperature phenomenon, so most of these junctions we are studying are stuck down at the bottom of a thermos which is usually about 6ft tall.

The principle point I want to make about this example though is that we study the quantum state in Josephson junctions by looking at curves relating classical voltages and classical currents and not at all, as the Born interpretation suggests, by looking at probability distributions of allowed values for so-called quantum observables. We are looking in this case neither at a probability distribution nor at quantities represented by linear Hermitian operators.

This is a case in which we use measurable quantities from other theories to probe the quantum state. We can see the same point by looking at the converse process, in which we use the quantum state to calculate a quantity from classical circuitry theory. Let us consider an example that I have been looking at lately in studying tunneling in Josephson functions, the so-called "normal resistance". This is the resistance of the junction when both metals are in the normal - as opposed to the superconducting state - which they will be when the junction is not cooled to just a few degrees Kelvin:

$$R_N = \hbar^3 / 4\pi \ 1/(e^2 N_r(0) N_l(0) <|T^2|>).$$

Here $N_r(0)$ and $N_l(0)$ are the densities of states at the fermi level at the right and left metal respectively and $<|T^2|>$ is the mean transition probability for an electron to move from left to the right.

Where does this identification come from? I follow here the treatment of Antonio Barone and Gianfranco Paterno. They first derive from fundamental microscopic theory an expression for the quasi-particle tunnelling current in a Josephson junction. (Quasi-particles are like single electrons as opposed to the Cooper pairs, which are responsible for superconductivity.)

$$I_{qp} = -\hbar / eR_N \int_{-\infty}^{+\infty} d\omega \ n_l(\omega) \ n_r(\omega - eV_o / \hbar) \left[f(\omega) - f(\omega - eV_o / \hbar)\right]$$

Here $n_l(\omega)$ and $n_r(\omega)$ are the quasiparticle densities as a function of frequency on the left and on the right $f(\omega) = (e^{-\omega/k_B T+1})^{-1}$, for k_B the Boltzmann constant, and V_o is the voltage across the junction. We now apply the formula for I_{qp} to the case when both junction electrodes are in the normal state. Then $n(\omega) = 1$, and since

$$\int_{-\infty}^{+\infty} d\omega \left[f(\omega) - f(\omega - eV_o / \hbar)\right] = -eV_o / \hbar$$

we get

$$I_{NN} = V_o / R_N,$$

which shows that the quantity R_N defined above is indeed the resistance of the junction when both metals are in the normal state.

The point here again is that nothing like the Born interpretation is to be seen. I like the example especially because R_N is a straightforwardly measurable quantity unlike most of the things that get labelled as "observable" in quantum mechanics. And we do not treat it by finding a linear Hermitian operator to represent it and then taking inner products of that operator.

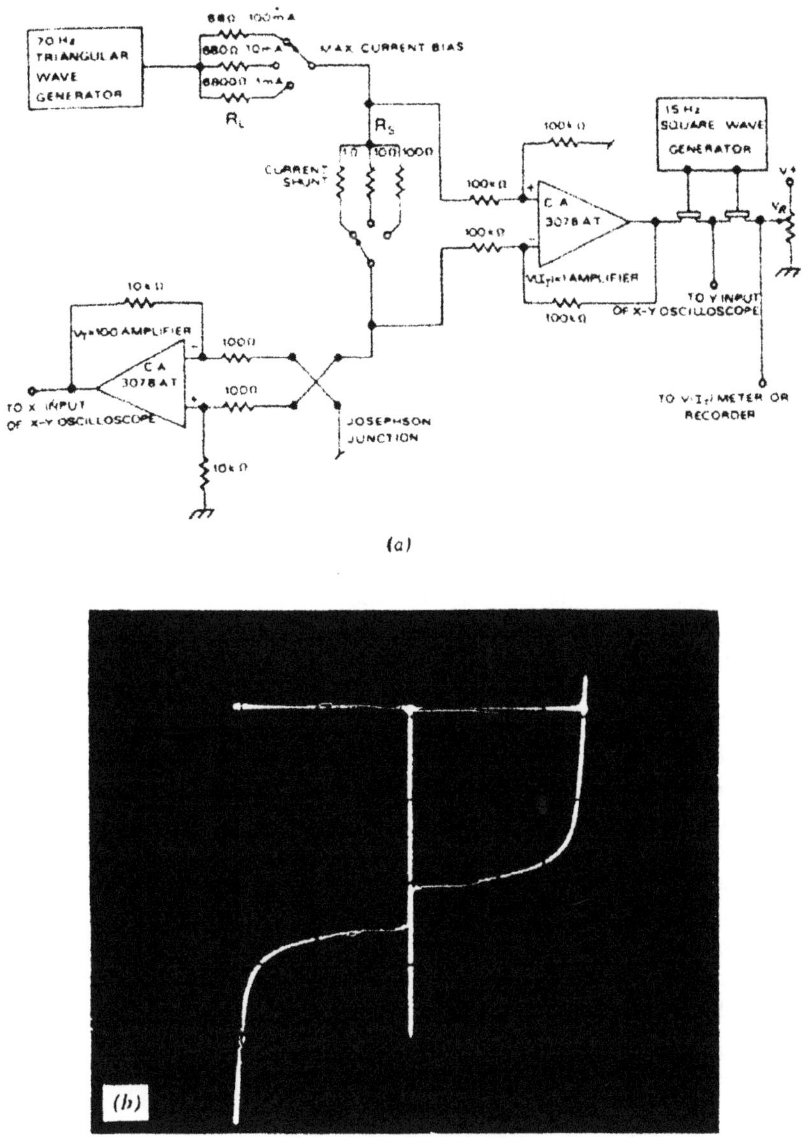

voltage

Figure 2

a) Schematic of the circuit used to measure junction *V-I* characteristic and Josephson current variation. (After Balsamo *et al.* 1974.)

b) Oscilloscope display. The horizontal line corrisponds to the reference voltage V_R.

We do have here of course in the expression for R_N some remnants of an operator: $<|T^2|>$ is a quantity that comes by ignoring the differences in a lot of matrix elements T_{kq} and pulling them outside the integral. These matrix elements are from the transition matrix that appears in the interaction Hamiltonian between the metals on both sides of the junction, along with electron creation and annihilation operators. And it's no surprise that we should find something like this because the mathematics of quantum mechanics is after all nothing but vectors and operators, so if we want to get numbers we are going to have to do something like look at the matrix elements or take their trace - which last we read, I think mistakenly usually, as calculating an expectation. (There is also the question of n_l and n_r which is yet another story.)

There is one last point to make with this example. The identification of an expression we can calculate in quantum mechanics with the quantity R_N which we can measure using Ohm's law with totally conventional volt meters and ammèters is not a once and for all identification as the usual account of the Born interpretation suggests. There you assume you have in quantum theory a lot of fixed rules of correspondence that assign fixed operators to represent observable quantities. In this case we have quite the opposite. The expression for R_N depends very much on the specifics of the situation studied: here one simple quantum junction driven by a constant dc circuit with no spatial phase variation. In other situations we will need a very different quantum expression for the normal resistance. Finding it is just the kind of job that theoreticians working in superconductivity spend their time doing. You can see then why I call the Born interpretation "quantum mechanics for the lazy theoretician".

It also serves *the lazy philosopher*. The relations of quantum to other theories are complex, rich, highly varied. The interrelations between various theories and how they must be deployed together to describe any real physical system are highly context dependent. Trying to put some order into the matter and understand how these kinds of interrelations work is a tedious job which we would rather not have to tackle - and for the most part we don't.

Let me give one further example of the interplay of quantum and classical theory in jointly producing a single account. In this case we see not only local identifications of quantum and classical quantities; we also picture a causal interaction between them. The example is from Willis Lamb's semi-classical theory of the laser. Lamb's theory assumes a classical electromagnetic field in interaction with the laser medium. The field introduces a 'dipole moment' ($\langle er_i \rangle$) in the atoms of the medium; the "dipole expectation" is identified with the macroscopic polarization of the medium; this in turn acts as a source in Maxwell's equations. Sargent, Scully and Lamb provide the following diagram[8]:

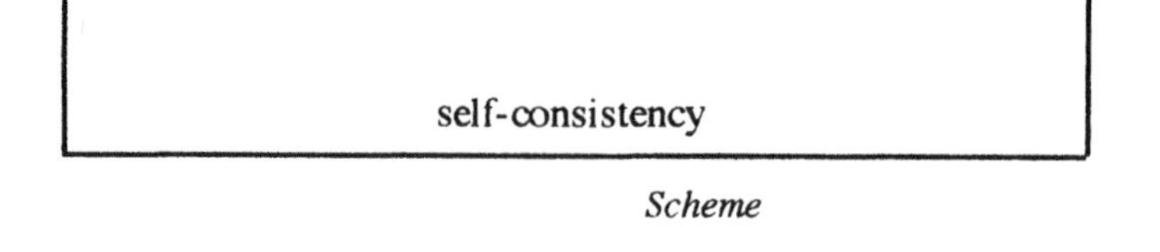

Scheme

In the semi-classical theory the first step, linking the field and the dipole expectation, appears as a causal interaction between classical and quantum characateristics. In the step marked "statistical summation" we see an identification of a quantum quantity, the suggestively named "dipole expectation", with a purely classical polarization. I put the term "dipole expectation" in scare quotes to highlight the fact that we need not think of it, ala the Born interpretation, as a genuine expectation. This identification with the macroscopic polarization is guided by a powerful heuristic: the quantum electric dipole oscillator. We are told by Sargent, Scully, and Lamb, as we are in many texts on laser physics, that quantum electrons in atoms "behave like charges subject to the forced, damped oscillation of an applied electromagnetic field." This charge distribution oscillates like a charge on a spring, as in the following pictures[9]:

a) P = 0 *b)* P ≠ 0

Figure 3

The discussions of the oscillator waver between the suggestive representation of $(\boldsymbol{er}_i)$ as a genuinely oscillating charge density and to my mind the mistaken interpretation in terms of time-evolving probability amplitudes. From the standard point of view this should be troubling; without a generalized Born interpretation looming in the background I see no reason to be troubled. The oscillator analogy provides a heurisic for identifying a quantum and a classical quantity in the laser model. The identification is supported by the success of the model in treating a variety of multimode laser phenomena - the time and tuning dependency of the intensity, frequency pulling, spiking, mode locking, saturation, and so forth.

Conclusion: In closing I should like to turn directly to the topic we are supposed to be addressing at this conference: the relationship between observation and realism. The Born interpretation is a kind of take-over devise for quantum mechanics. It takes all the other very different kinds of quantities that we study elsewhere in physics and says what they really must be: They must be these funny "observables" that we need in order to interpret quantum mechanics. Now that seems to make then epistemologically primary. But ontologically it does them in! It is quantum mechanics that is primary. Everything that nature fixes about these quantities is determined by the quantum state. They have no motive force of their own. All they can do is what the quantum state tells them to. Interestingly though, quantum mechanics is not the most thorough kind of take-over theory, for the classical properties that get relegated to the role of observables are in no way reduced to facts about the quantum state nor do they supervene on them. These properties must in a fairly robust sense be real and independent of the quantum state, even though everything about them that is nomologically determined is determined by the quantum state. It is to mark this peculiar independence/dependence relationship that we have in quantum mechanics a kind of distinction between states and observables that is not necessary elsewhere in physics.

Our successes in modelling the world suggests that nothing like this picture is correct, and that is so even in cases where quantum mechanics plays an absolutely central role - cases like Josephson junctions and lasers - which is why I've chosen them as examples. There we see very complicated kinds of relationships between all sorts of different kinds of properties governed by laws from appropriate theories. The current through a junction has a quantum state; but it has a lot of other characteristics as well, not least of which is that it is hooked into a classical circuit. Indeed it is only because it is part of a classical circuit that we can begin to make any local identification at all between the quantum state and other properties the current has - like the identification of the normal resistance that I showed earlier. We should beware of concentrating too much on that example, though, for the possibility of these kinds of local reduction may suggest that all these other properties really do after all depend entirely on the quantum state. And that is just what the evidence of modelling practice belies: First, because the identifications are local and different from case to case; second, because there are not enough; and third, because they are only one type among many relationships between the quantum state and other properties. Causal relations between the two of the kind we saw in the laser example are an especial problem for the take-over view.

But if this picture is correct, what does it say about quantum realism? Nothing. Whether quantum mechanics can be judged true or not depends on the evidence that things have quantum states and that the quantum states behave as they are supposed to in the relevant quantum theory. That question is independent of whether systems have other properties as well and even independent of whether these other properties influence the quantum state. The idea that quantum theory is a closed whole and that no concepts from outside can play any role if quantum states and quantum claims are taken realistically is in large part, I think, a consequence of the Hilbert-ization of theory that I criticized at the start.

It is in light of these remarks about realism that the real motivation for my attack on the Born interpretation can become clear. Of course the Born interpretation is bad for quantum mechanics because, as is notorious, it is hard to make sense of. What are the probabilities it talks about probabilities of? We know we get into trouble if we say they are the probability for systems to have the relevant value of the so-called "observable" in question. So we resort to their being probabilities for these observables to take the relevant values *on measurement*. And now we are saddled with the measurement problem.

But that is not the real source of my antagonism to the Born interpretation. My antagonism is grounded in the succor that the Born interpretation provides for the quantum take-over programme. If we really go along with the spirit of the Born interpretation, all the rich variety of properties studied by all the rest of physics - not to mention the rest of the natural and social sciences - get relegated to the role of observables, whose behaviour is dictated (up to a probability) by the quantum state. That is not the way I see science working - when it actually is working to treat real systems and not just sitting in the *Journal of Mathematical Physics*. The Born interpretation is not true to scientific practice. But this does not mean that we must in any way be non-realists about quantum mechanics. I can agree that quantum terms genuinely refer and that the bulk of quantum theoretical claims are true (or as true as most other claims we make). But that is not to accept that quantum terms can pick out all there is to talk about nor that quantum claims are all the truths there are. To assert that is not quantum realism but quantum imperialism. So the moral of this talk in slogan can be put: "Be a realist if you like, but resist quantum take-overs".

The London School of Economics and Political Sciences, London
United Kingdom

NOTES

1 See Collins 1995.
2 See Hacking 1983.
3 See Galison 1987.
4 See Wise and Smith 1983, 1989.
5 See Warwick, forthcoming.
6 See Barone and Paterno, 1982, p.13.
7 See Barone and Paterno, 1982, p.67.
8 See Sargent, Scully and Lamb 1974.
9 See Sargent, Scully and Lamb 1974, p.32.

REFERENCES

A. Barone and G. Paterno, *Physics and Applications of the Josephson Effect*, Wiley, New York 1982.

H. Collins, "Changing Order", in *Replication and Induction in Scientific Practice*. Sage, London 1985.

P. Galison, *How Experiments End*, University of Chicago Press, Chicago 1987.

I. Hacking, *Representing and Intervening - Introductory Topics in the Philosophy of Natural Science*, Cambridge University Press, Cambridge 1983.

M. Sargent III, M. O. Scully, and W. Lamb Jr, *Laser Physics*, Addison-Wesley, Reading MA, 1974.

A. Warwick, *Forthcoming book on Cambridge Wranglers*.

N. Wise and C. Smith, *Energy and Empire*, Cambridge University Press, Cambridge 1983; *Work and Waste, Political Economy and Natural Philosophy in Nineteenth Century Britain (I), (II), (III)*, in *Hist. Sci.* xxvii, pp. 263-301 and pp. 391-449 and *Hist. Sci.* xxviii, pp. 221-261, 1989.

BERNARD D'ESPAGNAT

OBSERVATION, CONTEXTUALITY, AND REALISM[1]

In the title of this paper the terms "observation" and "realism" refer to notions of which both the nature and mutual relationships are well known to philosophers. The middle term, "contextuality", is a new acquaintance and may sound enigmatic or out of place to philosophers.

The last term will be commented upon in a moment. For the time being let us consider the familiar terms of "realism" and "observation". Concerning these notions there is, as we all know, a choice between two possible standpoints. Let us call them, for short, that of the Aristotelians on the one hand, and that of the *mechanists* on the other. Of the two, the Aristotelians remain more faithful to sense-data. They take notice of the observational fact that a moving body not subjected to any force finally comes to rest, and they incorporate this fact in their system. They note that living beings have individualities that are qualitatively different from one another, and this also they incorporate as a basic element in their system.

Thence, in the end, the Aristotelians have a great wealth of concepts at their disposal without there being a rigidly organized hierarchy among them. Thence also, for them, a great openness to possible qualitative descriptions but a weakness in the area of the quantitative. For the mechanists, on the contrary, what matters is the hierarchical organization of concepts. In their eyes, there are the basic concepts and all the others, and these others must finally be explained in terms of the first, so that in the end the description of the physical world is formulated entirely in terms of just a few basic notions. This means that in many cases the link with observation becomes extremely indirect. As a reward, however, this attitude favors the quantitative. We do know, of course, that in classical physics and in other sciences it is this last attitude that has taken the upper hand; although, the recent "discovery" of complexity and the investigations of some mathematicians (such as René Thom) concerning morphogenesis did recently bring back a certain lustre to the Aristotelian approach.

This contrast between the two schools often attracts the attention of philosophers, but the point I would like to emphasize here is one idea that the schools have *in common*. This is the idea that the basic concepts are more or less obvious or, more precisely, are "clear and distinct" right from the start. These are concepts which "common sense" (as Descartes said) assures us are valid and irrefutable. Both schools admit, whether explicitly or not, that the description of physical reality must be made in terms of these very concepts or in terms of concepts derived from them. I call this conception "near realism": "realism" because what is aimed at is a description of reality itself, and not merely our perception of reality, and "near" because the basic concepts, those according to which others must be defined, are all familiar concepts (essentially position, form and motion) for which evidence guarantees validity.

E. Agazzi and M. Pauri (eds.), The Reality of the Unobservable, 251-255.

Now it is *not* with the aid of familiar concepts (even largely idealized) that one can explain that space transforms itself (partially) into time (and vice versa) when one changes the referential and that pure motion is changed into things (I refer here to the phenomenon of particle creation in high energy collision). Otherwise said, historically speaking, the advent of relativity stands out, in this matter, as a sort of rupture. The familiar concepts, the commonsense notions dear to Descartes, even idealized, are no longer sufficient and are often seen as faulty. In other words, near realism is proved false. We must go beyond it. As we all know, this transgression is possible because of mathematics. It is mathematics, for example, that allowed us to define such notions *as three-dimensional curved space.*

Let me now call your attention to something that is a consequence of this transgression. It concerns a (very wise) limitation that nineteenth-century physicists imposed upon their domain of competence. This limitation consists of considering that science is concerned with the *behavior* of the components of physical reality, and not with their existential reality. What I would like to emphasize is that, precisely because of the just mentioned necessary transgression of familiar concepts, this limitation rule - for which physicists have preserved a nostalgia - has, compared with what it was before, dramatically changed in nature. The point is that as long as physics was anchored in familiar concepts alone, this rule could be applied without stepping out of conventional realism. One could have "realism without ontology". A clear example of this is provided by Fourier's theory of the *propagation of heat.* In Fourier's time there was competition among a few rival theories concerning the nature of heat: that of *caloric*, that of *molecular agitation,* and presumably others as well. The point of view adopted by Fourier consisted in not taking sides at this level, otherwise said, in not exploring the *nature* of heat, and in focusing quite exclusively on its behavior. This is what made it possible for him to write down the equation for the propagation of heat. This equation is true and will for ever remain so, precisely because it is independent of any speculation concerning the fundamental nature of the object studied and deals exclusively with its behavior.

In view of what follows let it be noted that, notwithstanding this restriction, the *equation for the propagation of heat* is nevertheless considered as dealing, not just with collectively defined observational predictions but with an entity, heat, conceived as a reality somehow existing independently of us.

What should be noted at this stage is that the latter circumstance is made possible due to the fact that everyone has an idea of what heat is, of what cold is, and in consequence the theory is spared having to define heat. Unfortunately, when speaking of *curved space*, quarks etc..., this is no longer possible. It is impossible to claim to speak exclusively of the behavior of these entities without first having defined them, that is to say, without having said what they are. This necessitates, therefore, in their case, either giving up the rule of *behavioral limitation*, that is to say, accepting, in the end, to come to a decision about existential realities, or greatly modifying this rule. This second issue exists but it demands that we step out of traditional realism, meaning that we give up the idea that science aims at the description of reality in itself, "the real such as it truly is".

As we all know, this last option does not at all imply dropping the notion of scientific objectivity. It does imply, however, a redefinition of the latter.

An elementary but philosophically quite important (although quite often disregarded) point is that when it is claimed - and rightly so, of course - that science is objective, all has not been said because, at least for a realist, the word "objective" has two different meanings between which it is important to distinguish.

In order to make this distinction explicit, it suffices to observe that objective statements in physics are not all formulated in the same fashion. Some have a form allowing them to be interpreted as informing us, in a direct way, about the attributes of the thing under study. Statements such as "two bodies with opposite electric charges attract each other in such and such a manner" and, more generally, all statements in classical physics (with the notable exception of those of thermodynamics) are evidently of this type. It is not indispensable to understand these statements in this manner, and there are, in some cases, good reasons, extrinsic to science (philosophical, for example) not to do so. These statements, however, do have a form which at least renders such an interpretation possible. I call them "strongly objective".

However, even though, to repeat, all scientific statements are objective, not all of them are strongly objective. In, for example, *statistical thermodynamics*, as usually formulated, explicit reference is made (through the notion of coarse graining) to what it is possible to observe, or to the types of systems that it is possible to prepare. In the same way, in standard textbook quantum mechanics, the statements of some basic laws refer in an essential way to what will actually be observed in such or such a circumstance. It is entirely appropriate to call such statements objective since, by definition, they are true for anyone. Their very form, however, makes it impossible to interpret them as stating how things really are. Such statements I call "weakly objective".

The acceptance of weak objectivity is truly a sort of substitute for the classical rule of *restriction to behavior*. This is true at least in the sense that weak objectivity too makes it possible for us to avoid taking ontological stands: if I tell you that in doing this you will observe that, it is clear that I am not dealing with ontology. Indeed, I do not see any other substitute for this rule than this. It is consequently not surprising that the founders of quantum mechanics adopted the position of weak objectivity.

The fact that they adopted it is entirely clear as is shown, for example, in the following citation of Bohr: "The description of atomic phenomena has a perfectly objective character in the sense that no explicit reference is made to any individual observer, and that therefore no ambiguity is involved in the communication of information".[2]

I think the difference is clear. When, for example, Fourier speaks of heat, he does so in terms of strong objectivity: and he can legitimately assume that he is speaking of a reality that is "in itself", while refraining from speculating on its nature. On the contrary, in the areas of physics where new concepts must be formed from mathematics, we can no longer think this way. Rejection of ontology imposes weak objectivity. This closes the first part of my paper, which can be summed up in the form of a dilemma. *Either* we keep the standard realistic way of conceiving things and raise it to the level of a full-fledged ontology *or* we accept the view that our science is only weakly objective.

My second part concerns the first of these two options. This option is not easily put into practice because some axioms of quantum mechanics are only weakly objective. Consequently it is then necessary to modify this mechanics while preserving, of course, its verifiable predictions. This can, for example, be done by assuming that there exist, outside wave functions, parameters that are called "hidden".

It is at this stage that the third term of my title, "contextuality" becomes relevant. In the title I used this word in a general sense, to designate several closely related aspects of Reality, such as non-separability and non-locality. These notions essentially come into play when dealing with correlation at a distance. Let me briefly recall what is in question by using a simple example: Let us imagine a "source" that emits pairs of gloves, the two gloves in each pair going off in different directions one, say, towards the North, the other one towards the South. Only chance decides if it is a right glove that is sent off to the North, and a left glove to the South or the contrary. In the North an observer notes the hour of arrival of a glove and its nature "right" or "left". In the South the same thing happens (two gloves from the same pair reach observers at the same time).

At the end of the day the observers meet each other and compare their notebooks. They then note a strict correlation: each time "North" observed a right glove, "South" observed a left glove and vice versa. This is clearly explained by a correlation established at the source. Each pair is made of one right glove and one left glove. Each "system" (glove) leaves the source with a certain configuration (right or left) that is proper to it, that is not the same for all elements of the ensemble of gloves departing toward the North (sometimes it is a right glove, sometimes it is a left glove that goes) and this *determines the result of the particular observation* that will be made about this glove.

What I call "separability" is precisely this: it is that the result of the measurement made, let us say in the North, depends only on the configuration of the glove departing toward the North, and not on the observations of the other distant observer making his measurements in the South. Let us now replace the pair of gloves with a pair of particles. Concerning this pair, it is a priori completely natural to assume that separability applies as well. In other words it is a priori quite natural to assume that the result of the measurement made on the particle moving toward the North depends only on the parameters that fix the state of the particle moving toward the North. However, this proves to be false. The Bell theorem predicts from this hypothesis, which disproves separability, some statistically observable results. These predictions are contrary to those provided by quantum mechanics and, *above all, are invalidated by experience. Non*-separability holds true. Contextuality is a generalization of this.

This is of interest in connection with our subject. The reason why this is so has to do with a point which, although unusual, has been demonstrated: that this non-separability does not touch the level of the *"operational"*, that is to say, that when all is said and done, the level of "everyday" science. This means that there is necessarily an immense difference between *apparent order*, which is the order of both *operative* science and common sense (an order centered on the idea that objects lying at different places are separable, at least by thought), and the real order - or "implicit order" as Bohm said - in which non-separability is the rule.

It would, however, I think, be an error to conclude from this that physics is totally confined to the description of the *behavior* of things, that what touches existential questions is fully outside its realm, lies in the domain of pure philosophy, and that, therefore, the philosopher can come to a decision in all independence and liberty. Physics gives us, in any case, existential information on what reality *is not*, and consequently the philosopher must - which is very Popperian- take into account constraints originating in Physics. More precisely, the philosopher cannot demand a non-contextualistic realism.

This, of course, does not mean that the philosopher no longer has a choice. He still may choose between the following:

- a radical idealism (possibly camouflaged under some other noun), making it a principle that *the notion of knowledge is conceptually prior to the notion of existence*;

- a "traditional" realism, traditional in the sense that in it, Reality per se is considered as intelligible. In this framework, however, there exist *several* models, all non-separable, of course (because of Bell's theorem), which means that all these models meet with relativity troubles, are very much distant from the explicit order which is that of observation and everyday science, and are considered in general as exageratedly speculative for these reasons;

- and, lastly, there is the solution of a weak structural realism, which is the one I recommend. In it, phenomena are conceived - somewhat in Kant's manner - as something like rainbows, that is to say, as dependent for a large part on our sensorial apparatus and on our understanding. They are, all the same, conceived of as not emanating entirely from us. Otherwise stated, these phenomena, and more precisely, the mathematical laws which they obey, although they cannot constitute independent reality, are still vague reflections of something. This something lies, in its detail, above our cognitive possibilities and even seems to be conceptually prior to space and time. It is this something that I call "the Real".

If you happen to know of any solution quantitatively compatible with Physics (other than these), I would be most happy to be informed.

Paris-Orsay University
France

NOTES

1 Translated from the French text.

2 See Bohr 1963.

REFERENCES

N. Bohr, *Essays 1958-1962* on "Atomic Physics and Human Knowledge", A. Bohr ed., Suffolk, Bungay 1963.

MASSIMO PAURI

LEIBNIZ, KANT AND THE *QUANTUM*

A Provocative Point of View about Observation, Space-Time, and the Mind-Body Issue

1. INTRODUCTION

The concepts of space and time, both in regards to Newtonian space + time (non-relativistic theory) and Minkowski's spacetime (relativistic quantum field theory), remained essentially unchanged during the development of quantum theory. There are, however, several aspects indicating their provisional status. I leave aside the attempts to construct a quantum theory of gravity, given their lack of relevance, as far as they stand today, to the problem I intend to discuss here. Some features of the so-called secondary or covariant theories, such as the *string* theory, certainly give very interesting indications of a direction consistent with my own point of view. On the other hand, the non-perturbative theory still has to be constructed and the perturbative one is a 'background' theory. The non-perturbative canonical theory is not yet sufficiently developed and its overall conceptual structure is far from being clear. It seems peculiar that such a long stasis, nearly a century, occurred because, historically, an ontological correlation between the notions of object and process, on the one hand, and those of space and time, on the other, has always been present. Every time there is a change in the conception of 'thing', it has been accompanied, sooner or later, by a corresponding change in the conception of space and time, or *vice versa*. It suffices to recall the pressure exerted by the formation of the modern concept of space towards the overcoming of the Aristotelian ontology of substance, during the Renaissance. On the other hand, we have before us a modification of the notion of 'object' within the context of quantum theory, which amounts to a substantial decrease in ontological value. And this, after more than seventy years, still does not seem to have explicitly affected the conception of space and time.

On March 25 1995, sixty years had passed since the publication of the famous Einstein-Podolsky-Rosen paper in the *Physical Review*[1]. From then onward, foundational debates about quantum theory were essentially centered on two distinct issues. The first concerns questions like non-locality, non-separability, or the so-called micro-realism. The second is about the so-called *macro-objectification*, the formulation of a quantum theory of measurement, the issue of macroscopic decoherence and a resolution of related paradoxes (*Schrödinger Cat* and the like[2]). While the first group of problems has to do with counter-intuitive features of micro-processes (which, I believe, are unavoidable and should be incorporated into any future possible theory), the second group calls into play the fundamental problem of the objectivity of the

E. Agazzi and M. Pauri (eds.), The Reality of the Unobservable, 257-282.

macro-world as described by quantum theory. It is well-know that not only does quantum theory entail indefiniteness of property attributions to the micro-structure in certain circumstances (a fact which I deem unavoidable), but it also entails that, *if no additional assumptions are made*, this lack of definiteness should be propagated to the macro-realm. The received view of quantum theory (the so-called 'Copenhagen interpretation' advocated by Bohr, Heisenberg, Pauli and Born) introduces the *additional principle* that the very *measurement* procedure prevents the propagation of property indefiniteness to the macro-realm (*state-vector reduction* postulate). This is tantamount to asserting, in principle, the *classical* nature of any particular measurement apparatus, *qua* measuring device, and thereby to admitting the epistemic necessity of a dichotomic world description: the quantum and the classical. At the same time, one introduces a *shifty* sort of nature for the demarcation line (*cut* or *Schnitt* of von Neumann) in so far as its position along the chain of processes, which extends from the observed micro-system to the final measurement apparatus, is left completely undetermined by the very nature of the reduction postulate, provided that, to quote von Neumann, the chain is not extended beyond the point at which the result of the observation "is perceived by the observer"[3]. The *shifty* nature of the line of demarcation between the object and the subject is the basis of the so-called *complementarity* of descriptions. It is evident, on the other hand, that one cannot give up the introduction of a *cut* without literally depriving the theory of any physical meaning, as a consequence of the lack of objective attribution of properties to the *whole* macroscopic world.

Let us note, for the sake of comparison, that in statistical thermodynamics, the parameter 'number of particles' determines the scale, on the basis of which a collection of molecules may be considered (macroscopically) a gas. Admittedly, an approximation is implied here, but this is justified by the underlying fundamental theory of classical dynamics. In quantum theory, on the other hand, no such parameter allowing the introduction of a suitable scaling exists. It is only by assuming *ad hoc* that the reduction of the state vector takes place by a measurement or, alternatively, by a suitable approximate selection of states in the density matrix (the 'butchering' of the density matrix in the words of John Bell[4]), that the cut is introduced and thereby a physical meaning attributed to the theory. But, a variety of opponents, in regards to the first assumption, objects that it renders the theory subjectivist one way or the other, and, regarding the second choice, it finds it difficult to accept that a supposedly fundamental theory may acquire its meaning *via* an approximation, a solution which may entail an epistemic and logical circularity.

In spite of the manifest technical and philosophical relevance of the macro-objectification issue, decades of debate have not yielded any agreed upon and satisfactory solution. The inconclusive character of this debate seems to have led a growing number of physicists and philosophers of science to believe that this issue, by itself, contains the seeds of self-destruction of quantum theory as a fundamental theoretical structure not merely reducible to a set of recipes. But, of course, the macro-objectification stands as an open issue to the extent that one refuses Bohr's thesis according to which it is necessary, for the epistemic foundation of the theory, to assume the existence of a classical background including, for any kind of experiment, at least one ultimate observer. Another well-known weakness that critics point out, is that the purely epistemical interpretation of the wave function, maintained by the orthodox view, cannot be easily reconciled with the fact that "measurement-like processes are going on more or less all the time, more or less everywhere"[5]. I believe, nevertheless, that, although Bohr's standpoint may be judged unsatisfactory in its renunciation of the traditional program of mutual support between epistemology and metaphysics, it contains important elements of truth and should not be rejected in its entirety. The viewpoint that motivates the present essay is based on the conviction that the classical nature of the background should be related to more profound and more general arguments than merely taking notice of it, as in the Copenhagen interpretation. Far from being an issue having merely to do with *interpretations* of quantum formalism (as some people claim), or with technical modifications of the mathematical structure of the original theory (as others propose), I propound that the macro-objectification issue, which is essentially a problem of *composition*, has all the marks of a profound technical *and* philosophical issue, concerning the possible *forms* of human knowledge of physical reality. I argue that the radical denial of *naive atomism*, implied by the discovery of the *atomization of actions* (i.e. of 'processes' instead of 'things'), must entail far reaching consequences for the concepts of space and time in their relationship to the mind-body problem, which, so far, have been neither properly drawn up nor envisaged. Against this backdrop, I stress the surviving relevance of the debate about spatial 'wholes' and 'parts' that ideally took place between Leibniz and Kant, and I advance a *non-transcendental reinterpretation* (in a peculiar 'realistic' sense) of the role of intuition of the pure homogeneous extension of space (and time) as *form* of sensible perception.

My argument will be mainly of a philosophical nature. One could legitimately say that no philosophical argument can provide, by itself, solutions to problems in theoretical physics. Yet, as Feynman has remarked, every technical achievement is based on some intuition and it is difficult to say whether or not this last is philosophical. I believe that when a new, fully coherent, synthesis (of quantum theory and relativity) is reached, philosophical reflection will be one of the ingredients, of a renewed intellectual exchange between empirical science and philosophy. Thus, instead of commenting on the present theoretical situation, I will take the risk of looking ahead. Although I am quite serious about my thesis, I qualify my point of view as 'provocative' since my considerations will follow a problematic and speculative line and part of my claims will be rather venturesome. I will also present viewpoints that do not belong to current, mainstream philosophical thought and that conflict with what could be considered accepted wisdom in today's philosophy of science. I shall limit myself to presenting a new point of view from which to look at still unsolved problems. Also, my proposal implies certain theoretical consequences concerning the role of the *non-local formalism* in our physical theories and could even be exposed to empirical refutation in certain circumstances.

2. QUANTUM THEORY AND THE MIND-BODY ISSUE

There are many ways in which it is said that quantum theory is linked to the issues of observability and the mind-body relationship[6]. But it is mainly through the macro-objectification problem that this is postulated. I shall briefly summarize the many ways in which this problem is tackled by gathering them into a limited number of main strategies, each possessing different levels of philosophical commitment[7]. A first group of strategies is characterized by the thesis according to which all problems and paradoxes originate from erroneous *interpretations* of the formalism. Therefore, the various positions of this group stand out because of the new interpretations they propose. Many other strategies are justified by the necessity of 'closing the circle'[8]. Yet, the enforcement of this slogan hides very different philosophical and technical projects. The 'circle' to be closed by theoretical physics (quantum theory in particular) is often strongly intended as the 'Democritean circle' of perception *tout court*. These positions more or less saddle a particular, historically contingent, physical theory with the task of 'solving', as it were, the mind-body problem itself and almost all the other problems of knowledge. Therefore they implicitly attribute to quantum theory a sort of meta-empirical *status* since they tend to draw forceful and ultimate philosophical conclusions from the specific structure of a 'supposedly final' physical theory. This is always suspect, at the very least, because it is in any physical theory's nature to *not* be *final*. A subgroup of the 'formalistic' positions (which is worth pointing out) directly calls observability and the mind-body relationship into play through a merely implicit theory of perception. This is the method pursued by a variety of so-called 'many-worlds', or even 'many-minds', interpretations of quantum theory. These positions allow for an *indefinite macro-realm*, thereby bypassing the necessity for the *cut*, with the limited scope of securing only *definite appearances*, and thus denying that the macro-world of our experience is, in fact, what it appears to be in observation. The variety of actualizations of properties envisaged by quantum formalism is simply translated into as many evolutionary branches of the universe ('many worlds'). In a certain sense, one could say that these views realize a *sheer interpretation* of quantum formalism *as it is*, because, bypassing the necessity of von Neumann's Schnitt, the *same* quantum formalism is asserted in order to describe the dynamic evolution of *all* physical systems, whether or not they possess consciousness, or any other property. On the contrary, other interpretations require that certain systems or entities in the world (including conscious organisms in particular) evolve according to different laws of dynamics. Even technically, however, these programs do not keep their promises. Since the 'many worlds' interpretations must describe the *whole universe* quantum-mechanically, they should account also for *our perceptions as observers*. Yet, due to the well-known indifference of quantum theory with respect to the choice of bases in the state space, coherently superposed states of consciousness must be allowed. Since the 'many worlds' interpretations fail to explain why states of mixed perception never occur, the only alternative would be to assert that superpositions of distinct states of consciousness actually do occur and to claim that this is an experimentally testable statement altogether[9]. I believe that the 'many worlds' (or 'many minds') interpretations are insufficient for tackling the problem, in that they evade the basic issue through figures of speech.

A radical and rather paradoxical variant of 'closing the circle' is the original position held by Wigner, and even before by others[10], according to whom the impossibility *in principle* of fixing von Neumann's Schnitt necessarily leads to admitting that consciousness itself is responsible for state vector reduction, thus entangling the mind-body problem with quantum theory *as it is*. It is not clear, however, among the other things, which type of consciousness should play this role, if one considers the range of subjectivity that presents itself with evident *graduality* in the entire animal world.

Finally, another group of strategies pursues the 'closure of the circle' at the level of physical knowledge, as a program of solving the basic riddle of objectification, according to a macro-realistic perspective, by means of suitable *modifications* of the theory. These strategies aim at securing a definite and objective macro-realm by means of a unified dynamic description of the micro and macro-worlds, thus theoretically explaining why the macro-realm is as we *observe* it to be. These attempts explicitly resist any involvement of the mind-body issue with physical theory, according to a traditional self-restrained view of scientific practice.

Here, I shall argue in favor of a quite different perspective of considering the 'closure of the circle'. A definite implication of my proposal is that, at the level of our spatiotemporal experience, no mixed state of perception be possible in principle so that no coherent superposition of states corresponding to macroscopically distinguishable spatiotemporal localizations be allowed. This entails that the linear laws of quantum theory must somehow be violated[11].

3. OBSERVATION AND PERCEPTION

At this point, the direction of my discourse requires the recollection of certain points and distinctions of the different components of the issúe of observation in physics, namely: (i) the different roles of the *observer* as a conscious subject in the process of observation; (ii) the notion of *object of observation* and its necessary, fundamental relation, with *space and time*; (iii) the *mind-body problem*; and (iv) the basis of the distinction between *observational* and *theoretical* terms. According to Roberto Torretti[12], we must distinguish between: a) personal and *impersonal* observations, according to whether awareness is primarily present or not. "In personal observation the *receiver* is always the observer's body - while in impersonal observation the receiver can be a wide variety of physical things". A further distinction is between: b) *direct* and *indirect* observation. "In personal observation the observer always pays attention to something that is *before his eyes*, or within reach of his ears". This is the *direct object* of observation and it is being *directly observed.* On the other hand, the observer's main concern may be the object of an impersonal observation whose receiver he observes personally. Torretti calls the latter the *indirect* object of observation, and says that it is *indirectly observed* by the observer. Some other relevant points are the following: "The direct object of observation is *never simple*. Its parts and aspects stand to one another and to other *directly* observable *objects* in *relations of time and space*... . It is chiefly through the steady presence of this network that the world of possible objects of personal observation is held as a permanent background to whatever one happens to be actually observing". Concerning our understanding of the process of observation, Torretti says: "For us, as for Descartes, Boyle, and Huygens, all *physical action* boils down to a *transfer of momentum*, or as we now prefer to say: 4-momentum *in* spacetime. However, the modern idea of physical action has burdened us, since its inception, with the so-called mind-body problem. For transfer of momentum will neither account for, nor be explained by a change of mind. All the attention devoted to the problem since Descartes has not brought us any nearer to understanding how a man's decision can initiate a definite outward flow of energy and momentum across his skin, or how an inward energy-momentum flow across it can modify his state of awareness". Torretti concludes (and I share his view entirely): "It is unlikely that this rift between the *two sides of observation* can be closed without some *radical, incalculable innovations* in our understanding of physical action". Finally: "The distinction between direct and indirect observation is crucial to philosophers who wish to play down the importance of the intellectual factor in experience, or to deny it altogether.... But it is not easy to say just at what point observation becomes indirect, and hence dependent on memory and reason".

Actually, the real issue at stake in the traditional debate is whether or not it is possible to classify entities into *observable* and *unobservable* ones in a reasonably well-founded way, and to take a position on the ontological relevance of this distinction. On the realist side, Grower Maxwell, in a well known paper[13], maintains precisely that, first of all, the *observable-unobservable* distinction necessarily implies that the cognate *observational-theoretical* distinction be sharp and ontologically crucial (a fact that he maintains to be easily falsifiable); and, secondly, that, once this is shown to be untenable, the former distinction is undermined by the existence of a nearly continuous series of observational situations corresponding to *higher and higher resolutions in time and space*. The core of the argument is that, in these conditions, we are left without ontologically significant criteria that would enable us to draw a *non-arbitrary* line between acceptable and non acceptable observationality and, thus, between a claim to real existence for the relevant objects and lack of it. According to Maxwell, we cannot say that what we see through spectacles is a "little bit less real" than what is observed by unaided vision. Therefore, although there is certainly a *continuous transition* from observability to unobservability, any talk of such continuity from full-blown existence to nonexistence is nonsense. In conclusion, the arguments usually given by the realists about observability are generally of two sorts: one directed precisely against the very possibility of drawing the distinction, the other against the ontological relevance of the distinction to be drawn; and the realist argumentation is carried out to the effect that both must be denied. Thus the realist holds that, even if a feasible distinction could be made, which is not the case, this distinction would be irrelevant to the attribution of existence or reality to unobservable entities, since the latter is independently guaranteed by the global truth-content and general super-empirical virtues of scientific theories, and cannot depend on the questionable distinction between what is and what is not "humanly observable".

Before concluding my discussion of these premises, let me observe that, independently of the specific context in which these sentences are found, what seems to count here in the various cases is precisely the *difference of experience* in relationship to the *spatiotemporal macro-physical intuition*. In what follows, I shall try to argue (on the side of the realist) that the details of the traditional debate are strongly overshadowed by quantum theory in that the latter entails a peculiar new meaning of what must be understood as 'humanly observable' (meaning *anschaulich* or, so to speak, intuitively observable[14]), a meaning that bypasses the usual vague and controversial notion of what is observable *simpliciter*. I also maintain that this is a rock-bottom and wholly intrinsic notion. It is technology independent, and far from introducing a discontinuity between full-blown existence and non-existence, it distinguishes between different levels and modalities of reality in a profound sense. Thus, I will jump into the realism/anti-realism debate by sidestepping it in a rather provocative way. I shall take an unwarranted position in favor of the first alternative, namely the realist one, by exploiting the macro-objectification problem of quantum theory with a strong ontological commitment, relating it to spacetime and the mind-body issue. In other words, I propose that the whole discussion of observability depends on a most radical philosophical issue: on one hand, that of the structural conditions of our sense intuition and of the relationship they may have with certain preconditions (in particular with the inexorable framing of every sense perception inside the pure intuitions of space and time); and on the other hand, that the quantum macro-objectification issue, itself, plays a crucial role in determining the *epistemic priority* of our experience of spacetime as intuited homogeneous extension.

Before expounding my view on quantum theory, let me observe that the very *terminology* of observation and perception might be deeply affected by the desired realistic solution to the objectification problem. According to the above definitions, a direct object of observation is something that is before the observer's eyes, and this something is supposed to be an 'object'. Now let me cite, as a mere conceptual possibility, what would happen if we look at the consequences of a quantitative model of spontaneous state vector reduction that is now well-known[15]. This dynamic reduction model (GRW) belongs to the last group of strategies listed above and tries to describe state vector reduction on the basis of a unified dynamic governing all natural processes. The underlying philosophy is quite simple: one introduces a 'small' stochastic modification of the standard dynamics, that has a negligible impact on micro-systems, but which nevertheless leads to a dynamic suppression of macroscopic superposition. Note that reductions occur even at the level of a single electron. Although this model is merely phenomenological and cannot easily be rendered relativistic, it is nevertheless very interesting and instructive. According to this model, a simple and realistic (micro-macro) experimental situation is envisaged, in which a coherent superposition of two beams of a few photons, emerging from a single-atom Stern Gerlach experiment, is dynamically reduced by the triggering of a potential action in a human retinal neuron. This is possible because the number of ions involved in the process along the optic nerve fulfills the model's conditions for decoherence within the standard temporal limits of perception. At the same time, it can easily be demonstrated that reduction by one observing subject implies reduction by any other subject and the rest of the universe. In such a situation, there is no objective *spatial property attribution* (as to which screen originated the photons) before the dynamic reduction of the coherent superposition of the beams takes place in the observer's optical nerves. And it would be difficult to assert that the observer is *directly* or *indirectly* observing an 'object' at all. For it is his own act of observation (or, better, his optical process) that *constitutes* the objective, external reality of the spatial location of one of the emerging beams of photons as a possible *direct* object of observation[16] for another subject (assuming, for the sake of argument, that other coherently superposed photons are simultaneously directed to a second observer). Let me remark that no specific role is attributed here to the observers' consciousness. According to the authors of the dynamical reduction model, the observer's brain is the only place in the universe in which there is a difference of mass density, for the two superposed states, sufficient to spontaneously induce the objectification. This theoretical experiment (based on a rather neutral philosophical framework) should warn us of the possibility that an eventual clarification of the macro-objectification issue may substantially transform our understanding of the relationship between subjectivity and perception of spatiotemporal objective reality, well beyond the epistemological relevance of the distinction between direct and indirect observation. Consider again the GRW model and assume for instance that, for some reason, the reduction does not take place within the optical nerve or the lateral geniculate body of the observer's brain (surely an empirically admissible hypothesis). What kind of perception should we envisage for the coherently superposed states?[17]

4. THE ANTINOMIES OF ATOMISM: LEIBNIZ, KANT, AND THE CONTINUUM

Just one month after the Einstein-Podolsky-Rosen paper was published, in a letter to Heisenberg, Pauli[18] wrote: "Quite independently from Einstein, it seems to me that in a systematic foundation of quantum theory, one should start more from the issues of *composition* and *separation* than has been done until now". It is here asserted, with crystal-clear intuition, that the issue of *divisibility* and *composition*, hence the *part-whole relation*, should play a fundamental conceptual role at the very basis of quantum theory. Surely, the cultural community at large is not properly aware of the radical implications of quantum theory. In fact, it constitutes an empirical response to the two thousand year-old, philosophical question about continuity, discreteness, and divisibility of matter. But, surprisingly, it is an answer that constitutes a definitive blow to the atomistic conception of Nature. Aside perhaps from relativity and Kant, this historical event represents the first empirical refutation of a general philosophical view about the world. It is beneficial to my aim that I now briefly recall certain historical passages concerning the issue of divisibility and composition.

The paradoxes of the notion of 'atom', construed as an indivisible element of a *homogeneous extension*, have been widely known since Zeno, through Descartes, and up until Kant. It is important to recognize that this paradoxical issue constantly dealt with the nature and intuition of space (possibly intertwined with mathematical structures). In the paradox of 'The plurality', the force of the argument is geometric: if extended things exist, Zeno argues, they must be composed of parts; thus there is a plurality of parts. Furthermore, these parts must themselves have parts (since they are extended). Since the process of subdivision is indefinitely repeatable, there must be an infinity of parts, and so on, given the implications of the measurability of the putative parts. I am not interested here in following Zeno's argument through to its end, but I would like to make the following remarks: It is clear that the atomic constitution of physical matter has no direct bearing upon this argument. It is not necessary for the parts in question to be physically separable from one another; it is sufficient that they be conceptually distinguishable. Even if there are atoms, which cannot be split, there are still spatial extensions for which we can distinguish different parts. Yet, I believe that the argument is strong enough to weaken the idea of the physical existence of atoms as well. For, if atoms are spatially extended, the very possibility of conceptually distinguishing between their parts confers upon them, as it were, some kind of secondary qualities, some differentiating factors which are properly distinctive features of the phenomenal things which atoms are supposed to explain. In other words, although it does not constitute a logical difficulty, it turns out that, if spatially extended, then physical atoms cannot be those very *simple* entities which they are imagined to be. It is no surprise then, to find that Descartes, who assimilates matter with space, argues against the existence of atoms precisely in this way[19]. Historically, the most thorough philosophical analysis of the part-whole relation, and of the issue of divisibility can be found in the Kantian Second Antinomy of Reason. In both the proof of the thesis as well as the antithesis, Kant refers to the debate on atomism that took place during the XVII and XVIII centuries, due mainly to Descartes, to English empiricists, to Wolff, Euler, Leibniz and Clarke. With the remarkable exception of Leibniz, the whole historical debate concerned what I would like to call *naive* or *spatiotemporal atomism*. The thesis of the Kantian Antinomy[20] states: "Every compounded substance in the world is constituted by simple parts, and nowhere something exists which is not simple or compounded by simple parts", while the antithesis states: "In the world, no compounded thing consists of simple parts; and nowhere something simple exists in it". In other words, the thesis asserts that an arbitrary *part* of a *whole*, when analyzed thoroughly, is *simple*, (i.e. it cannot be analyzed further). The antithesis states instead, that any part of a whole in the world is composed of *non-*

simple parts, which can be analyzed *ad infinitum*[21]. This thesis and its antithesis, however, are not contradictory propositions in the simple logical meaning of the term (i.e. such that they could be proven under the same hypotheses). According to Kant, they are in fact 'contrary' propositions (i.e. such that both follow from a self-contradictory concept). Actually, the antinomy derives its force from the inconsistent concept of a 'real compositum of parts'. On the one hand, the whole must be composed of primary (and thus supposedly *simple*) parts having *external relations* with each other (i.e. relations which are contingent or accidental). And on the other hand, *qua* real (i.e. possible contents of experience), the whole must be given in space. But space, like time, has no simple parts and the extension of a spatial whole propagates downwards to all parts of it. In other words, any putative 'real compositum' of simple parts should be, at the same time, a *phenomenal* substance (i.e. a possible object of experience, necessarily given in space) and a *noumenal* entity, a pure object of thought, since it is supposedly constituted of *actual parts* in an ontological sense (for Kant, not reachable in experience).

The logical aspects of this Antinomy have recently been analyzed by Brigitte Falkenburg in terms of the mereological relations of part and whole[22]. Falkenburg points out an asymmetry between the logical structures of the thesis and the antithesis. As she clearly demonstrates, if one scrutinizes the logical structure of the proofs, one sees that, in proving the antithesis, Kant cannot limit himself to excluding that compounded substances have simple parts, but he must also exclude the reality of *simple* substances which are *not* parts of a whole. To this end, Kant resorts to an additional transcendental demonstration relying on the relational character of all sensible experiences. What Kant needs to exclude, essentially, is Leibniz's solution of the problem of substance, since Leibniz's *monads* are precisely conceived as simple substances (*viz.* the unique substances) which share no external (i.e. spatial) relationships and, therefore, are not constituent *parts* of some compounded thing (they do not stand among themselves according to the part-whole relation with respect to the compounded thing). The *monads* can be considered as parts only in the elliptical sense of being principles of composition which cannot be spatialized a-posteriori since their relationships (essentially non-spatiotemporal) are without causal efficacy and are *implicit* in their very nature, hence non-contingent. Not being spatial parts of a *compositum*, they cannot be identified with the phenomenal and spatially extended parts of material things. Kant clearly refuses any element of spatial reality beyond homogeneous extension experienced in the phenomenon and, correspondingly, he asserts an exhaustive *transcendental priority* of extension. In conclusion, *noumenal* entities are not possible objects of knowledge, while entities as phenomena, in that they are spatially extended, can be subdivided *ad infinitum*. Whether or not they are actually subdivided cannot be asserted or denied. It is interesting (and impressive for the modernity of intuition) to quote here, by contrast, the following passage of Leibniz[23]: "Space, like time, is something *not substantial*, but *ideal*, and consists in possibilities, or in order of co-existents, that is, in some way, *possible*. And thus there are no divisions in it, but such as are made by the mind, and the *part* is *posterior* to the *whole*. In *real things*, on the contrary, *units* are *prior* to the *multitude*, and multitudes exist only through units. The same holds for *changes*, which are *not* really *continuous*". As Falkenburg notes, there is another important asymmetry. In the thesis, space plays no role and the argument depends entirely upon the logical meaning according to which the part-whole relation is understood. In the antithesis, however, the part-whole relation is considered a spatial relation from the beginning. This relation being an external relationship among different substances (and as such, accidental and conceptually removable), must exist in the form of an external sense, hence can only be thought of in space. Therefore, the argument of the antithesis is synthetic *a priori*, to the extent that it implies the conjunction of the concept of compounded substances (exploited in the thesis) and the spatial interpretation of the part-whole relation. Eventually, Kant can conclude that a real substance in space cannot have simple parts because, in the spatial interpretation, these last could only be point-like (but *points* do not actually have a true spatial nature, and so on). As Falkenburg emphasizes, the self-deception into which Reason falls (according to Kant), when moving from one side of the Antinomy to the other, is precisely determined by the implicit and incorrect assumption that the interpretation of the part-whole relation be the same. For, were one to spatially construe even just the statement of the thesis, one would have to explain how a 'real compositum' could have constituent parts that *fill* a space and are, at the same time, indivisible.

This is a crucial point. A possible answer to this question is clearly given by the Cantorean structure of the mathematical continuum, even if it is difficult to convince oneself, on the basis of the axiomatic notion of measure, that the real numbers 'fill' a space. As Weyl has stressed[24], this answer entails overcoming a philosophical 'abyss': precisely the jump from the notion of space as an external continuum, understood as real, sensed intuition (the 'Raum' of experience), to the abstract set-theoretical notion of it, by means of which theoretical physics reconstructs the continua of space and time. Certainly, one cannot assimilate the structure of the Cantorean continuum to a morphological representation or to a mathematical 'imitation' of the perceptual wholes of sensed experience, or lived temporal flow. While perceptual wholes, as such, cannot be broken down into simple and ultimate parts, the Cantorean continuum precisely amounts to a breaking down into simple parts (real numbers) beyond any threshold of sense perception. Yet, I claim that we should not confuse the vague continuity of perceptual wholes (let me call it the *psychological continuum*) with the ideal continuum of homogeneous extension, which is a presupposition of our notion of the continuum in general, before its bifurcation in spatial and temporal continuity. Surely, we cannot think of space and time, hence of the external continua, as 'objects'. As a matter of fact, we don't perceive space[25] or time; we perceive and think of inhomogeneities and variations over an extended homogeneous background. This latter - as singular intuition - realizes the idea of a homogenous spatial extension with no body posited *in it*. The pre-Cartesian *synthetic* continuum of geometry is just a conceptual abstraction of this intuition. Admittedly, many classical problems and paradoxes of divisibility, can find a solution within the framework of the Cantorean continuum. As Grünbaum[26] argued some twenty-seven years ago, the modern theoretical notion of spacetime based on the Cantorean continuum has the strength to overcome many of the Zeno paradoxes. But the essential point in my analysis is the possibility of paradoxes, or at least counter-intuitive conclusions, having to do with the notion of the synthetic continuum, rather than with the set-theoretical notion of it.

Surely, we cannot re-propose Kant's notion of intuition today with *all* of its implications, in particular in relationship to its connection with *intellectual schemata*, as generating *a-priori* synthetic judgments, which, as such, are apodictic and constitutive of the mathematical structure of space and time[27]. Nevertheless, I do believe that the core of the Kantian notion of pure intuition of homogeneous extension, as an essential feature of the real (sensible) notion of space-and-time should not be jettisoned altogether. I mean here a primary, non-inferential, grasp of pure extension, which is neither based nor reducible to perception, let alone to memory or introspection: pure extension as a singular *substratum* for all intuitions and perceptions of particular objects. Any attempt at constructing the space and time of our experience from completely isolated elements, clashes with the particular nature of spatial and temporal relations. In order to be able to speak about *real* spatial relations among entities it seems necessary for these last to already be individuated as *spatially distinct*, in the specific (sensible) meaning of the term so that the intuited extension must be presupposed. Thus, the relational conceptions of space cannot avoid some form of logical circularity. The substantivalists, however, avoid the possible circularities of the relationists (concerning the notion of spatial location) at the expense of having to require an account of what it is that confers individuality on the points of space. On the other hand, the notion of spatial extension as a form of sensible intuition is a natural starting point for the formalization of *structural* concepts of space like those of Newton and Einstein[28].

Leaving aside, for the moment, the philosophical *status* of the synthetic continuum, we can say that its points are not *individuals* and lack any intrinsic localization. They are pure, ideal entities immersed in the perfect background of qualitative homogeneity over which all multiplicities, discontinuities, variations and the points themselves only become thinkable. The 'points' ideally introduced into the homogeneous extension are preliminary elements for the axiomatic characterization of space. It is only after the axiomatization that they become monadic referents of symbols and objects of quantification, hence individual *qua* numbers, even if nearly all of them turn out to be non-computable in the algorithmic sense of the term. It has correctly been said that the continuum possesses an intrinsic and fundamental *epistemic bi-modality*[29]: that of the homogeneous intuition and that of the axiomatic categorization. The *bi-modal* nature of the continuum holds a central position in gnoseology. First of all, it is the source of the irreducibility of mathematics to logic, since, even after axiomatization, the mathematical content of the continuum stands as a horizon that cannot be completely determined or exhausted by means of a logical syntax. Secondly, while the axiomatizability of the continuum renders the scientific description of space possible, it is its synthetic modality that, by filling the so-called Platonic Gap (i.e. the 'abyss' between mathematical entities on the one hand and perceptual experience on the other), provides the grounds for the distinction between physics and mathematics, as related to the distinction between concrete sensed intuition, and intellectual construction and understanding. In particular, the continuum grounds the 'dimensionality' of physics by forcing any measurement to be a ratio with respect to a conventional unit[30]. In addition to this, one should not forget that the adoption of the Cantorean continuum, as the basic mathematical layer of the theoretical description of physical space and time, cannot be justified *factually* but only on the basis of an induction principle meant in a broad sense, at most. Finally, let us note that, even disregarding the constraints of spatiotemporal intuition, any putative discretization of space and time runs into serious difficulties. It's worth recalling that, in discussing the paradox of the 'The Stadium', while correctly arguing that current quantum theory is not at all challenged by this paradox, Grünbaum issued an important cautionary lesson against any imagined program of *direct atomization* of space and time, which would imply, at the very least, a dependence of the *event-status* on relative motion. I am not interested in pursuing a deeper analysis of Kant's Antinomy in this study. And I am actually more interested in the arguments put forward in the proof of the Antithesis than in the fact that an Antinomy exists at all. So let me now give some attention to the essential features of quantum theory in what concerns the *part-whole relation*.

5. THE ATOMIZATION OF PROCESSES

The ontological core of quantum theory, well beyond the specific structure of its formalism, is an extraordinary *fact of the world* that disproves the very terms in which the historical tradition of atomism developed. Namely, this fact is the discovery that *action* is what is really made up of *indivisible* elements or *quanta*. But action is a theoretical entity that cannot be cut out in space (or spacetime). In other words, what is actually atomic is not a spatial or temporal extension but an algorithmic quantity (of the classical description) which intrinsically encodes both *spatiotemporal* and *dynamic* elements. The atomization involved in quantum theory is therefore neither that of a *thing* nor that of a *homogeneous extension*, but rather that of a *process*. I take this *process-atomization* to be a fact of the world that has absolutely radical consequences. Not only can processes corresponding to exchanges of action which are smaller than the Planck constant not exist, but the *elementary quantum act*, which corresponds to the exchange of a single *quantum* of action among any parts in which a physical system can be *ideally subdivided*, cannot be described in *any possible local way* within, nor literally *belong to*, space and time, since no continuity of real physical states is physically conceivable in between the initial and the final states of parts. For, were any putative local description possible, one could always reconstruct processes corresponding to arbitrary exchanges of action. On the other hand, no restriction whatsoever is put by this fact alone on the mathematical description of the underlying spacetime structure even if, as already remarked, the conceptual *status* of the latter seems to be jeopardized[31].

Yet, restrictions on ordinary spatiotemporal language necessarily emerge. 'Heisenberg inequalities' (unhappily defined as 'uncertainty relations'[32]) reflect precisely the conceptual tension arising between the *symbolic structure* that *replaces* the impossible spatiotemporal description, on the one hand, and the unavoidable utilization of causal, spatiotemporal language, for the description of the theory's empirical façade, on the other hand. Consider, for example, a single atomic transition between two nearby energy levels of an atom. Since its spatiotemporal description is *onto-logically forbidden* by the quantum principle, the process is represented in the following way: The atom's state-vector evolves in a coherent superposition of the atom's state-vectors corresponding to the relevant energy levels until the observation is eventually completed. During the intermediate period, no objective, energy attribution property is possible and the energy values can be referred to only potentially[33]. The transition is treated by means of a linguistic circumstantial compromise in terms of a so-called 'uncertainty' ΔE and Δt of the 'energy values' and of the 'moment of transition', in the form of the quantitative limitation $\Delta E \cdot \Delta t \geq \hbar$. In other words, instead of describing the transition, one expresses the probability that the transition has taken place at a certain time, under the constraints of action atomization. The implications of *process atomization*, as well as the relationship between the latter and the *principle of coherent superposition*, are shown here in an exemplary manner[34]. It is significant that Heisenberg inequalities do not confine themselves to expressing the limits of applicability of macroscopic spacetime language. They also play the active role of expressing the spatiotemporal extension of the actualization of potentialities under given empirical conditions: so, for example, they say that a micro-entity 'photon', characterized by an 'emission time interval' of $\approx 10^{-8}$ sec., has a 'coherence length' of ≈ 3 m.

Since quantum theory reveals that matter (in the *latu sensu* of matter-energy) possesses neither a continuous nor a discrete structure, but a peculiar quantum structure, we are faced with the task of *composing simple processes* instead of simple spatiotemporal *continuants*, or *things*. Process atomization confronts us with conceptual problems, which, although quite different from those hidden in the traditional Zeno paradoxes, and discussed by Kant and Leibniz, are not totally unrelated to them. As Brigitte Falkenburg notes[35], all conflicting positions about atomism, until the XVIII century (Leibniz aside), were determined by their relying both on the *spatial interpretation* of the parts-whole relation and on the traditional concept of substance, conceiving the composition of material things in terms of a spatial relation of substances. On the other hand, in both classical physical theories and in quantum theory, the prevailing point of view (at least at the pragmatic level) has always been based on a characterization of the extension of a compounded system, in terms of internal interactions among the 'elementary' constituents, which are first identified, and then grouped together. According to these interpretations, the part-whole relation is treated as partly spatial and partly dynamical. And while the 'elementary' constituents of matter are pragmatically held to be indivisible at a given level of approximation, space itself, as an extension *within which* the composition is described, is a presupposed background (in the traditional Newtonian form of a 'real compositum') and is therefore divisible *ad infinitum*. So, it would seem that the antithetical statements about composition, typical of Kant's analysis, should be irrelevant for any modern theory of matter. However, I believe things to be radically different, and far more complex in quantum theory because the transformation of the concept of composition, induced by the quantum principle, is by far more profound than it may appear at first.

A process is essentially an *interaction among parts*. If the elementary process is universally atomized, the very concept of *part*, in its spatial interpretation, is radically altered. Were it not, one could always exploit the spatiotemporal description of the evolving parts, in order to reconstruct processes with arbitrary exchanges of action. Thus, action atomization has a direct effect on any conceivable distinction of parts of a whole, by denying, in definite circumstances, individuality and objective attribution of properties to the putative parts. Consequently, any subdivision of a whole into parts can only be thought of in a metaphoric sense, and described in a purely symbolic way.

The subtlety of the conceptual prohibition of conceiving parts in a conventional sense, which is implied by action atomization, should be explicitly emphasized. Various (methodologically distinguishable, though not independent) mechanisms are relevant for composition and separation within the quantum framework: 1) the principle of superposition and the consequent role of correlations among putative parts of compounded systems, with or without explicit mutual interaction[36]; 2) indistinguishability of identical nomological individuals; 3) the fact that any *ideally conceivable* separation of a whole system into parts must be such that *all possible relative changes in time* of these last correspond to exchanges of action that are multiples of the elementary quantum; 4) the fact that from action atomization - via Heisenberg inequalities and relativistic requirements - follows a regime of *virtual entities* that radically transform even the *dynamic* aspect of the description of composite systems in spacetime.

An entirely new interpretation obtains for the concept of 'force' among constituent parts, as exchange of virtual entities (I would say the first historically significant explanation of 'force').

This new concept, however, undermines the very distinction between the 'parts' and the 'forces' that link them together. Indeed, although the various nomological species of entities that enter the description are characterized by symmetries and are asymptotically defined in spacetime, they lose their own individuation within the compound. This is so, not only because they may be indistinguishable but also because any of them - irrespective of its belonging to the *fermion* or to the *boson* species - is 'virtually' made up of every other one. Even if the classical limit of a theory of *fermion fields* is a particle theory, while the classical limit of a theory of *boson fields* is a classical field theory which describes the 'forces' in the conventional formulation, away from this limit the difference between *fermions* and *bosons* reduces to a difference in statistical properties and, therefore, has no special ontological relevance. I take this unification to be one of the most important and unique achievements of quantum theory[37]. It crucially depends on the synthesis of the quantum principle and special relativity and shows that any claim for a particle-ontology is definitely naive: this is true in particular for the De Broglie-Bohm re-interpretation of quantum theory, quite apart from its contextualism. Finally, even stronger is the *duality* emerging in *string* theory, where *elementary* and *compounded* entities appear to be interchangeable, a fact that makes the very distinction between the *whole* (as compounded entity) and its *parts* (as composing entities) entirely relative[38]. It seems as if, at a certain level of analysis, Nature could easily *count* without *individuating*. The 'whole' cannot be conceived as being composed of independently existing parts. Not only are the properties of the whole unable to be explained in terms of parts connected by external relationships of a mechanist nature, but, what is more, the properties of the whole logically contradict the assumption that parts do exist at all, except in a potential modality. This peculiarity leads to the predominant role of symmetry in the definition of nomological properties, since a potential structure can possess *a priori* symmetries which are then broken, either by removing a certain approximation (as in the case of isotopic spin), or by the internal dynamics of the fields (Goldstone bosons). To conclude, the ontological properties described by the symbolic structure of quantum theory are neither *spatiotemporal* nor essentially *atomistic* in a naive sense.

6. QUANTUM THEORY AND THE PRESENT STATUS OF SPACETIME

All this said, one should ask how it happens that relativistic quantum field theory (RQFT) is able to exploit spacetime structure so successfully. My claim, however, is that even if the mathematical structure of micro-spacetime has not been modified until now, its theoretical *status* has been substantially altered. There are many symptoms of illness and of the provisional character of the concept of spacetime in RQFT, but the cure is far from apparent. Let me enumerate some *pros* and *cons* of local spacetime description. The *pros*: On the one hand, the extraordinary empirical success of RQFT seems to rely heavily upon the notion of background spatiotemporal *locality*, mathematically represented in terms of the Cantorean continuum. Both the basic condition of micro-causality, and certain regularization procedures, require a topological and metrical Minkowski microstructure. What is more, Minkowski's micro-spacetime constitutes an *absolute* spacetime in the constitution of micro-entities in an even stronger sense than that of Newton's space since it selects, at least kinematically, the *a priori* possible forms of quantum nomological objects through the so-called 'unitary-representations' of its symmetries (*Poincaré group*). Finally, a spectacular *pro* in favor of locality is provided *prima facie* by the empirical success of *gauge* theories, whose symmetries crucially depend on the points of micro-spacetime. This might not be a real *pro*, however, given the growing consensus among quantum field theorists that the only consistent quantum field theories are the so-called asymptotically-free non-Abelian gauge theories. If we interpret 'physical observability' in this context as possibility of probing micro-structures by means of interactions, we must conclude that such *gauge* theories are, in fact, consistent to the extent that they are insensitive to short-scale spacetime behaviour. This conclusion is supported by the fact that theories which do not satisfy the above condition (like gravitational field theory, for example) are not consistent RQFTs. In this sense, each point of the manifold, on which the symmetries of *gauge* theories crucially depend, should be understood, as it were, as a representative *'compendium'* of something spatial but non continuously extended, as lattice theories themselves tend in turn to suggest[39].

This leads us to the *cons* of local spatiotemporal description. The instrumental interpretation of any point-event as a representative *'compendium'* seems to be further strengthened by the following facts: a) physical observables are not defined on individual points, but rather over extended regions of the micro-spacetime; b) interacting quantum fields are essentially quantities that interpolate asymptotically defined states; c) the *status* of the micro-spacetime of RQFT is quite peculiar. From the chrono-geometric point of view, this is a universal, classical, deterministic macro-spacetime, actually Minkowski's space of the special theory of relativity, utilized without any scale limitation from below. Actually, it is introduced into the theory through the group-theoretical requirement of relativistic invariance of the statistical results of measurements with respect to the choice of *macro reference frames*. Therefore, this micro-spacetime is *anchored* to the macroscopic medium-sized objects that asymptotically define the experimental conditions in the laboratory. It is, in fact, in this asymptotic sense that a physical meaning is attributed to the classical spatiotemporal coordinates on which the quantum fields' operators depend as parameters. Thus, the spatiotemporal properties of the micro-Minkowski manifold, including its basic causal structure, are, so to speak, projected on it *from outside*.[40] Furthermore, in classical field theories spacetime points play the role of individuals. The recent reinterpretation of general relativity, prompted by the quantum gravity program, claims that they are thus individuated by the fields they carry, and primarily by the basic *individuating metric field*. But no such possibility is left open with quantum fields. From this point of view, Minkowski spacetime is in a worse position than general relativistic spacetime. Since the Riemann tensor vanishes in this case, no invariant of the Riemann tensor can be used to effect the individuation[41]. On the other hand, all the additional elements which are being used for the individuation of the points of Minkowski macro-spacetime, like rigid rods and clocks or *genidentical* world-lines of free particles and so forth, cannot be instrumental here.

Finally: d) a further sign of the uncertain *status* of spacetime in RQFT comes from the so-called 'peaceful coexistence' between this latter and special relativity (but I will not elaborate on this point here); and: e) the limitations posed by quantum theory upon the operational procedures for measuring microscopic spacetime intervals (easily reducible to time intervals) are very strong indeed. It can be shown[42] that every possible kind of clock is necessarily a non-microscopic object and, in particular, that even the so-called atomic clocks can be considered microscopic devices only on the basis of a rough idealization which conflicts with the fundamental concepts of measurability. Actually, there is no way of measuring microscopic intervals (e.g. intra-atomic) in a direct operational manner. While nearly every kind of physical measurement is based on indirect procedures that involve a theoretical interpolation, the unique aspect of the microscopic case is that the theory involved is precisely quantum theory with all its peculiar features concerning potentiality and spatiotemporal representation. We should then distinguish between physical Minkowski spacetime, on the one hand, and *analogical* Minkowski micro-spacetime, on the other. The former, as operationally founded on the behaviour of (macroscopic) *clocks* and *rods*, and the latter as a mathematical structure whose empirical meaning rests on a theory that theoretically prevents an exhaustive representation within the homogeneously intuited spatiotemporal extension. Far from worrying about a direct operational justification, what I want to emphasize here is that the role of micro-spacetime in RQFT is *analogical* because of the lack of any physical individuation of its points so that its empirical meaning is conferred on it solely by the macroscopic (classical) level of control of quantum theory, and in particular by its being extrapolated from the macro-spacetime of the special theory of relativity, which is a well-defined physical concept. Thus, the role of this micro-spacetime seems essentially that of an instrumental *'external' translator* of the symbolic structure of the theory into the language of the macroscopic irreversible traces, which constitute the experimental findings within macro-spacetime. All this tends to emphasize that the main quantum role of the classical representation of spacetime, as a homogeneously extended continuum, is that of being an epistemic precondition for the formulation of RQFTs.

A final remark is necessary. The provisional and instrumental nature of the microscopic utilization of Minkowski spacetime within RQFT is also revealed by its peculiar conflict with the spacetime concept of the general theory of relativity. There is a profound theoretical gap in our understanding of spacetime at the macro and the micro levels. We can say the following: on the one hand, the topological and metric structures of general, relativistic, macroscopic spacetime depend, at least partially, on the distribution of matter in bulk. But, their theoretical formulation is conditioned by the fact that general relativity describes matter in a purely phenomenological way (in terms of classical fields or fluids), so that it is not able to account for the medium-sized macroscopic objects that provide the basis for its empirical foundation (the *clock* above all). On the other hand, Minkowski macro spacetime, which has to be understood as a local approximation (in the proper sense of 'tangent structure') of the general-relativistic concept, plays the role of an absolute structure for the constitution of nomological micro-entities which, from a naive logico-ontological point of view, are considered to be primary constituents of macroscopic material structures. The medium-sized macroscopic objects, which constitute the empirical operational basis of physics, possess - as Einstein himself remarked - a mere plausible existence within general relativity and, until a consistent solution of the macro-objectification problem is found, they will continue to share only a formal limited status with respect to the microscopic theory of matter. We can say that the two fronts are controlled by theories which both lay claim to an unlimited validity that requires mutual ignorance because they conflict at a deep theoretical level. This conceptual clash could be reduced if the very nature of quantum objectification could explain away the conception of material aggregation as naively taking place *in* an independently existent spacetime (i.e. if the composition in extension of ontologically prior extended elements could be blocked).

7. KANT REVISITED

"Est aliquid praeter extensionem, imo extensione prius"[43]

The atomistic nature of processes thus has the effect of radically altering the concept of composition or aggregate: the very distinction between the whole (as a compound) and its parts becomes relative. It should no longer be permitted for one to consider aggregates of parts, which interact *in* spacetime, if this is not done in a pragmatic sense, useful for actual calculations. I believe that we have here a fundamental basis for an intrinsic notion of unobservability as spatiotemporal *Unanschaulichkeit*, which directly follows from action atomization. We have to admit a kind of reality, which cannot be exhaustively brought within the realm of spacetime, if this realm is simply meant as an extensional order. In other words, our intellectual knowledge is not limited to an *anschaulich* modality: as Quine has remarked, quantum ontology is one of abstract entities, though not of mental ones[44]. It seems to me that the discovery of the quantum principle implies a further profound detachment from the Kantian conception of space and time, in addition to that induced by the discovery of non-Euclidean geometries and by the formulation of macroscopic relativistic theories. Although, for Kant, space and time are pure *intuitions* and not *concepts*, he nevertheless allows for empirical conceptual determinations of space and time, namely the scientific concepts as constructed by the analysis of the behaviour of sensible objects. These concepts, however, cannot be the original and *primary* representations of space and time, and are only real to the extent that they belong to a context of possible (according to the empirical laws) past, present, or future perceptions, within the progress or regress of experience, precisely in the sense of macrophysical intuition. We have already learned, from the possibility of non-Euclidean geometries and relativity theories, that the intellect does not *schematize* the apprehension of space and time in terms of *a priori* synthetic judgments, which, as such, are univocal and apodictic. But now RQFT requires the utilization of concepts of space and time which, though surely empirical, cannot be traced to any synthesis of sensible perceptions. Furthermore, the analogical status of these concepts is qualified by the circumstance that the underlying intuition of extension cannot be supplemented by a physical individuation of the point-events.

With this in mind, I shall try to revisit the objectification problem of quantum theory. We can say the following: (a) our ability to reach a symbolic representation of the ontic substructure described by quantum theory tends to prove that our conceptual knowledge is not confined to an *anschaulich* modality of description (in terms of spatial and temporal extension), but, as it were, 'goes beyond it'. Of course, I am not willing to say that the kind of reality, which is symbolically represented in quantum theory, has a *noumenal* character in the sense of Kantian terminology. For to hold that we can know *'things in themselves'*, or even to know that they somehow exist, would contradict Kant's critical doctrine in a radical way[45]. I nevertheless believe that we are entitled to say (exploiting Kantian terminology) that the quantum domain appears to be, as it were, *ultra-phenomenal*, and that quantum theory, even more than relativity theory, forces us to tear up the Kantian straitjacket of spatiotemporal pure forms, as an insuperable boundary of reality. Thus, with respect to Kant's dictum, the *'quantum noumenon'* is no longer confined to meaning the mere intellectual conceivability of the 'thing in itself', but it rather exhibits an articulation of the representation which, although symbolic and non-spatiotemporal, is nevertheless connected to an empirical context within a nomological network; (b) this ontic substructure is such that it is logically impossible to think of it - at least in certain conditions - in terms of *parts* of a *whole*, if not potentially; (c) on the other hand, our perception and observation of the macro-realm is objective and allows us to distinguish - with a certain degree of approximation (precisely that of classical theories) - *separable parts* which can be represented *in extension* of space and time. Therefore: (d) the basic question is: How is it possible to reconcile and integrate a basic ontology of *simple processes* - which denies full reality to the parts of aggregates - into a derivative ontology of *genidentical* world-lines or continuants in space[46].

I suggest that, far from being a challenge to technical ingenuity, the objectification issue of quantum theory (as a problem of *composition*) has all the distinctive marks of a profound ontological and epistemological issue that cannot be solved without some radical modification of our notions of space and time, likely to have direct implications also on the mind-body conundrum. I submit that the solution of the objectification problem should be structured in such a way that *the objectivity of the extensional properties of spacetime itself be a consequence of a fundamental quantum process and not a presupposition of it*[47]. On the other hand, we are transparently reminded of the unique way in which the conception of space is strictly connected to the issues of mind-body and subjectivity by the following passage in the *Critique of Pure Reason*[48] (*italics mine*): "In order that certain sensations be referred to something *outside me* (that is, to something in another region of *space* from that in which *I* find myself), and similarly, in order that *I* may be able to represent them as *outside* and *alongside* one another, and accordingly not only as *different* but as in *different places*, the *representation of space must be presupposed*". Thus, not only is the presupposition of the representation of space necessary for recognizing things as spatial, distinct and next to each other but, above all, for recognizing them as *outside the subject*. The representation of space is therefore essential even to the *self-individuation of the subject* and for recognizing the *plurality of conscious subjects*. Surely, I am not committed here to following Kant in what regards his transcendental innatism. Still, as I have said already, I am not willing to accept the complete suppression of the role of spatiotemporal intuition, which has been under way since the middle of the last century. Likewise, I am not willing to concede that the intuition of homogeneous extension is entirely parasitic of the empirical process of vision, since this would involve the issue of the causal theory of perception in a circular way. Thus, I claim that the pure forms of intuition of the homogenous extensions of space and time maintain a fundamental and special role in the mind-body issue, in relationship to our intentional constitution and individuation of phenomenal objects. On the other hand, if space and time were lacking any ontological root beyond extension, as for Kant, knowledge would be confined to phenomenal appearances and the empirical basis of subjectivity would be put under strain, with the consequence of giving up any ontology of subjectivity, and of being doomed to a fatal idealistic drift.

Pure intuition of extension and, more generally, the way in which the subject is modified by external objects, is a perpetual riddle, *on par* with the relationships of the qualitative homogeneity of experience of *qualia* and experience of temporal flow to the physical description of material processes[49]. I claim that, far from being a mere technical problem, this is a fundamental philosophical point: the challenge we are faced with (as Kant himself had foreseen in his *Opus Postumum* vertiginous reflections) is the understanding of the material preconditions of pure intuition of extension. As it should be clear, my thesis is not committed to the existence of any formal structure having the role of grounding synthetic *a priori* judgments about space and time[50]. Rather, we have to clarify the processes bringing from the quantum structure to our cognitive apparatus as living organisms and, eventually, to the generation of the phenomenological intuition of the continuum in his peculiar essence of 'non-propositional formatting'[51]. This looks as a 'new' methodical task about the mind-body issue having to do with both physics and the cognitive sciences.

In his book on Leibniz, Russell[52] says that "There is evidently in space something more than relations". I am suggesting that there must be, in the reality of space, something which is even *more than extension and relations* and is *beyond them*. If this is sound, there must be, in the material constitution of humans (and, with suitable graduality, living organisms as well), something innate, in the empirical sense, that causes our perceiving things in extension, and this something should be the *same something* that brings about the composition of elementary processes in spatially extended and inter-subjective macro-reality. In this sense, vision would be parasitic on the constitution and representation of space and not the other way around, and, in the same sense, quantum decoherence would not be an effect of the intervention of consciousness, but rather a specific and necessary *precondition* for our observational experience in spacetime as it is. Well beyond the known fact that, due to indistinguishability, we cannot rigorously speak of, say, the electrons of *this* table or *this* chair, there might be a profound and important sense in which we would *not* be entitled to say literally that tables and chairs are *compounds* of their elementary constituents *in* an extensional spacetime background[53, 54].

To conclude, I suggest that space, if understood as the homogeneously extended content of our sensible intuition, is, as it were, *epistemically prior* but not *ontologically prior* (as for Kant), and that there is something in the reality of space (and time) which is indeed *ontologically prior* with respect to the intuition of homogeneous extension. If extension is not ontologically basic, but only epistemically prior, in the specific sense of being a presupposition of sensible perception and therefore of observation, the ontic structure cannot be intelligibly decomposed into parts in extension (i.e. understood in the *anschaulich* modality which is unique to appearances or phenomena) as perceived and observed. In this sense, and in this sense alone, extension would be *ideal* and not real.

At the same time, the epistemic priority of homogeneous extension would justify the peculiar resilience of scientific imagery against the attempts to renounce the mathematical continuum in our physical theories[55] or even vindicate an appeal to a primary classical reference in the sense advocated by Bohr. Therefore, from a theoretical viewpoint, the conjunction of the obligatory epistemological priority of intuition of the synthetic continuum in representation, on one hand, and the ontological priority of non-extensional elements of quantum structure, on the other hand, would imply that *the role of 'non-locality' in our formalisms should be constitutive, hence programmatic and by far more extensive and intensive than used so far*[56]. The content of non-locality of the current theory does indeed seem to be insufficient for both the issue of so-called 'peaceful coexistence' of quantum theory and relativity, as well as that of quantum gravity and macro-objectification. Furthermore, it seems likely that rather than modify the quantum temporal evolution, one should try to *modify the structure of the Hilbert space of compounded quantum states*. Here, of course, I can do nothing more than note that we already know of elements which should have something to do with space (or spacetime) but are not extension-dependent, like: incongruous counterparts, difference of signature between space and time, spectral dimensions in graph theory, knots, spinor connectivity, and the like[57]. Finally, let me remark that this way of looking at the objectification problem links it to the quantum gravity issue[58], although it is clear that, according to what I have suggested above, any program simply based on 'quantizing geometry' should appear as a weak strategy. As early as 1963, in his Pasadena Lectures on Gravitation, Feynman was the first to suggest that gravity could be, as it were, a residual effect of quantum decoherence[59]. If the atomic nature of action has something to do with space (and time), it should be with those elements of it which are ontologically prior to extension. The latter, being a phenomenal derivative notion, could always be 'ideally reconstructed' in terms of the Cantorean continuum. In this sense, whatever quantum gravity may mean, it would be connected to something built into the composition mechanism and not something presupposed or obtained by quantizing general relativity in a conventional sense. As I noted in the Introduction, in history of thought there has always been a continual, though implicit, ontological correlation between the notion of space and time on the one hand, and the notion of things on the other, a correlation which has nourished the relational/absolutist debate. For this reason, every change in the conception of objects has eventually induced a corresponding transformation of the concepts of space and time. The transformations in the concepts of space and time, corresponding to the profound ontological *diminutio* of the concept of 'object' brought about by quantum theory, has not, as yet, been completely accomplished. If and when that is achieved, we will perhaps be able to say that 'the circle tends to close'.

Physics Department - Parma University
Italy

NOTES

1 See Einstein *et al.* 1935.

2 See Schrödinger 1935.

3 See von Neumann (1973), who adds: "The intellectual inner life of the individual ... is extra-observational by its very nature (since it must be taken for granted by any conceivable observation or experiment)".

4 See Bell, 1990.

5 See Bell, 1990.

6 See for example Butterfield, 1995 and 1998; Lockwood, 1989 and 1993.

7 A thorough classification of the various positions about the macro-objectification issue can be found in Ghirardi 1995.

8 The slogan 'closing the circle' has been coined by Abner Shimony: "... to understand the knowing subject as an entity in nature and to assess claims to knowledge in the light of this understanding. Such a program aims at the integration of epistemology with the natural sciences and metaphysics. It intends to show how claims to human knowledge of the natural world can be justified, and in turn how the resulting view of the world can account for the cognitive powers of the knowing subject. For brevity I shall refer to this program as 'closing the circle'", see Shimony 1993.

9 See, for example, Deutsch 1985, 1990. An attempt to escape these difficulties, called 'many minds' has been proposed by David Albert (see Albert 1992). I will not elaborate on this point further; see Butterfield 1995.

10 For instance London and Bauer, already in the thirties, write (see London *et al.* 1939): "The observer has an entirely different point of view [as compared to a measurement apparatus]. For him it is only the [micro]object x and the apparatus y which belong to the external world - to that which he calls 'objective'. By contrast, he has with himself some relations of a completely special character: He has at his disposal a characteristic and quite familiar faculty, which we may call the 'faculty of introspection'. He can thus give an account of his own state in an immediate manner. It is in virtue of this 'immanent knowledge' that he claims the right to create for himself his own objectivity, that is to say, to cut the chain of statistical quantum correlations ... by certifying: 'I am in the definite state [such and such]'". Of

course, the putative transition induced by introspection must be non linear and stochastic in nature, even if it is assumed that quantum consciousness be represented by an ordinary quantum state. It seems rather awkward, however, to say the least, to relate the superposition principle to, for example, perceptual vagueness.

11 As is well-known, Penrose has repeatedly suggested that a fundamental link should exist between the possible violation of the linear Schrödinger evolution and *perception* processes. See Penrose 1990, 1994.

12 See Torretti 1986 (*italics mine*).

13 See Maxwell 1971.

14 This concept will be clarified in the following.

15 See Ghirardi *et al.* 1986; also Aicardi *et al.* 1991.

16 Provided, of course, that one does not mean by 'observer' the abstract, disembodied consciousness of the empirical observing individual.

17 *Note added in proofs*: I have been informed by professor G.C. Ghirardi (*private communication*) that he is now suggesting a quantum optical experiment which might discriminate between *conscious perceptions* of pure photon states and *conscious perceptions* of coherently superposed photon states. According to him, this should be experimentally feasible in the near future by exploiting the possibility of explicitly testing that quantum coherence persists, between its terms, up until the moment in which the conscious observer 'looks' at it (*quant-ph/9810028v2*). If we concede, for the sake of argument, that blurred or intermediate *color perceptions* are concretely possible (for instance superposing green and red photons), it would seem that a corresponding blurred or 'opalescent' *spatial perception* (as to which screen originated the photons) should also be possible. This result, however, would appear astonishing and nearly unbelievable.

18 See K. von Meyenn *et al.* 1985, p.404: letter [412] (*italics mine*).

19 "D'autant que si petites qu'on suppose ces parties, nèanmoins parce qu'il faut qu'elles soient ètendues, nous concevons qu'il n'y en a pas une entre elles qui ne puisse ètre encore divisée en deux ou en un plus grand nombre d'autres plus petites...", see Descartes 1724.

20 See Kant 1963.

21 Of course, by 'analyzable in parts', I mean here 'analyzable by means of an iterative process any step of which is characterized by elements homogeneous to those of the previous step (for example *intervals*)'.

22 See Falkenburg 1995.

23 G.W.Leibniz, VI letter to B. de Volder, August 19, 1702 (*italics mine*), see Loemker 1976.

24 See Weyl 1932.

25 See, e.g., Gibson 1950.

26 See Grünbaum 1968.

27 Of course, once disentangled from the categorical function of the *intellectual schemata*, the form of pure intuition does not suffice to ground the apodictic and universal synthetic judgements of geometry. In this situation, the geometric axioms could not be synthetic *a priori* and, as Kant himself declares in his 1770 *Dissertation*, one should have to admit the possibility that real space be endowed with geometrical properties different from the Euclidean ones (see Kant, 1912).

28 That Newton's concept of absolute space is in fact a *structural* concept (neither relational nor substantivalist, even if its chrono-geometric properties are obviously non-dynamic) is transparently shown by the following passage: "Thus the parts of space have their individuality from their positions, so that if any two could exchange their positions, they would thereby exchange their individualities, and each would be converted numerically into the other. By their mutual order and positions alone (*propter solum ordinem et positiones inter se*) are the parts of duration and space understood to be just what they are in fact; nor have they any other principle of individuation besides that order and those positions, which therefore they cannot change", see Hall and Hall 1962 (*italics mine*). I am indebted to Professor Roberto Torretti for having brought this crucial passage to my attention. For the modern general-relativistic concept, see Stachel 1993.

29 See Petitôt 1993.

30 Consider for instance a *graph*. We can say that a set of points endowed with a binary, symmetrical, anti-reflexive relation, realized by links connecting the points pairwise, establishes a *space* in the modern sense of the word, that is the sense in which Riemann firstly spoke of 'spatial manifoldness'. A metric in such a space can be defined *intrinsically* in terms of the number of links in the minimal path(s) connecting two points. In order to build up the whole construction, we only need *distinguishability* of 'points' (entities) and *degree* of their relations. Also a *spectral dimension*, for example, can be defined provided the graph is infinite. There is a fundamental difference, however, between this 'space' and a *lattice* of the kind used by quantum field theorists for the sake of regularization procedures, since the *lattice* is constituted by introducing a *reticular spacing*. In this way an *extrinsic metric, physical dimension* (a length) and, consequently, the *continuum* are brought in. Then, the 'lattice space' is necessarily *thought of* as immersed in a

multidimensional manifold and, intuitively, in the synthetic continuum. On the other hand, were physics - to quote Wheeler's slogan - essentially "it from bit", any measurement would be a mere absolute counting. Therefore, unless one be ready to accept the informationally unorthodox notion of an 'actual infinity of bits', physics would be a-dimensional. A well-known distinction between 'physical continuum' and 'mathematical continuum' is that made by Poincaré (see Poincaré 1968).

31 Of course, what I am saying in regards to spatiotemporal limitations should not be confused with the fact that the wave function of a compounded system is a function defined in terms of configuration space and not in physical space. It is also interesting to recall what Erwin Schrödinger wrote in a letter on August 1926: "[Concerning] Bohr's standpoint, that a space-and-time description is impossible, I reject a *limine*" (see Moore, 1994, p.16: letter to Willy Wien). Of course, it is difficult to disagree with Schrödinger, for the very notion of a spatial (spatiotemporal) *limine* is self-contradictory, provided space is meant as the intuited homogeneous extension that theoretical physics mathematizes. It seems, however, that Schrödinger - unlike Pauli - failed to recognize the fundamental fact that in quantum theory there is no spatial (spatiotemporal) *limine* at all and that all spatiotemporal limitations are essentially indirect. From Planck constant and other physical constants, it is certainly possible to derive spatial or temporal limitations, such as, e.g., the Compton wavelength. These limitations, however, have a mere pragmatic meaning and do not set a theoretical boundary to furthering spatiotemporal analysis. This analysis can even bring about qualitative changes in the nature of the putative 'parts' of the composition (as in the case of *quarks*) and it is far from clear what kind of limitations it may meet. It is not at all obvious, for instance, that what matters with the so-called Planck fundamental units is their role of being *absolute* limitations rather than their possible effective role as *differences* of values. This alternative may be of paramount conceptual relevance because - due to the enormous scale differences it entails - the *absolute interpretation* tends to decouple the quantum gravity and the macro-objectification issues. Finally, it is worth noting that David Bohm had a deep awareness of the crucial ontological role played by action atomization, independently of the specific technical structure of quantum theory. In his *Wholeness and the Implicate Order* (see Bohm, 1980: Ch.4, Sections 12,13), he argues to the effect that a (so-called) 'realistic' view of quantum phenomena entails the existence of a 'sub-quantum level' of reality in which action is *not* at all atomized in units of Planck constant. This would mean that action atomization of the current theory could only be approximate. But, of course, no empirical finding whatsoever supports this conjecture, which, on the other hand, would open a Pandora's Box of conceptual and physical problems that quantum theory has already explained.

32 I use the term 'unhappily', because this terminology suggests the idea that quantum entities are 'objects' in the sense of macro-physical, spatiotemporal intuition, i.e., entities 'having' a definite individuation and characterization in terms of parameters such as 'position', 'momentum' and the like, which then would become blurred. By now, we know that the basic ontology of the theory is one of (relativistic) quantum fields, and that the notion of 'particle' is approximate and asymptotic, and furthermore it loses its invariant meaning for accelerate observers or in curved space-times. In general, one should no longer talk about 'disturbances' caused by the measuring apparatus on a putative, spatiotemporally located (micro)object. Bohr's assertion of the inseparability of the apparatus from the measured micro-entity can be coherently understood in terms of the effect of the objective spatiotemporal restrictions imposed by the experimental setting on the *a priori* possible manifestations of the quantum domain.

33 It is well known that Heisenberg was the first to assert that quantum theory requires a modality which is situated between logical possibility and actuality, which he calls '*potentia*' (see Heisenberg 1962; see also Margenau's notion of 'latency' in Margenau 1949). Of course, the quantum 'potentiality', in its being linked to nomologically determined probabilities, has nothing to do with the classical Aristotelian concept of 'potentia'.

34 I would like to remark in connection with this, that speaking of energy quantization as a general characteristic of the quantum domain, is quite misleading. Energy quantization is not universal, it depends upon specific experimental conditions and, above all, it is derivative with respect to action quantization (there are continuous energy spectra and, what is more, energy is not, unlike action, a relativistic invariant). Also, it is not by chance that all the observables that are universally quantized have physical 'dimensions' of action, like spin and angular momentum, or are non-spatiotemporal, like charge and 'internal' quantum numbers.

35 See Falkenburg 1995.

36 The EPR phenomenon is a typical example of correlation without explicit interaction between 'parts'.

37 This unification will result even stronger if the so-called super-symmetry (between *fermions* and *bosons*) could be empirically verified.

38 "Elementary objects now seem to be made of the very particles they create. More specifically, *duality* makes *elementary* and *composite* objects *interchangeable*: whether a particle or other entity is irreducibly fundamental or is itself made up of even more fundamental entities, depends on one's point of view. Either perspective ultimately yields the same physical results. For example, *tangles* of *quarks* may give rise to *solitons* that are *monopoles*, *tangles* of

monopoles may give rise to *solitons* that are *quarks ... The fundamental scale associated with quantum theory (Planck constant) is intimately entwined with duality*", L. Susskind, quoted by M. Mukerjee, Scientific American, January 1996 (*italics mine*).

39 It may be interesting to note that a peculiar relation between mathematical localization and non-local interactions emerges even in the framework of classical dynamic systems (non-quantum, non-relativistic) by applying an adapted *gauge* methodology to systems with dynamical symmetries (see De Pietri et al. 1995, 1996).

40 The standard Minkowsky structure of this micro-spacetime has been probed down to the scale of $\approx 10^{-18}$ m., yet only from the point of view of *scattering* experiments, involving a limited number of *real* 'particles'.

41 See, for example, Stachel 1993.

42 See Salecker and Wigner 1958; also Peres 1980.

43 See Leibniz, in Gerhardt 1850.

44 It should be clear that my notion of *Anschaulichkeit* is akin to the Kantian *Anschauung* rather than to the notion spelled out by Heisenberg in his famous 1927 paper on quantum theory (see Heisenberg, 1962). See also Miller (1984) who revisited Heisenberg's original notion.

45 As Nicholas Rescher has remarked, "A thing in itself whose nature is brought within the reach of the categories of understanding is *ipso facto* unable to do the job of endowing appearances with the intentionality of indicating something that stands altogether outside the phenomenal order, to assure that appearances are appearances of something... . A 'cognitively domesticated' thing in itself would (ex hypothesis) not be able to perform the key mission assigned to such a thing in the Kantian framework, viz. to provide a basis of 'externality' for the objects of our knowledge (where the 'externality' at issue is not a matter of spatiality, but one of grounding)".

46 That is, following Jonhson, "individuals with changing properties but lasting identity over time, interacting with other individuals in their environment", so that the notion of spatial separation of parts evolving in time is appropriate (see Johnson, 1921-24).

47 Among other things, this would justify the privilege of the spatiotemporal basis that is explicitly asserted in some recent attempts to solve the macro-objectification problem. For example, in the GRW model of spontaneous dynamical reduction, the *preferred spatial basis* only guarantees that reduction take place at the macroscopic level without modifying the usual microscopic quantum behavior. In my view, of course, we can expect to uncover semi-macroscopic coherent super positions of, say, momentum or angular momentum but never of position states.

48 See Kant 1963.

49 This is, in particular, the source of the so-called 'grain objection'; see Sellars 1971; also Meehl 1966, and Lockwood 1993.

50 My position concerning the relationships between homogeneous intuition and space may be reminiscent of Helmholtz's famous assertion that "space can be transcendental without the axioms being so", see Kahl 1971.

51 See Petitôt 1993.

52 See Russell 1900-1937.

53 It is obvious that the lasting *spatial isolation* of single subatomic 'particles' like that obtained by, e.g., Brown and Gabrielse or Dehmelt (see Brown *et al.* 1986, Dehmelt 1990), has to be meant only FAPP (viz. 'for all practical purposes').

54 *Note added in proofs:* In this connection, it is interesting to see that, according to the picture suggested by Prosperi and coworkers (see Prosperi's essay in this volume, *italics mine*), "...the so-called *elementary constituents* and their related observables would appear simply as a *description* of the *deviations* of the basic macroscopic quantities from a deterministic behaviour".

55 It is interesting to recall what Einstein wrote in a letter to one of his former students: "The problem seems to me how one can formulate statements about a discontinuum *without calling upon a continuum (space-time)* as an aid; the latter should be banned from the theory as a supplementary construction not justified by the essence of the problem, [a construction] which corresponds to nothing 'real'. But we still lack the mathematical structure, unfortunately. How much have I plagued myself in this way !...... *The continuum is more ample than the things to be described...*" (Einstein to Walter Dallenbach, November 1916, [*italics mine*], see Stachel 1986). Furthermore : "Perhaps, the success of Heisenberg's method points to a purely algebraic method of description of nature. Then, however, we must also give up, on principle, the utilization of the space-time continuum ... Meanwhile, however, *this project resembles the attempt to breath in an airless space*" (see Einstein 1954, [*italics mine*]). Finally: "The alternative continuum-discontinuum seems to me to be a real alternative; i.e., there is no compromise... [In a discontinuum theory] *space and time cannot occur*... Physics up to now is naturally in its essence a continuum physics, in spite of the use of the material point, which looks like a discontinuous conceptual element, and has no more right of existence in field description. Its strength lies in the fact that it posits *parts which exist quasi-independently, beside one another*... On the other hand,

dimensionality (as four-dimensionality) lies at the foundation of the theory" (letter to H.S.Joachim, August 1954, [*italics mine*], see Stachel 1986).

56 Maybe Wolfgang Pauli had something similar in mind when writing (*italics mine*): "It appears that it will be possible to overcome the difficulties only by a new generalization of the concept of complementarity of present-day quantum theory which will necessitate still more far-reaching renunciations of *Anschaulichkeit* in the narrow sense", see Pauli 1992.

57 Let me also recall that Penrose's attempts in the direction of twistors and spin networks, as well as the so called 'loop representation' in canonical quantum gravity, head precisely in this direction.

58 See, however, footnote 31.

59 See Feynmann 1962-63; for different proposal about the link between quantum gravity and the macro-objectification issue see Karolyhazy 1966, 1990, Penrose 1990,1994. Another basic issue which is often connected to macro-objectification is the arrow of time. It should not be indeed by chance that - in a way similar to the ultimate classical background in quantum theory - the arrow of time plays a role of *epistemic precondition* for the empirical control of all (time-reversible) fundamental theories which, on the other hand, do not explain it.

REFERENCES

F. Aicardi, A. Borsellino, G.C. Ghirardi, and R. Grassi, "Dynamical Models of State-Vector reduction: do they ensure that measurements have outcomes ?", *Foundations of Physics Letters*, **4**, 1991, 109-128.

D. Albert, *Quantum Mechanics and Experience* , Harvard University Press, Cambridge, Mass., 1992.

J. Bell, "Against 'Measurement'", in *Sixty-two Years of Uncertainty: Historical, Philosophical, and Physical Inquiries into the Foundations of Quantum Mechanics* , Arthur I.Miller ed., Plenum Press, 1990.

D. Bohm, *Wholeness and the Implicate Order*, Ark Paperbacks, London 1980.

L.S. Brown, and G. Gabrielse, "Geonium theory: Physics of a single electron or ion in a Penning trap", *Rev. Mod. Phys.*, **58**, 1986, 233-311.

J. Butterfield, "World, Minds and Quanta", *Aristotelian Society Supplementary* Volume 69, 1995, 113-158.

J. Butterfield, "Quantum Curiosities of Psychophysics", in *Human Consciousness and Identity,* John Cornwell ed., Oxford University Press, Oxford 1998.

H. Dehmelt, "Experiments with an Isolated Subatomic Particle at Rest", *Rev.Mod.Phys.*, **62**, 1990, 525.

R. De Pietri, L. Lusanna and M. Pauri, "Standard and generalized Newtonian gravities as gauge theories of the extended Galilei group: I. The standard theory", *Class. Quantum Grav.* **12**, 1995, 219-214; "Standard and generalized Newtonian gravities as gauge theories of the extended Galilei group: II. Dynamical 3-Space theories", *Class. Quantum Grav.* 12, 1995, 255-272; "Gauging kinematical and internal symmetry groups for extended systems: the Galilean one-time and two-times harmonic oscillators", *Class. Quantum Grav.*, **13**, 1996, 1417-1450.

R. Descartes, "Qu'il ne peut y avoir aucuns atomes, ou petits corps indivisibles", *Commentaire à la deuxième thèse de la seconde partie des Principes de la Philosophie*. Ecrits en Latin, par René Descartes. Et traduit en Français par un de ses amis. Nouvelle Edition, revue & corrigée. A Paris. Chez Denis Mouchet MDCCXXIV. Avec Privilège du Roi.

D. Deutsch, "Quantum Theory as a Unversal Physical Theory", Int. J.Theor.Phys., **24**, 1985, 1-41; 'Three connections between Everett's interpretation and experiment', in *Quantum Concepts in Space and Time*, C.Isham and R.Penrose eds., Oxford University Press, Oxford 1986.

A. Einstein, "Physics and Reality", in *Ideas and Opinions*, Crown, New York 1954.

A. Einstein, B. Podolsky, and N. Rosen, "Can quantum-mechanical description of physical reality be considered complete ?", *Phys.Rev.* **47**, 1935, 777.

B. Falkenburg, "Kants zweite Antinomie un die Physik", *Kant-Studien*, **86**, 1995, 4-25.

R.P. Feynman, *Lectures on Gravitation 1962-63*, Lecture Notes by F.B. Morinigo and W.G. Wagner, California Institute of Technology, 1971.

J.J. Gibson, *The Perception of the Visual World*, Houghton Mifflin, Boston 1950.

G.C. Ghirardi, A. Rimini and T. Weber, "Unified Dynamics for Microscopic and Macroscopic Systems", *Physical Review D34*, 1986, 470-491.

G.C. Ghirardi, "An Attempt at a Macrorealistic Quantum World-View", in *Proceedings of the Symposium The Interpretation of Quantum Theory: Where Do We Stand ?*, L.Accardi ed., Istituto della Enciclopedia Italiana, Academy for Advanced Studies USA at Columbia University, April 1992, Fordham University Press, New York 1994.

A. Grünbaum, *Modern Science and Zeno's Paradoxes*, George Allen & Unwin Ltd., London 1968.

R. Hall and M.B. Hall, *Unpublished Scientific Papers of Isaac Newton*, Cambridge University Press, Cambridge 1962.
W. Heisenberg, *Physics and Philosophy*, Harper, New York 1962.
W.E. Johnson, *Logic*, Cambridge University Press, Cambridge, England 1921-1924.
R. Kahl, *Selected Writings of Hermann von Helmholtz*, Wesleyan Press, Middletown 1971.
I. Kant, "De mundi sensibilis atque intelligibilis forma et principiis", Dissertation 1770, *Kants Gesammelte Schriften*, hrsg. von der Königlich Preussischen Akademie der Wissenschaften, 1912, vol. II. pp. 385-419.
I. Kant, *Kritik der reinen Vernunft*, Johann Friedrich Hartknoch, Riga, 1st edition 1781, 2nd edit. 1787; English translation by Norman Kemp Smith: *The Critique of Pure Reason*, Macmillan, London 1963.
I. Kant, *Prolegomena zu einer jeden künftigen Metaphysik die als Wissenschaft wird auftreten können*, 1783, Felix Meiner Verlag, Hamburg 1976; English revision by L.W. Beck of the Carus translation: *Prolegomena to Any Future Metaphysics*, Bobbs-Merrill, Indianapolis 1955.
F. Karolyhazy, "Gravitation and Quantum Mechanics of Macroscopic Objects", *Il Nuovo Cimento*, A**42**, 1966, 390.
F. Karolyhazy, "The Breakdown of the Superposition Principle", in *Sixty-two Years of Uncertainty: Historical, Philosophical, and Physical Inquiries into the Foundations of Quantum Mechanics*, A.I.Miller ed., Plenum Press, 1990.
G.W. Leibniz, "Specimen dynamicum", in C.J. Gerhardt, ed. *Leibnizens Mathematische Schriften*, Halle 1850-63, VI, p.235.
G.W. Leibniz, *Philosophical Papers and Letters*, L.E. Loemker ed., D. Reidel Publ. Co., Dordrecht, Holland 1956.
M. Lockwood, *Mind, Brain and the Quantum*, Blackwell, Oxford 1989.
M. Lockwood, "The Grain Problem", in H. Robinson ed. *Objections to Physicalism*, Clarendon Press, Oxford 1993, pp.271-291.
L.E. Loemker, ed.: *Leibniz's Philosophical Papers and Letters*, D. Reidel Publ. Co., Dordrecht, Holland 1976.
F. London and E. Bauer, *La Théorie de l'observation en mécanique quantique*, Hermann & Cie., Paris 1939.
H. Margenau, *Philosophy of Science*, **16**, 1949, 287.
G. Maxwell, "The Ontological Status of Theoretical Entities", in *Minnesota Studies in the Philosophy of Science*, vol.III, *Scientific Explanation, Space and Time*, H. Feigl and G. Maxwell, eds., Minnesota University Press, Minneapolis 1971.
P.E. Meehl, "The Compleat Autocerebroscopist: A Thought-Experiment on Professor Feigl's Mind-Body Identity Thesis', in *Mind, Matter, and Method - Essays in Philosophy and Science in Honor of Herbert Feigl*, Paul K. Feyerabend and Grover Maxwell, eds., University of Minnesota Press, Minneapolis 1966.
K. von Meyenn , A. Hermann and V.F. Weisskopf, eds.: *Wolfgang Pauli, Scientific Correspondence with Bohr, Einstein, Heisenberg* a.o., vol. II: 1930-1939, Springer, Berlin-Heidelberg 1985.
A.I. Miller, *Imagery in Scientific Thought - Creating the Twentieth Century Physics*, Birkhäuser, Boston 1984.
W. More, *A Life of Erwin Schrödinger*, Cambridge University Press, Cambridge, England 1994.
J. von Neumann, *Mathematical Foundations of Quantum Mechanics*, translation from the German edition (1932) by R.T.Beyer, Princeton University Press, Princeton N.J. 1973.
W. Pauli, 'Space, Time and Causality in Modern Physics', in *Wolfgang Pauli. Writings on Physics and Philosophy*, Ch.P. Enz and P. von Meyenn eds., Springer, Berlin-Heidelberg 1994.
R. Penrose, *The Emperor's New Mind - Concerning computers, minds, and the laws of physics*, Oxford University Press, New York 1990; "Précis of The Emperor's New Mind - Concerning computers, minds, and the laws of physics", in *Behavioural and Brain Sciences*, vol.13, 643-705, Cambridge University Press, Cambridge 1990: with Open Peer Commentary by other authors; "The nonalgorithmic mind", author's response to criticisms of the Précis, in *Behavioural and Brain Sciences*, ibid., 692-705.
R. Penrose, *Shadows of the Mind - A Search for the Missing Science of Consciousness*, Oxford University Press, Oxford 1994.
A. Peres, "Measurement of Time by Quantum Clocks", Am.J.Phys., **48**(7), 1980, 552.
J. Petitôt, "Continu et Objectivité. La Bimodalité Objective du Continu et le Platonisme Trascendental", in *Le Labyrinthe du Continu (Colloque de Cerisy)*, J-M. Salanskis et H. Sinaceur eds., Springer-Verlag, Berlin 1993.
H. Poincaret, *La science et l'hypothèse*, Flammarion, Paris 1968.
B. Russell, *The Philosophy of Leibniz*, George Allen & Unwin, London 1900-1937.
H. Salecker and E.P. Wigner, "Quantum Limitations of the Measurement of Spacetime Distances", *Phys.Rev.*, **109**, 1958, 571.
E. Schrödinger, "Die gegenwärtige Situation in der Quantemmechanik", *Die Naturwissenschaften*, **23**, 1935, 807-812; 823-828; 844-849; English translation by J.D.Trimmer, "The present situation in Quantum Mechanics", Proc.Am.Phil.Soc., **124**, 1980, 323-338.

W. Sellars, 'Time and the World Order', in *Minnesota Studies in the Philosophy of Science*, vol.III, *Scientific Explanation, Space and Time*, H. Feigl and G. Maxwell, eds., Minnesota University Press, Minneapolis 1971.

A. Shimony, "Reality, Causality, and Closing the Circle", in *Search for a Naturalistic World View - Natural Science and Metaphysics*, vol.I, Cambridge University Press, Cambridge, Mass., 1993.

J. Stachel, "Einstein and the Quantum: Fifty Years of Struggle", in *From Quarks to Quasars: Philosophical Problems of Modern Physics*, R.G.Colodny ed., University of Pittsburgh Press, Pittsburgh 1986.

J. Stachel, "The Meaning of General Covariance - The Hole Story", in *Philosophical Problems of the Internal and External Worlds - Essays on the Philosophy of Adolf Grünbaum*, J. Earman, A.I. Janis, G.J. Massey and N. Rescher eds., University of Pittsburgh Press, Pittsburgh, Pa, 1993.

R. Torretti, "Observation", *Brit.J.Phil.Sci.*, **37**, 1986, 1-23.

H. Weyl, *Das Kontinuum. Kritische Untersuchungen über die Grundlagen der Analysis*, Walter de Gruyter & Co. Berlin 1932.

OLIVIER COSTA DE BEAUREGARD

EFFICIENT AND FINAL CAUSE AS *CPT* RECIPROCALS[1]

1. INTRODUCTION

Aristotle's final cause is little respected in the hard sciences where operationality of concepts is required. Three "common sense" reasons have dictated this rejection. *First*, "Is not the idea that a cause may operate from a not yet existing future obviously absurd?" *Second*, "Everybody can see that a stone thrown in a pond emits a diverging wave, but to visualize the reversed sequence one has to run a film taken from the reality backwards; retrocausation is nonsense." *Last but not least*, "Who believes that an idea can move matter?" Common sense may be fooled by these sorts of arguments. Today a billiards player or a car driver easily accepts that if there were no friction, motion would persist indefinitely; in ancient Greece the flying javelin did not convince Aristotle or anyone else that no force is needed to sustain uniform motion. Galileo did not prove the law of inertia by high precision measurements; he made it plausible by simple observations and experiments. High precision verifications came later, first from celestial mechanics, including the most impressive one from general relativity: advance of Mercury's perihelion.

In quantum mechanics, the "consistent histories of particles" theory, termed by Omnès[2] as "logical interpretation", aims, in spite of the time reversibility of the equations, at justifying *as best as one can* the common way of thinking and speaking concerning (quote) "intelligibility (i.e. having a visual representation of an object), locality (a definite object has a definite place), causality, and distinguishibility." True, physicists do use such wording when describing their experiments, but then it is full of unsaids.

According to Omnès, this common sense conceptualization is an acceptable approximation since it derives logically from the normal use of quantum mechanics. Thus, he finds that (quote) "the formalism of logic is not time reversal invariant." *But logic is atemporal.* In fact Omnès subordinates his logic to finding (quote) "the difference between past and future," *a presupposition shared by common sense and by "the normal use of quantum mechanics."* Retarded causation has been assumed, which in Mehlberg's wording, is "factlike, not lawlike"[3].

In his nonrelativistic approach, Omnès confers values "at a definite time t" to the "properties of a system", a fact that, due to the fourth uncertainty relation, raises questions in physics. Also, the idea of a thicknessless present is one more preconception of common sense, which experiments by Libet[4] disprove in neurophysiology.

E. Agazzi and M. Pauri (eds.), The Reality of the Unobservable, 283-291.

So, on the whole, aiming at justifying the common sense view may not be the good option, especially if some of its presuppositions are assumed, which imply circularity. Also, there is a risk that some unknown but interesting fish may escape through the net.

More radical than Omnès, I demand that both the interpretation and the use of any theory be unambiguously derived from its formalism.

Objecting that "the normal use of quantum mechanics" violates CPT invariance, I replace the state vector by the transition amplitude as a "sesame opener". Being time-extended and time-reversible between preparation and measurement, the transition amplitude formalizes the paired concepts of an efficient cause in the past and a final cause in the future, acting (this is a metaphor) like "sources" and "sinks" in hydrodynamics by pressure from upstream and suction from downstream.

In physics, a symmetry is understood either passively or actively with mutually implicative meanings. Thus, the cognizance-organization symmetry is implied in the preparation-measurement one, termed preselection-postselection in the frequential jargon, or coding-decoding in cybernetic parlance[5]. In a world conceived as twin-faced, *reality-and-representation*, the negentropy-information transition continuously goes on. Rather than "distant-and-veiled"[6] *reality is not self-sustaining, being the objectivation of intersubjective information, the actualization, so to speak, of a shared prejudice.*

Wigner[7] also assumes from symmetry arguments "a direct action of mind upon matter," termed today as psychokinesis, conceived as *reciprocal to gain-in-knowledge*. Carnot's factlike-not-lawlike principle, stating forward easiness and backward uneasiness of the negentropy-information transition, then *defines* psychokinesis as "paranormal".

Also, *non-separability* and *time extendedness* of matter, both formalized in the *transition amplitude* concept, do postulate, and thus rationalize, the phenomena of precognition and of telepathy.

In *Physical Review A* of July 1994, Stapp[8] discusses an experiment in psychokinesis by Schmidt[9], in a context of interpretation of quantum mechanics. Earlier, Jahn[10] had presented his experiments and views on this matter at international meetings, where I[11] had argued that cognizance-organization reciprocity is tied with reversibility of the joint probability.

So, rather than (possibly alternative) "consistent histories of particles," I uphold Bohr's concept of "a history of preparations and measurements," of what one *has done* and one *has learned* about particles. Nothing has changed in the conversations between physicists, but there is a radical change in the metaphysics.

A particle (or a system of particles) *transiting* from preparation to measurement is neither in the retarded state emitted by the preparation, nor in the advanced state going into measurement; Wheeler [12] says it is a "smoky dragon" living so to speak above the empirical world in the complex plane - or the Hilbert space. Regrettably, Omnès does not discuss the spectacular "delayed choice experiments" which retro-activate at will one or another of mutually inconsistent histories of particles...

From *delayed choice* to *retropsychokinesis* the veil is thin; Schmidt's experiment, alluded to earlier, was in retropsycho-kinesis. *Psychokinesis is retropsychokinesis*: influencing a random event generator is *influencing the hidden transiting states before amplification and display* of the outcome. Unbelievable as this may seem, Schmidt[13] has consistently done this sort of experiment, *perhaps the strongest proof of the views presented here*. Psycho-kinesis combined with an EPR style correlation *is* faster than light telegraphing[14] via a relay in the past, a feasible experiment disproving Omnès' locality criterion.

So, *the metaphysics opened by the "sesame" of transition amplitude is more radical than that of a "veiled reality."* Linking *reality* with *representation* as the twin faces of a medal, and *cognizance* with *organization* by turning the medal over, strips the verbs *to have* and *to be,* as understood by common sense, from reality. Assuming with Weyl[15] *extendedness of time*, it reshapes the concept of existence in a way which will now be examined.

2. SPATIO-TEMPORAL EXISTENCE

The first argument for excluding the final cause concept from physics, "non existence of the future", falls with the advent of the relativity theory. Who would have guessed in 1887 that Michelson's result, isotropy of the velocity of light in *any* inertial frame, entails time extendedness of matter ? This came as the "surprise du chef" in Poincaré's 1906 paper, and with great solemnity, in Minkowski's 1908 paper as well.

Disconnecting the concepts of *existing* and *now*, relativity sees past, present, and future as *coexisting - not now* of course, that would be self-contradicting!

The *fact* that had so long hidden Einstein's relativity of time was an existential one: largeness in practical units of the conversion coefficient c between space and time, evidencing a disproportion between our subjective feeling and objective perception of the "time flow." Conversion of time into space by relative motion was obvious to Galileo; the reciprocal conversion of space into time was Einstein's discovery, a consequence of the finiteness of c.

Still today, the full philosophical import of this momentous discovery is largely underestimated outside the physics community.

3. EFFICIENT AND FINAL CAUSE SYMMETRY

The second objection against the idea of final cause was past-future asymmetry: "advanced actions are not observed in physics." A burning meteorite emitting a retarded ballistic and thermal wave illustrates that *an interaction limited in time produces after- and not before - effects*. Thermodynamics and statistical mechanics formalize this situation as entropy and probability increase. The latter is now tied with wave retardation by the wave-particle dualism.

So, based on "fact-but-not-law"[16], the second objection is just as unfounded[17] as was the first; moreover, experimental facts, such as *spin echo* or recovering of *phase coherence*, refute it.

Let us show on a basic equation of the calculus of probabilities that rejection of final cause is a prejudice similar to Aristotle's ignorance of inertia.

Bayes' principle states: "The joint probability of two correlated occurrences A and C equals the product of the conditional probability of A-if-C times the prior probability of C or, *inversely*, the product of the conditional probability of C-if-A times the prior probability of A": a proposition which is symmetrical in A and C not only grammatically, but *also geometrically*, as explained presently.

A correlation of physical occurrences, termed spacetime events, is an *interaction*; the *joint probability* is its *expression*. The symmetry of it thus formalizes action-reaction equality for a space-like separation, cause-effect symmetry for a time-like one. A *physical* joint probability is "CPT invariant"

So *the calculus of probabilities is symmetrical in efficient and final cause*; Loschmidt's reversibility argument would have been more radical if enunciated this way.

The *factlike* preponderance of retarded causality in physics (or, as Fantappié[18] contends, of advanced causality in biology) is formalized via the prior probabilities, as van der Waals[19] explains. This calculational recipe is similar to the one selecting retarded or advanced solutions of a wave equation via the boundary conditions. Retarded causation and statistical prediction, and advanced causation and statistical retrodiction, respectively, have identical formalizations. That "one cannot see in the future nor act in the past" is a radical - too radical - exclusion of finality.

The Bayesian approach may not be the best one in physics, as researchers resemble less poker players than mushroom seekers in a wood. The probability they are using is *intersubjective* rather than subjective; its appropriate expression is not relative but *geometrically covariant*.

I therefore replace[20] the joint probability $|A)\&(C|$ by a *joint number of chances* $|A)\cdot(C| = |A)(A|C)(C|$, the product of a *reversible intrinsic conditional probability* $(A|C) = (C|A)$ times *both* the *a-priori probabilities* $|A)$ and $(C|$. When completed by the composition law $(A|C) = \Sigma\,(A|B)(B|C)$, this scheme is identical to that of *transition probabilities* in statistical mechanics, which are Lorentz and PT invariant, and also topologically invariant for chained collisions pictured in spacetime. *It is a geometrized calculus of intersubjective probabilities.*

4. PROBABILITY, ENTROPY, INFORMATION

It is clear today that Pascal and Fermat's discovery of the calculus of probabilities, and Maxwell and Boltzmann's formulation of statistical mechanics, were anticipating information theory. *Chance* for the former and *disorder* for the latter were lack of both *knowledge* and *control*. These twin faces of *information* are exchanged by the cybernetical *coding* and *decoding*. *Entropy* was Clausius' name for Shannon's[21] *missing information*, Boltzmann's *k Ln 2* denoting the change rate.

The third objection to final cause, that "an idea cannot move matter," is dispelled by the information-negentropy equivalence: turning *knowledge* into *control does* precisely this, which fundamentally is *psychokinesis*. Inside the cybernetics operator, psychokinesis is at the origin of coding, as it operates inside anyone moving his body by will. Descartes, in a 1648 letter to Arnauld, writes that "an indisputable everyday experience shows that our soul *does* move our body, this in a way differing entirely from the one by which a body moves another body." Eccles[22] confirms this by neurological experiments.

What delayed the momentous unveiling[23] of information-negentropy equivalence (implicit in Boltzmann's work) was the disproportion in the change rate, *k Ln2* # 10^{-16}: a "bit" of information is extremely cheap in Clausius entropy units, but a "Clausius" costs a lot of bits. This is why so many advertisements go straight to the wastebasket, while the cost of an object rewards the insight or craftsmanship of the discoverer, the maker, the conceiver, the manager. A non zero *k* means that "one cannot get anything for nothing, not even an observation", as Gabor puts it; smallness of *k*, Brillouin emphasizes[24], makes the negentropy cost of information easily forgotten. Cybernetics asks *consciousness-the-spectator* to pay a few cents for her ticket, but grants exorbitant wages to *consciousness-the-actor*. Thus, many spectators and few actors is one more wording of factlike irreversibility.

As for the lawlike mutual reversibility of negentropy and information implying existence of psychokinesis, Wigner[25] says, tongue in cheek, that "every phenomenon is unexpected and most unlikely until...discovered, and some...remain unreasonable a long time after [being] discovered."

Psychokinesis, as now repeatedly evidenced in the laboratory *via* random event generators[26], brings us back to the cradle of probability theory. In a November 1648 letter to Elisabeth, Descartes states that he routinely observes psychokinesis in chance games.

The paradigmatic implication is that our world is a twin faced one, *reality-and-representation*, and that a needle goes back-and-forth between the tapestry's weft and displayed face. More than one biologist or philosopher has expressed this sort of idea, for example, Claude Bernard[27] with his "idée directrice" of an organ. Some truly wonderful facts in mimetism do suggest even more in the direction of telepathy.

Retropsychokinesis deserves special consideration; this almost unbelievable but repeatedly displayed fact[28] speaks strongly in favor of time extendedness of both mind and matter.

5. WAVELIKE CALCULUS OF PROBABILITIES AND NON-SEPARABILITY

So, the general principle we call upon is that *both the use and interpretation of quantum mechanics should be uniquely derived from its Lorentz and CPT invariant formalism*; has not the subject of the verb "to undulate" been dismissed by the relativity theory promoting its spacetime paradigm ?

The 1926 Born-Jordan *wavelike probability calculus* radically invalidates the verbs *to be* or *to have* as applied to the *now* of particles, or their *present* "properties", thus reducing "reality" to a "trompe l'oeil". It does this by the calculation recipe of adding partial and multiplying independent amplitudes rather than probabilities. Dirac's[29] *bras* and *kets* scheme, Feynman's[30] Lorentz-style and Lüders[31] CPT-style dressing of it, have put on the final touch.

The manipulation rules "correspond" to the rules of the statistical transition probabilities. The *Hermitian reversible complex transition amplitude* $\langle A|C\rangle = \langle C|A\rangle^*$ corresponds to the *real reversible intrinsic transition probability* $(A|C) = (C|A)$; the *dressed transition amplitude* $|A\rangle\langle A|C\rangle\langle C|$ corresponds to the *dressed transition probability* $|A)(A|C)(C|$, implying both *a-priori* probabilities; the *composition law* of the amplitudes $\langle A|C\rangle = \sum \langle A|B\rangle \langle B|C\rangle$ corresponds to the composition law of the transition probabilities $(A|C) = \sum (A|B)(B|C)$.

Conditionality is essentially implied: the transition amplitude holds *if* each and every preparation and/or measurement as expressed in the formula is performed. Disregarding this warning has caused[32] some serious misinterpretations in *(inter alia)* the non-separability problem.

Dismissal of the verbs "to be" or "to have" is displayed by the cross-terms in the transition probability $(A|C)=|\langle A|C\rangle|^2$ via the composition $\langle A|C\rangle= \sum \langle A|B\rangle\langle B|C\rangle$ of amplitudes; these are *interference* terms, entailing *space or time non-separability*.

In Dirac's conceptualization, the *transition amplitude* between the *prepared* and the *measured* (*retropared* says Hoekzema[33]) states of an evolving system formalizes "inversely" either prediction and retrodiction (cognitive interpretation), or preselection and postselection (organizing interpretation).

Hermitian invariance $\langle A|C\rangle = \langle C|A\rangle^*$ of the transition amplitude expresses CPT invariance in a nutshell: the $\langle A|C\rangle \Leftrightarrow \langle C|A\rangle$ reversal symbolizes the geometric PT symmetry, while the $\langle A|C\rangle \Leftrightarrow \langle A|C\rangle^*$ one symbolizes particle-antiparticle exchange, C. Actively interpreted, the CPT symmetry implies cognizance-organization symmetry, or efficient-final cause symmetry, or cognizance-psychokinesis reciprocity.

Timelike non-separability consists of this: being neither in the retarded state issuing from the preparation, nor in the advanced state going into retroparation, an evolving particle or system is *not* in a "real hidden state", but (1927 wording) in a "virtual state". Wheeler[34] says it is a "smoky dragon", living, so to speak, above empirical spacetime, in the complex plane, letting its tail hang down here, on the real axis, grabbed as preparation, and its teeth biting as measurement. Whether tail and teeth "really are" down here, is an interesting question to be left aside here.

Each vertex of a Feynman graph hides a smoky dragon. The simple V- and Λ- shaped graphs, respectively picturing an EPR[35] and a reversed EPR correlation, display the spacelike non-separability.

The point is that in an EPR correlation the distant measured states $\langle A|$ and $|C\rangle$ of correlated amplitude $\langle A|C\rangle = \sum \langle A|B\rangle \langle B|C\rangle$ *cannot* have preexisted as such in the source B, due to the cross terms. Whence, mutually implicative[36] *retro-causation* and *spacelike non-separability*.

One often reads that "finding at A the state $\langle A|$ instantaneously collapses at C the distant state $|C\rangle$". *This is all wrong*, relativistically non-invariant, and A&C non-symmetric. As the correlation amplitude is A-and-C Hermitian symmetric, "which measurement collapses the other one"? Last but not least, *both* measurements *can be* and *indeed are* performed, their probabilities being $(1|1)=(0|0)$ and $(1|0)=(0|1)$ for the coupled answers *yes* or 1, *no* or 0.

Retrodictive non-separability shows up in the inverse EPR correlation, a Fresnel style interference, where the "paradox", known since the twenties, consists of this: one can neither retrodict nor find out experimentally which source any one of the detected photons came from.

In a famous EPR experiment by Aspect[37], the measuring devices at A and C were optical polarizers oriented *after* the correlated photons had left the source, *thus evidencing retrocausation.* New tests are now in progress with *both* orientations made randomly. Reliability of the detectors will be increased. Possibility of making the AC distance future-timelike, by folding the BA path, is envisaged; this would imply telegraphing backwards in time, from C to A. There is no reasonable doubt that the (Lorentz-and-CPT invariant) transition amplitudes scheme will be completely vindicated.

So, the "paradoxical" paradigm of the wavelike probability calculus keeps on unfolding - even more mind stretching than Kepler's heliocentrism was in its days. Even professional physicists get headaches when thinking upon what they successfully calculate.

6. PSYCHOKINESIS AND FINAL CAUSE

Is it *rational* in physics what is *First* "Deducible from the accepted equations"; *Second* "Experimentally verified"; *Third* "Sometimes 'paradoxical' in terms of the previous paradigm".

Psychokinesis is predicted as reciprocal to gain-in-knowledge by the negentropy-information equivalence, and by CPT reversibility of the transition amplitude. *Formalized by ponderation of the final a-priori amplitude, it is final cause operating.*

Schmidt[38], an engineer, initiated in 1976 the electronic random event generator for testing psychokinesis; Jahn, also an engineer, followed in 1980. Electronic noise in a circuit was the usual chance generator. Anybody can play the game. Small, the effect is evidenced statistically. The protocols have been found correct by external experts; one voiced criticism has been that "this sort of inquiry is unusual in physics" - an implicit definition of physics, and a restrictive one.

I have suggested a specific test which, in view of the many previous positive ones, should work[39]. A low intensity laser beam is separated by a semi-transparent plate. Normally, say, half and half; one emerging beam A is received by a photodetector, the other C is observed by an "agent" wishing to increase or decrease "paranormally" its intensity. The assumption is that the agent, retroacting at the plate, will attract or repel photons, correspondingly decreasing or increasing the other beam's intensity. This is "faster than light telegraphing" via correlation plus psychokinesis through a relay in the past, violating *not* Einstein's 1905 prohibition, but the taboo against retrocausation.

In 1994, *Physical Review A* published an article by Stapp[40] discussing an experiment in retropsychokinesis by Schmidt, in relation to the interpretation of quantum mechanics. Unobserved "pre-recorded targets" were set aside together with copies; some time later they were played before an agent invited to bias the outcome at will. "Believe it or not", the result was just as good as if psychokinesis were performed here and now. If confirmed, such results are no less revealing of the secrets of Nature than was Galileo's inertia.

7. VEILED REALITY, OR VEILED INFORMATION?

Bernard d'Espagnat[41] has shown in detail that reality as displayed in microphysics differs extremely from the idea one gets at the macro-level via common sense experience; his analyses confirm the more global ones made in the twenties by the founding fathers.

What is reality ? The latin *res* is heavy, opaque, passive. But the negentropy-information dialogue suggests a more dynamic view. Today d'Espagnat rather speaks of *Being*, with a capital letter. The Bible's Being, as revealed at the Sinaï, is Living Spirit.

This may be aiming too high; epistemology, not theology, is our business. An essay by Paul Claudel in *Figures et Paraboles* refers to biological evolution. He submits that God's responsibility is here indirect, as He has delegated powers to Prakriti, the goddess of life. Her inspiration is of variable quality, sometimes superb, as with the albatross, sometimes funny, as with the rhinoceros. This is Claudel's version of a twin faced universe, *reality-and-representation*. Today it should be extended to cosmology, as preferable to the anthropic (anthropocentric?) model. When one sees life's obstination to occupy incredibly hostile niches (at least they seem so to us) it seems that "information-as-organization" is playing an energetically low level game everywhere all-the-time, thus quietly countering irreversibility.

Disorder, for Bergson[42] the philosopher and Shafroth[43] the physicist, is *order differing from expectation*. Thus *physical irreversibility may well be more subjective than we feel*, meaning lack of *either* side of *information*.

Coming back to physics, it so happens that recent theories of the quantum measurement process impute state-vector-collapse to deliberate ignorance of the parameters of the measuring device uncoupled to the measured properties. The question is, *how could deliberate ignorance generate a reality*, that is, anything else than *illusion of a reality*? *Why* the state vector collapses this or that way we do not know, but this does not imply that no one knows (or does not do) about the collapse.

Who would have guessed, at the eve of this century, that the two theories then born, the quantal and the relativistic, would invite such questions? No subsequent step forward, important as it may have been, has suggested such radical reorientations.

Fondation Louis de Broglie, Paris
France

NOTES

1 Translated from the French text.
2 Cf. Omnès 1995.
3 Cf. Mehlberg, in Feigl and Maxwell 1961.
4 Cf. Libet 1985.
5 Cf. Brillouin 1967 and Rothstein 1958.
6 Cf. d'Espagnat 1995.
7 Cf. Wigner 1967.
8 Cf. Stapp 1994.
9 Cf. Schmidt 1993.
10 Cf. Jahn *et al.*, in Papanicolaov and Gunter 1987.
11 Cf. Costa de Beauregard, in Papanicolaov and Gunter 1987, pp. 315, 318.
12 Cf. Miller & Wheeler, in Kamafuchi *et al.* 1983.
13 Cf. Schmidt 1976.
14 Cf. Costa de Beauregard, in Papanicolaov and Gunter 1987, pp. 117,124.
15 Cf. Weyl 1949.
16 Cf. Mehlberg, in Feigl and Maxwell 1961.
17 Cf. Lewis 1930.
18 Cf. Fantappié 1991.
19 Cf. van der Waals 1911.
20 Cf. Costa de Beauregard, in P. Bush *et al.*, 1993.
21 Cf. Shannon 1948.
22 Cf. Eccles 1986.
23 Cf. Lewis 1930 and Shannon 1948.
24 Cf. Brillouin 1967.
25 Cf. Wigner 1967.
26 Cf. Jahn & B. Dunne, in Papanicolaov and Gunter 1987.
27 Cf. Bernard 1865.
28 Cf. Schmidt 1976 and 1993.
29 Cf. Dirac 1947.
30 Cf. Feynman 1949.
31 Cf. Lüders 1952.
32 Cf. Costa de Beauregard 1985.
33 Cf. Hoekzema 1992.
34 Cf. Miller & Wheeler, in Kamafuchi *et al.* 1983.
35 Cf. Einstein, Podolsky and Rosen 1935.
36 Cf. Costa de Beauregard 1983 and 1984.
37 Cf. Aspect 1982.
38 Cf. Schmidt 1993.

[39] Cf. Costa de Beauregard, in Papanicolaov and Gunter 1987.
[40] Cf. Stapp 1994.
[41] Cf. d'Espagnat 1995.
[42] Cf. Bergson 1907.
[43] Cf. Shafroth 1960.

REFERENCES

A. Aspect *et al.*, *Phys. Rev. Lett.*, **49**, 1982, 91.
H. Bergson, *L'évolution Créatrice*, Paris 1907, Chap. 3.
Cl. Bernard, *Introduction à l'Etude de la Médecine Expérimentale*, Paris 1865, 2ème Partie, Chap. 2.
E.I. Bitsakis and C.A. Nicolaides eds., *The Concept of Probability*, Kluwer Academic Publishers, Dordrecht 1989.
L. Brillouin, *Science and Information Theory*, 2nd ed., Academic Press, New York 1967, Chap. XIV, XV, XVI.
P. Bush *et al.* eds., *Symposium on the Foundations of Modern Physics*, World Scientific, Singapore 1993, p. 153.
O. Costa de Beauregard, *Phys. Rev. Lett.*, **50**, 1983, 867.
O. Costa de Beauregard, in S. Kamefuchi *et al.* eds., *Foundations of Quantum Mechanics in the Light of New Technology*, Phys. Soc. Japan, Tokyo 1984, pp. 233-241.
O. Costa de Beauregard, *Found. Phys.*, **15**, 1985, 871.
B. d'Espagnat, *Veiled Reality: An Analysis of Present-Day Quantum Mechanical Concepts, Frontiers in Physics*, vol. 91, 1995.
P.A.M. Dirac, *The Principles of Quantum Mechanics*, 3rd ed., Clarendon Press, Oxford 1947.
J.C. Eccles, *Proc. Roy. Soc.*, **22**, 1986, 411.
A. Einstein, B. Podolsky and N. Rosen, *Phys. Rev.*, **47**, 1935, 777.
L. Fantappié, *Principi di una Teoria Unitaria del Mondo Fisico e Biologico*, Di Renzo, Roma 1991.
H. Feigl et G. Maxwell eds., *Current Issues in the Philosophy of Science*. Holt, Rinehart, Winston, New York 1961, pp. 105-138.
R.P. Feynman, *Phys. Rev.*, **76**, 1949, 749; 769.
D. Hoekzema, *Found. Phys.*, **22**, 1992, 487.
M. Kafatos ed., *Bell's Theorem, Quantum Theory and Conceptions of the Universe*, Kluwer Academic 1989, p. 285.
S. Kamefuchi *et al.* eds, *Foundations of Quantum Mechanics in the Light of New Technology*, Tokyo 1983, p. 140-152.
G.N. Lewis, *Science*, **71**, 1930, 570.
B. Libet, *Behavioral and Brain Sciences*, **8**, 1985, 529.
G. Lüders, *Zeit. Phys.*, **133**, 1952, 325.
H. Messel ed., *Selected Lectures in Modern Physics*, Macmillan, New York 1960, p. 268.
Nicolaides eds., *The Concept of Probability*, Kluwer Academic, Dordrecht 1989, p. 167.
R. Omnès, *Found. Phys.*, **25**, 1995, 605.
A. Papanicolaov and P.A. Gunter eds, *Bergson and Modern Thought*, Harwood Academic, London 1987, p. 271.
J. Rothstein. *Communication, Organization and Science*, The Falcon's Wing Press, U.S.A 1958.
C. Shannon, *Bell Syst. Tech. J.*, **27**, 1948, 379, 623.
H. Schmidt, *Journal of Parapsychology*,**57**, 1993, 351.
H. Schmidt, *J. Amer. Soc. Psychical Research*, **70**, 1976, 267.
H. Stapp, *Physical Review* ,A **50**, 1994, 18.
J.D. van der Waals, *Phys. Zeits.* , **13**, 1911, 547.
H. Weyl, *Philosophy and Natural Science*, Princeton Univ. Press, Princeton 1949, p. 116.
E.P. Wigner, *Symmetries and Reflections*, M.I.T. Press, 1967, pp. 171-184.

CHARLES PAUL ENZ

OBSERVABILITY AND REALISM IN MODERN EXPERIMENTS WITH CORRELATED QUANTUM SYSTEMS

1. INTRODUCTION: THE REAL QUANTUM WORLD

If I have chosen to discuss correlated, i.e. interacting, quantum systems it is because I feel that these quite involved situations have been neglected for too long in the philosophy of physics. Indeed, the epistemological discussion of quantum mechanics has so far been limited essentially to an analysis of free particles and the particle-wave duality including, of course, all its ramifications with nonlocality, complementarity, uncertainty and the Copenhagen Interpretation. It is true that this artificial limitation to free particles or fields is not simply a free choice but rather the only issue available to philosophical researchers who are ignorant about the *physics of correlated quantum systems*.

The main reason for this ignorance, I believe, stems from the fact that those researchers in general come from a background of elementary particle or high energy physics. Now this type of physics operates with particle beams directed on a target (which in modern experiments often is another beam) and the reactions are registered far away from the target. Thus what is really observed are free particles or fields obeying Einstein's special relativity theory, while the interactions happen in a space-time region much too small to be accessible to direct observation. This is why I suggest looking for the real quantum world in condensed matter physics which, in spite of its mainly nonrelativistic nature, I believe, should not be ignored any longer by philosophers. It does not mean, of course, that correlated quantum systems do not exist in elementary particle physics, e.g. in the form of confined quarks.

The relation between condensed matter physics and elementary particle physics has long ago been discussed by one of the outstanding representatives of the former: Philip Anderson (Nobel Prize 1977) who in an article entitled "More is Different" wrote in 1972:

> "The ability to reduce everything to simple fundamental laws does not imply the ability to start from those laws and reconstruct the universe. In fact, the more the elementary-particle physicists tell us about the nature of the fundamental laws, the less relevance they seem to have to the very real problems of the rest of science, much less to those of society". [1]

In the last decade condensed matter physics has witnessed impressive breakthroughs. The most spectacular of them all of course is the discovery in 1986 of the first *High-Temperature Superconductor* (HTS) which won the authors K. Alex Müller and J.

E. Agazzi and M. Pauri (eds.), The Reality of the Unobservable, 293-310.

Georg Bednorz the Nobel Prize of 1987. In their pure form these substances are insulators, and it is only by doping with metal or oxygen atoms that mobile charge carriers (in most cases positive holes instead of electrons) are liberated. These then form pairs that condense into the superconducting state at a transition temperature of up to –140 *C* which is very high compared to that of the "old" superconductors which never reached more than –240 *C*.

These new superconductors exhibit an unexpected new complexity, namely strong correlations in the form of Coulomb repulsion between neighboring holes, which rendered the understanding in terms of quantum theory quite difficult. Thus the first calculations of the excitation spectrum of the pure substances which ignored these correlations showed ideal metallic behavior instead of an insulator. Much of this difficulty persists still today. In addition, these materials turned out to be layered structures in which conduction occurs predominantly within these layers, i.e. in two dimensions.

Strong correlations and *reduced dimensionality* ($d < 3$) have become the characteristics of a new era of condensed matter physics. Already before the advent of the new superconductors, planar electron systems in semiconductors subject to a transverse magnetic field had given rise to another spectacular phenomenon, the *Quantum Hall Effect* (QHE) which earned the principal discoverer, Klaus von Klitzing, the Nobel Prize of 1985. In this effect the electrons move along the one-dimensional edge of a planar metallic domain and the Hall conductivity has plateaus at values proportional to a number ν called the *filling factor* which is an integer in the "integer QHE" but is a rational fraction in the "fractional QHE". Such domains are of the order of a few hundred nanometers (*nm*) in size, i.e. a few tenthousandths of a millimeter and are obtained artificially by electron - or X-ray-lithography on semiconductor surfaces, like the chips in a computer. This so-called *nano-technology* is one of the important experimental tools of today's research in condensed matter physics, besides the solid-state chemistry and metallurgy supplying the crystals for HTS research.

Nano-technology is also used in the fabrication of so-called *artificial atoms* in which a fixed number of electrons is trapped within artificially prepared boundary conditions. These "zero-dimensional" structures working like transistors have become an extremely important area of semiconductor research. A quite different type of trap is obtained with the aid of electric and magnetic fields. Such wall-less traps allow one to perform precision measurements on single electrons.

Critical examination of experiments in the mentioned fields of research will be the main subject of this paper. However, in order to establish the connection with the present-day philosophical discussion of quantum phenomena revolving mainly around the particle-wave duality and non-locality, yet another field of activity has to be mentioned, namely *modern optics* whose main tools are the *laser* and *nonlinear optical media*. It is here that much of the debate over the Copenhagen Interpretation is taking place: the artificial world of single photons.

Consider the problem of a single photon impinging on a semi-transparent mirror, a so-called beam splitter. The incident photon obviously conforms with the particle picture, and so does the reflected or transmitted photon impinging on a counter, while reflection and transmission are typical notions of the wave picture. What is artificial here is the assumption of an initial *one-photon state*. Indeed, the output of a laser may be characterized as a wave having about equal uncertainty ΔN in the number N of photons and in the phase of the wave. It is possible to produce a low ΔN, either by "squeezing" or by using a decay cascade of an excited atom. $N \sim$

1 may then be obtained by lowering the intensity of the incident photon beam. A pure one-photon state, however, is an ideal limit which is practically unreachable. But this residual compositeness of the initial photon state makes the use of a *hidden auxiliary wave* accompanying the photon, such as de Broglie's guiding wave (onde guide) or Selleri's ghost field (Einstein's Gespensterfeld) or Popper's propensity wave[2], a quite hopeless enterprise.

More recently beam splitting experiments have been performed with massive neutral particles, first with neutrons[3] and now also with atoms[4]. In these cases one-particle states are better defined than for photons. The last reference is particularly remarkable since it proves explicitly the wave nature of complete atoms or, more precisely, of their *center-of-mass motion*. The consequence of the internal degrees of freedom, on the other hand, is that, passing a beam of neutral atoms through a gas of other atoms or molecules, the index of refraction of the gas depends in a complicated way on the internal dynâmics.

With these introductory remarks the problem is, I hope, sufficiently circumscribed: At stake is the very notion of the *single free particle*, both because of *particle-number uncertainty* and because of *correlations*. But before turning to the *real quantum world* let me review more in detail the notion of reality as it appears in the literature.

2. THE NOTION OF REALITY

Since Einstein's general relativity, we know that all the space-time coordinate systems are equivalent so that heliocentrism is only the most economical way to describe the planetary motions while geocentrism represents best the reality of everyday life. But philosophically, the reality that counts from the point of view of the scientific, but also the artistic or religious, endeavors may be called *homocentrism* in the sense of Kant's "Copernican Revolution", i.e. Kant's realisation that it is the human intellect which creates the laws of Nature and prescribes them to Her so that the *human consciousness* becomes the center of the world[5].

In the everyday world reality derives from our sensual perceptions. However, the smaller the scale of the object to be observed the more we must rely on instruments. (The symmetrical situation holds for astronomical observation.) The following quotation of Pauli - which emphasizes the necessity of the link with everyday reality - may be considered a guide in the quest for reality in microphysics:

> "Science is a systematic refinement of the concepts of everyday life revealing a deeper and, as we shall see, not directly visible reality behind the everyday reality of colored, noisy things. But it should not be forgotten either that this deeper reality would cease to be an object of physics, different from the objects of pure mathematics and pure speculation, if its links with the realities of everyday life were entirely disconnected"[6].

Quantum theory has taught us that subdivision of matter has a limit, a fact that the ancient Greek atomists intuitively felt must be true. The dilemma they encountered, however, was their total ignorance of the size of these indivisible objects, the Greek atoms. Historically this dilemma was overcome by Planck's discovery in 1900 of the *universal quantum of action h* and de Broglie's hypothesis of 1924 of *general particle-wave duality*. This may be seen as follows: According to the particle-wave duality, a particle of energy E has a frequency $\nu = E/h$ associated to it and a wavelength $\lambda = c/\nu$, c being the velocity of the wave, here that of light. But according to Einstein the energy at rest of a particle of mass m is $E = mc^2$ which gives $\lambda = h/mc$. Now, again making use of the particle-wave duality, the wavelength must be considered as a minimum uncertainty of the particle's position. But from our everyday notions of reality we conclude that in order for a particle to be divisible the uncertainty of its position cannot be larger than its extension represented by its radius R; hence: $R > \lambda$. Thus h/mc gives a *limit of subdivision*. Now this limit may be shown to perfectly allow the division of the chemical atoms - this is what chemistry is all about - but sets a qualitative limit of divisibility at the size of the proton and the neutron[7].

What this simple argument teaches us is that in observing smaller and smaller objects a limit determined by Planck's constant h is reached below which the behavior of objects changes qualitatively. In this way quantum theory gives meaning to the Greek atoms. But it also indicates that we cannot expect that the notion of reality we may agree upon for small but non-quantum objects such as bacteria, genes or dust particles should be transposable to truly quantum objects. Now the example of a truly quantum object most often considered in the epistemological discussion of quantum theory is the single photon impinging on a beam splitter considered in the Introduction. Leaving aside the difficulty mentioned there to produce a single photon state, Pauli, for example, says with respect to the particle-wave duality in such an experiment,

> *"The difference in the consequences of the two pictures is thus just as irreconcilable as the analogous difference between the two logical relations 'either-or' and 'both-and'."*[8]

This statement of Pauli expresses very succinctly the *Copenhagen Interpretation* accepting both, the particle and the wave picture, as *complementary*, i.e. neither contradictory nor identical, notions. This attitude of the Copenhagen Interpretation emphasizes the *contextual nature* of the meaning of physical concepts, a point which has also been stressed by Agazzi[9]. But such an *ad hoc* choice depending on the experimental situation has met with strong opposition from those who maintain the ontologically contradictory nature of particle and wave and led them to postulate the existence of a *hidden auxiliary wave* accompanying the real particle, as mentioned in the Introduction.[10] However, not only are these auxiliary waves becoming rather hopeless constructs when one is confronted with correlated quantum systems, but the notion of free single particle itself loses meaning, as I hope to show in the following sections discussing real experiments.

The other important aspect of quantum theory is *non-locality*. While it follows quite naturally from the wave nature of quantum objects, non-locality has gained particular importance through the discussion around the Einstein-Podolsky-Rosen (EPR) Gedanken Experiment. It was in the original EPR-paper of 1935 that the notion of *element of reality* was introduced in the following terms:

> *"If, without in any way disturbing a system, we can predict with certainty (i.e., with probability equal to unity) the value of a physical quantity, then there exists an element of physical reality corresponding to this physical quantity"*[11].

This problem arises for *composite systems* obtained by separating a quantum system such that there are correlations between the parts. I will call this type of correlations *EPR-correlations* in order to distinguish them from the correlations which are due to interactions between particles. For such EPR-type composite systems there are physical quantities to which, because of these correlations, no element of physical reality in the sense of EPR may be ascribed. Pauli had immediately seen the importance of this aspect in the EPR paper. On 15 June 1935 he writes to Heisenberg (my translation):

> *"Quite independently from Einstein it seems to me that in a systematic foundation of quantum mechanics one should start more from the composition and separation of systems than it happens till now (e.g. with Dirac). - This in fact is - as Einstein felt correctly - a very fundamental point in quantum mechanics, ... "*.[12]

Composition and *separation* are deep philosophical and psychological notions. The first aspect was analyzed in great detail by Kant who in his "Second Antinomy" proved that both statements: *"any composite system in the world consists of simple parts"* and *"no composite system in the world consists of simple parts"* are true.[13] The psychological aspect of composition and separation, on the other hand, was discussed in depth by Pauli in his "Hintergrundsphysik" where he gives as examples the isotope separation and the process of giving birth[14].

3. INDIVIDUAL ELECTRONS

Besides the photons considered in the Introduction, electrons are the particles of universal utility due to their charge and spin, their stability, and their small mass. They inhabit condensed matter in multiple appearances as we shall see. To look at them individually we may localize them in various ways. Conceptually, if not experimentally, the simplest way is to put them individually into a box. However, because of the electron's charge, walls are not suitable and must be replaced by electric and magnetic fields. Such "Penning traps" have been developed by Hans Dehmelt and Wolfgang Paul whose work was rewarded by the Nobel Prize of 1989.

In one experiment Dehmelt stored a single electron during 10 months. He observed that the electron, which oscillates in the trap, makes random quantum jumps[15]. This shows that causality manifestly does not hold. It has been established in other experiments that quantum jumps are sudden, i.e. it is not possible to associate an energy uncertainty to the jump duration[16]. In Dehmelt's experiment the jumps are caused by the coupling of the electron's charge to the ever present vacuum fluctuations of the electromagnetic field. Therefore the individual electron is not really a free particle but interacts with its environment; it is an *open system* showing the same difficulties with respect to a description as particle with a hidden auxiliary wave accompanying it, as was mentioned in the Introduction for the photon. For the electron, the consequence of this coupling is that its magnetic moment is not one unit of Bohr magneton but, as Dehmelt has measured, 1.001159652188(4) units[17]. This is one of the most accurately measured *renormalization effects*.

In plotting this result as function of the extension R together with the results for heavier particles, Dehmelt was able to deduce by extrapolation an upper limit for the electron's extension; he found $R < 10^{-20}$ *cm* which is 10 million times smaller than the extension of proton and neutron. Therefore one may safely conclude that *the electron is point-like and non-composite*. Dehmelt, however, speculates about an infinite hierarchy of composition of the electron. But if, for simplicity, one assumes that the electron is composed of N bodies of mass M each, the value for R requires that N is 2 or 3 and M is of the order of the mass of a proteine molecule. And the reduction of the composite mass NM to the electron mass would require an equally huge binding energy[18].

Individual bound electrons may be realized in a so-called "quantum dot" (a potential hill) or in a "single electron transistor" which are various forms of *artificial atoms* of a size some 100 times that of chemical atoms[19]. They are produced by the *nano-technology* mentioned in the Introduction. In a single electron transistor one may add or subtract individual electrons by overcoming the energy barrier e^2/C, called the Coulomb blockade, where C is the capacity between this electron and the rest of the system. This experiment really demonstrates *charge quantization* in units of the elementary charge e. While the capacity C is an effect of electron-electron interaction, the charge e is not renormalized by interaction, in distinction, e.g. to the magnetic moment mentioned before, and also contrary to the mass of the electron. The observability of the universal value e of the electron's charge in condensed matter is a remarkable fact. As in elementary particle physics, it is the consequence of a symmetry called gauge invariance. The "single electron transistor" is also noteworthy because successive addition of electrons is analogous to moving through the periodic table of the chemical elements[20].

4. ELECTRON CORRELATION AND DIMENSIONALITY

This idyllic picture of integer charge and individual electrons ceases to hold when correlations between them become important. As mentioned in the Introduction, electron correlations became the main theoretical problem of the *high-temperature superconductors* (HTS), $La_{2-x}Sr_xCuO_4$ (see Fig. 1a) discovered in 1986, and $YBa_2Cu_3O_{6+x}$, etc. that followed. Correlations are responsible for the fact that the pure substances with zero doping, $x = 0$, are insulators. For each of these substances, x has a critical value at which there is a so-called *Mott transition* from an insulator to a metal which then becomes superconducting. It was confirmed quite early that the charge observed in the superconducting state is indeed $2e$, not e, i.e. the electrons form pairs. This pair formation again is due to an interaction which must be attractive but for which there is a satisfactory explanation only for low-temperature superconductors, not yet, however, for HTS's.

While correlations in the new superconductors are three-dimensional, although with a strong dominance in the CuO_2 planes (see Fig 1b), the *fractional quantum Hall effect* is a strictly two-dimensional phenomenon in which correlations are important. In the integer QHE, on the other hand, correlations are negligible. Consider instead the fractional QHE corresponding to a filling factor $\nu = 1/(2k+1)$, $k=1,2,...$ This occurs in a transverse magnetic field so large that, measured in units of the magnetic flux quantum hc/e, the field is $2k+1$ times the density of the electrons. The effect is observed as a plateau in the Hall resistance (reciprocal conductivity) at the value $h/\nu e^2$ (see Fig. 2a). One knew since the discovery of the effect in 1982 that the *quasi-particles* moving along the edges must somehow have fractional charge. However, it is only very recently that direct experimental evidence of a *fractional charge* $e/3$ has been obtained[21].

So what happened to the electrons? Does context here favor the wave picture over the particle picture? The answer necessarily had to come from a careful analysis of the many-particle wavefunction of all the correlated electrons together. Indeed, in 1983 Robert Laughlin proposed a highly nontrivial *condensate wavefunction* ψ_k whose excitations are *quasi-particles of fractional charge* $e/(2k+1)$. For its construction Laughlin was guided by the condition that the 2-dimensional electrons form an incompressible liquid. ψ_k then can be represented by a function of the positions of all the electrons such that ψ_k vanishes as δ^{2k+1} where δ is the distance between any two electrons.

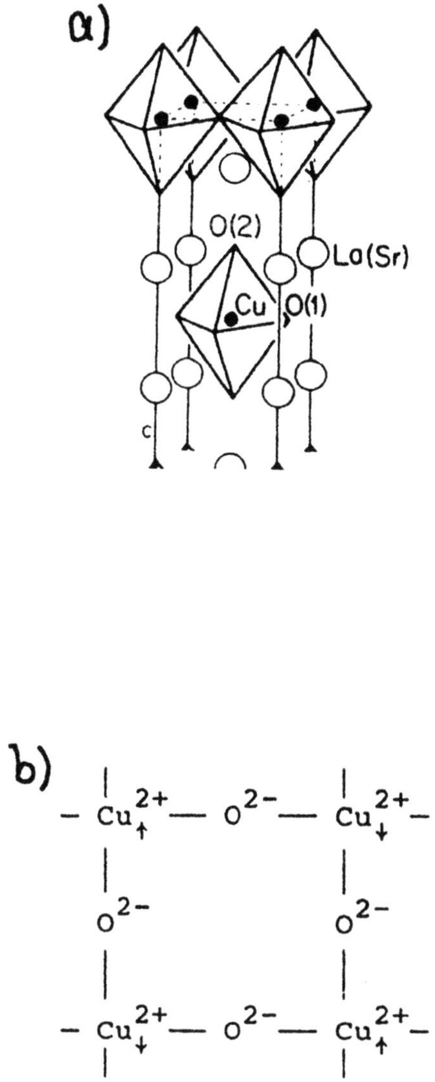

Figure 1 a), b)

a) Unit cell (bounded by the vertical lines) of $La_{2-x}Sr_xCuO_4$. The octahedra have O as corners ($O(1)$ and $O(2)$ are inequivalent positions) and Cu at the center.
b) The CuO_2- lattice in the horizontal symmetry planes of the octahedra, with associated valences and spin values.

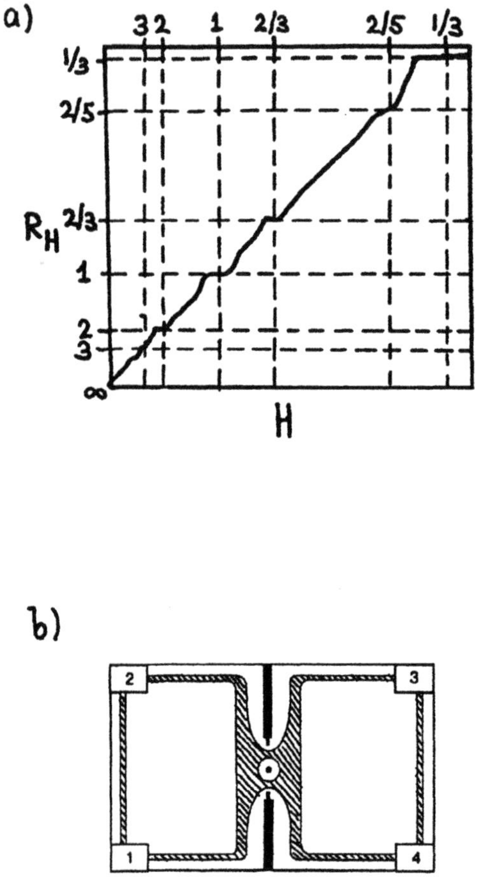

Figure 2 a), b)

a) Hall Resistance $R_H=h/\nu e^2$ as function of the magnetic field, parametrized by the filling factor ν *(H=hcn/νe)*, showing plateaus at integer values and at rational fractions of ν (*n* is the electron density).
b) Nano-Structure used in the experiment of Goldman and Su. The hatched areas represent the Quantum Hall condensate whose boundaries are the edge states. Quasi-holes tunnel from the inner edge at the anti-dot (center) to the outer edge and to the ohmic contacts 1,...,4.

But just to confuse the issue of the context dependence even more, Jainendra Jain succeded in 1989 to describe the same situation in terms of *local composite particles*, namely fermions which are bound states of an electron and 2*k* flux quanta. In this picture the flux quanta assume the role of the correlation, i.e. of the interaction between the electrons. Remarkably, this then allows a unified description of both, the integer and the fractional QHE, by defining the former to have zero flux, i.e. $k = 0$ in the particular case $\nu = 1$.

In the experiment, a "quantum antidot", i.e. an artificial atom binding quasi-holes instead of electrons was lithographically produced inside the incompressible fluid of the QHE-electrons. Measuring now the conductance due to tunneling of the quasi-holes from the antidot to the outer edge of the incompressible fluid (see Fig. 2b), the experimenters observed peaks as function of both the magnetic field and the gate voltage. Combining both results, they directly determined the charge q of the tunneling quasi-holes. The experiments were performed both in the integer ($\nu = 1,2$) and in the fractional ($\nu = 1/3$) regimes of the quantum Hall effect, yielding, $q = e$ and $q = e/3$, respectively[22].

If we lower dimensionality still further, other surprises await us. Indeed, in *one dimension* correlated electrons *dissociate* into a charge-carrying part and a spin-carrying part, and the two parts move quite independenly from each other. This so-called *Luttinger-liquid* description applies in particular to the edge excitations in the quantum Hall effect. It is interesting that this same dissociation which is rigorous in one dimension has long been suspected by Philip Anderson to hold true also for the 2-dimensional holes in the CuO_2 planes of the HTS's; he calls the parts *holons* and *spinons*. However, to this day, charge-spin separation has not been observed.

With this discussion I hope to have given convincing evidence that correlated electrons may behave in a way in which neither of the naive particle or wave pictures applies and that the only trustworthy guiding principle remains the *realistic many-particle wavefunction*. In the case of the above QHE-experiment the difficulty of interpretation is quite obvious from the description given there. But the difficulty of the experimenters in designing this experiment is also evident. This shows that *understanding is a give and take of interpretation.*

5. EPR-CORRELATIONS

50 years after the EPR-paper, Nathan Rosen, one of the original authors, wrote:

> *"The fact is that, in order to describe the results of new experiments, the trend seems to be toward theories that are more abstract, with less significance assigned to the role of space-time description and with a smaller degree of separability. Hence it is hard to believe that a theory will be found that will be complete, based on the criterion of an element of reality used in the [EPR-] paper. It may also be that in the future physical theories will describe reality in different terms from those to which we are now accustomed"*[23].

While I agree with these considerations, it is not in terms of more abstract theories that I wish to present EPR-correlations but, as before, by describing a real experiment. One of the cleanest EPR-type experiments I know uses two photons of different colors ω and ω_0-ω and perpendicular polarizations $\boldsymbol{o}$ and $\boldsymbol{e}$. They are produced from initial laser photons of color ω_0 in a BaB_2O_4 crystal as nonlinear response in the form of an ordinary (o) and an extraordinary (e) ray[24]. This procedure gives rise to an initial beam described by a state ψ_0 containing both photons. Passing this beam through a non-polarizing 50 : 50 beam splitter, a composite system in the form of an *entangled state* ψ_{12} consisting of two spatially separated beams 1 and 2 is produced such that the transverse polarizations $\boldsymbol{x}_1, \boldsymbol{y}_1$ of the two photons in beam 1 and similarly $\boldsymbol{x}_2, \boldsymbol{y}_2$ in beam 2 form a right-handed coordinate system with the respective beam direction. "Entanglement" means that ψ_{12} cannot be written as a product $\psi_1\psi_2$ of states of each beam 1 and 2 and physically implies *EPR-correlation.*

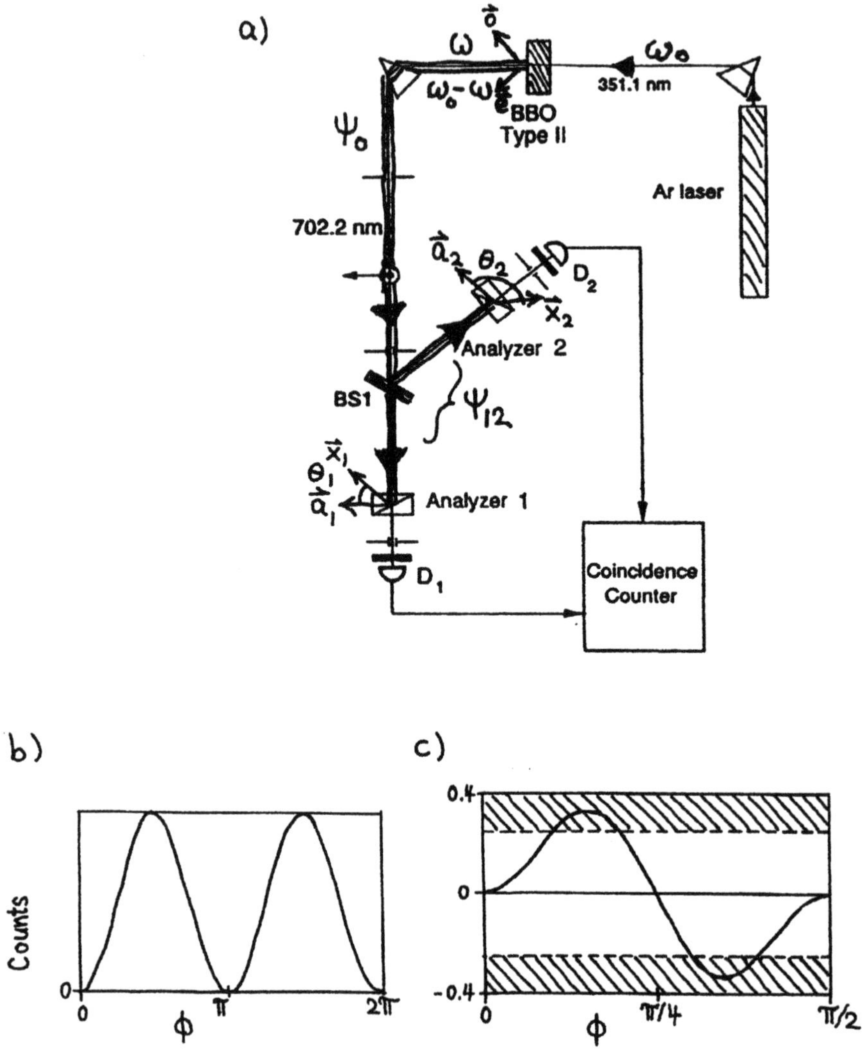

Figure 3 a), b), c)

a) Experimental setup of Kiess *et al.* which shows the down-conversion at the BBO crystal of the initial photon of frequency ω_0 into two photons of frequencies ω and $\omega_0 - \omega$ and polarizations $\boldsymbol{o}$ and $\boldsymbol{e}$, respectively, forming the initial state ψ_0. The entangled state ψ_{12} is formed at the beam splitter BS1, and the analyzers measure the angles Θ_1 and Θ_2 between the $\boldsymbol{o}$-polarizations $\boldsymbol{x}_1$ and $\boldsymbol{x}_2$ and arbitrary directions $\boldsymbol{a}_1$ and $\boldsymbol{a}_2$, respectively.

b) Coincidence counts as functions of $D = \Theta_2 - \Theta_1$

c) Violation of a Bell inequality represented by data in the hatched strips.

In order to see this, let me describe the experiment in some more detail: Each of the beams 1 and 2 now passes through an analyzer which projects the polarization on an arbitrary transverse direction $\boldsymbol{a}_1$ and $\boldsymbol{a}_2$, respectively, such that the angles θ_1 between $\boldsymbol{x}_1$ and $\boldsymbol{a}_1$, and θ_2 between $\boldsymbol{x}_2$ and $\boldsymbol{a}_2$, may be varied independently. The experiment which consists in counting coincidences between beams 1 and 2 behind the analyzers (see Fig. 3a) now shows that *the counting rate depends only on the difference* $\theta_2 - \theta_1$ (see Fig. 3b), and not on both angles separately. This establishes the EPR-correlation since the coincidences associate (within experimental error) to an arbitrary value θ_1 of analyzer 1 a unique value θ_2 of analyzer 2.[25]

This is quite remarkable since it appears as if information has been transmitted instantanously from analyzer 2 to analyzer 1, in contradiction with Einstein's postulate that information, i.e. energy, cannot travel faster than the speed of light. The remark that saves the situation of course is that no energy transfer is involved in the described measurement. We also see again that it is the complete wavefunction ψ_{12} rather than a single-particle or single-wave picture that technically solves the problem. I mention only in passing that this experiment also provided one of the strongest violations of a *Bell inequality*[26] and hence one of the best confirmations of the validity of quantum mechanics (see Fig. 3c).

6. OBSERVABILITY OF THE WAVEFUNCTION

Let me consider again electrons. In a perfect crystal at zero absolute temperature they will not be scattered by the atoms forming the crystal lattice. The electrical resistance is zero and the electron's wavefunction ψ extends over the whole crystal such that $|\psi|^2$ has the periodicity of the lattice. Thus one might think that taking as a crystal a piece of thin copper wire one should be able to cut electrons in two just by cutting the wire with scissors. This, of course, is not possible; what happens when cutting the wire is that new boundary conditions are established so that one part of the electrons will have wavefunctions ψ_1 extended over piece 1 of the wire while the remaining electrons will have wavefunctions ψ_2 extending over piece 2.

In practice a crystal is not perfect. Most important among the imperfections are impurity atoms as in the HTS materials. At zero temperature impurities scatter electrons elastically, i.e. without energy loss to the electrons. If there are many impurities, disorder in their positions does give rise to an electrical resistance, even at zero temperature. The reason is quite subtle: disorder implies that the *phase* ϕ *of the wavefunction* defined by the relation $\psi = |\psi| e^{i\phi}$ is *averaged incoherently,* while coherent averaging produces a zero-result. What is even more remarkable is that very strong disorder may localize the electrons completely so that the metal becomes an insulator. Thus disorder has an averaging effect on the electron's wavefunction. This so-called *Anderson transition* is analogous to the Mott transition mentioned in Section 4. But while in the latter localization is due to correlations, in the former it is due to disorder. In both cases it is an interaction that modifies the appearance of the electrons from extended to localized.

Since, as I tried to show all along, the wavefunction ψ is such a prominent tool for the understanding of the quantum reality, one may ask whether it is possible to observe ψ directly. As we shall see, the answer is affirmative as concerns the amplitude $|\psi|$. But again, the question is more subtle concerning the phase ϕ. The reason is that the particle number N and the phase ϕ approximately form a canonically conjugate pair of variables so that there is an uncertainty relation $\Delta N\,\Delta\phi > 1$. This implies that for a fixed number N of electrons, $\Delta N=0$, so that the phase ϕ is completely undetermined and hence unobservable.

Very recently the observation of $|\psi|$ has indeed become possible thanks to the invention of the *scanning tunneling microscope* (STM) by Gerd Binnig and Heini Rohrer who were awarded the Nobel Prize of 1986 for it. Remarkably, both were working at the same IBM-laboratory in Rüschlikon near Zürich as Bednorz and Müller who got the Price for the HTS one year later. The STM is able to probe surfaces on an atomic scale: If a very fine tip ending in a single atom is held sufficiently close to a metallic surface, say at a distance z, a voltage V applied between the tip and the surface will give rise to a tunneling current I. I then is a function of z, V and the local electron density $\rho = |\psi|^2$. At normal operation one scans a surface, parametrized by coordinates x,y, at constant V, I and ρ so that one measures directly the topography $z(x,y)$ of the surface (see Fig. 4a).

The same instrument also allows one to position single atoms on a metallic surface. In this way Michael Crommie, Christopher Lutz and Donald Eigler of IBM's Almaden Research Center in California succeded in constructing a "quantum corral" consisting of 48 iron atoms arranged on a circle with a diameter of 14.3*nm* on an oriented copper surface[27]. Then the authors put an electron into the 2-dimensional box formed by the corral and determined the local electron density ρ by scanning tunneling *spectroscopy*, i.e. by measuring the slope dI/dV as function of V. The result of their measurement, namely the topography $|\psi(x,y)|^2$ of the electron's wavefunction appeared on the cover of *Physics Today* in November 1993 (see Fig. 4b). The result also agrees well with the theoretical wavefunction for zero angular momentum and an effective mass of 0.38 electron masses. From the point of view of quantum reality this is, in my opinion, one of the most spectacular experimental results.

A detailed quantum theory of the STM measurement process of course is quite complicated. Measurement always involves damping due to statistical fluctuations which gives rise to the celebrated process of *wavefunction reduction*. The statistical aspect of the problem is taken into account by generalizing the wavefunction to the *density matrix*. Wavefunction reduction then becomes *decoherence* of the density matrix to diagonal form which yields the result of the measurement[28].

The particular case of an STM measurement has recently been analyzed in terms of decoherence by Milburn, Jacobs and Walls for an STM-like model in which the tip of the STM is attached to a cantilever. Instead of the distance z, it is then the force on the tip that is measured in such an "atomic force microscope" (AFM). The cantilever now is modeled as a perfectly reflecting wall of a strongly damped cavity. Comparing with recent experiments, the authors conclude that current AFM's are not quantum limited, i.e. may still be described, as Bohr claimed, as a classical instrument.[29] But this may not continue to be the case for all future measuring devices!

Let me now turn to the question of the observability of the phase ϕ. Evidently, we have to look for situations where the particle number N is not fixed. This is the case in *superfluidity* and *superconductivity*. An interesting candidate is a *Josephson junction* which consists of two superconductors S_1 and S_2 separated by a thin non-superconducting interface. Superconducting electron pairs described by pair-wavefunctions $\psi_j = |\psi_j| e^{i\phi_j}$ in S_j, $j = 1, 2$, are able to tunnel between S_1 and S_2, and the phase difference $\phi = \phi_2 - \phi_1$ obeys the *two Josephson laws* for which the author received the Nobel Prize in 1973.

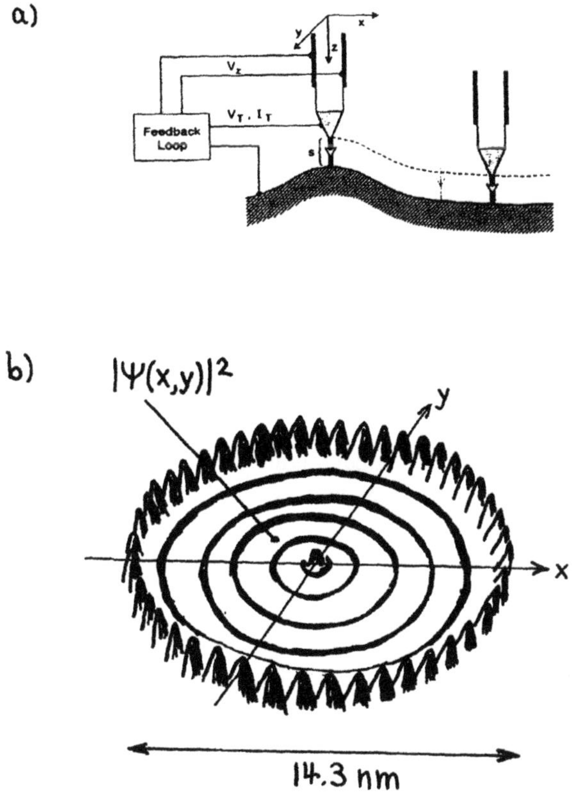

Figure 4 a), b)

a) Principle of Scanning Tunneling Microscope (STM).
b) $|\psi(x,y)|^2$ for a single electron confined in a "quantum corral" of 48 Fe atoms positioned on a *Cu*-surface in a circle of radius 14.3 *nm*, as obtained by scanning tunneling spectroscopy in Crommie *et al.*
[Figure 4a) is reprinted from G.Binnig and H.Rohrer, *Nobel Lecture*, December 8, 1986, p.396, in G.Ekspong ed., *Nobel Lectures*: Physics 1981-1990, World Scientific, Singapore 1993]

The first law relates ϕ to the tunneling current: $J_s = J_o \sin \phi$ while the second relates the time variation of ϕ to the applied voltage: $d\phi/dt = 4\pi eV/h$. Taking into account the capacity of the junction and applying an external current (see Fig. 5), ϕ satisfies the same equation as the elongation of a pendulum and hence may easily be observed[30].

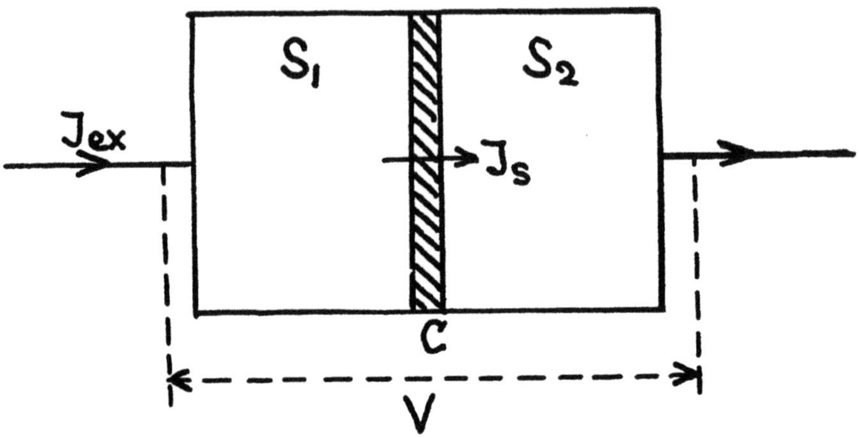

Figure 5

Josephson junction consisting of two superconducters S_1, S_2 separated by a thin normal interface characterized by a capacitance C, with tunneling current J_s and applied voltage V and current J_{ex}.

7. CONCLUSION: THE REALITY OF SYMBOLS

It had been stated some time ago by Evandro Agazzi that in quantum mechanics one needs new physical concepts; he says (my translation):

> 'The attempt to render the theorization about the quanta more satisfactory should be directed towards the quest for some *new* concepts not only, as already mentioned, for the simple fact of resulting in the new kind of combination of the classical concepts but, more directly, for the fact of replacing in total or in part those classical 'components' by something really new'[31]

In view of the above discussion of real experiments, I am tempted to say that this new concept Agazzi wants is nothing else than the old wavefunction but taken in its realistic correlated many-body form.

Notions which one has coped with for a sufficiently long time in order to understand some physical phenomenon, gradually acquire the status of ontological reality. This is in perfect agreement with what Selleri defines as *empirical realism*, meaning that the reality of properties precedes the reality of the object[32]. The wavefunction is a good example: We do not know what the object behind its properties really is. In this sense the object has less reality than its properties summarized in ψ. Or, expressed in the terminology of Bernard d'Espagnat, "there is something beyond the phenomena" which, very much like Kant's thing-in-itself, we cannot grasp. Therefore, calling this something "independent reality", d'Espagnat then concludes that this reality is "veiled"[33].

Practising quantum physicists and chemists have befriended the wavefunction in a way that makes it indeed this kind of *reality of properties*. In order to give a taste of the attitude typical among practitioners, be they experimentalists or theorists, let me quote the introductory passage of one of the papers discussed in Section 6. The authors Milburn and Walls are among the outstanding representatives of quantum optics; they write:

> 'If there is a problem of measurement in quantum mechanics, it must surely be this: why is quantum measurement theory apparently so irrelevant to experimental physics? More than 60 years of unresolved debate on the interpretation of quantum mechanics and measurement has not in the least slowed the progress of experimental physics. Perhaps the answer to this question is, until very recently, no measurement has been limited by quantum noise. Thus it has simply not been necessary to consider in detail the action of a measuring device at the quantum level".[34]

However, in spite of the observable properties of the wavefunction discussed in the last section, ψ is a complex-valued function in a multi-dimensional space and hence has a quite abstract and even symbolic meaning - it stands for the veiled something. Now ontological reality of symbols is not such a strange thought. After all, even the world of our everyday sensual perceptions is full of it, just think of the strange reality of dreams or, more generally, of the unconscious. Interestingly, ψ also serves as a shorthand symbol for psychology.

Pauli was quite familiar with this world and therefore it had a high degree of reality for him. In a letter to Markus Fierz, dated 12 August 1948 he writes (my translation):

> "The layman usually thinks that when he says 'reality' he is speaking of something obviously known; while it seems to me just to be the most important and exceedingly difficult task of our time to work for the establishment of a new idea of reality. This is also what I mean when I always emphasize that science and religion 'must' have something in common. (I do not mean religion within physics nor also physics within religion -since both would be onesided - but inclusion of both into a whole.) What appears to me to contain the new idea of reality I would like to tentatively call: the idea of the reality of the symbol. A symbol is, on the one hand, a product of the efforts of man and, on the other hand, a sign for an objective order in the universe of which man is only a part. It owns something of the old notion of God and also something of the old notion of object. (An example within physics: 'the atom'. The primary qualities of space-filling are obviously lost. If it were not a symbol, how could it be at the same time wave and also particle?)."[35]

For a Chinese or a Japanese the reality of symbols is exemplified by the Chinese character representing his name. Indeed, for a Chinese (my translation) "*a caligraphic character thus is, after all, like the cipher of his identity*" [36]. In the same way as the Chinese characters have to be appreciated not as static pictures but by recreating in one's mind the dynamics of the caligrapher's mind, body and brush drawing the successive strokes, the understanding of a specific quantum phenomenon - which, as I tried to show, is most accurately described in terms of the wavefunction ψ - is a *dynamic process*. This suggests that the quest for quantum reality may be best characterized as *active realism*. Indeed, in order to understand the *symbolism of* ψ one has to recreate in one's mind the *properties expressed by* ψ much in the way Chinese characters have to be read. This then may be called "*interpretation*"!

Admittedly, to a practising quantum physicist[37] labels like "realism" do not come easily. He may in general accept the existence of a mind-independent reality but will find the large choice in "varieties of realism" rather confusing because "*the philosophical views of real-life physicists are usually 'mixed' cases of different influences and doctrines, and they cannot be reduced to the 'pure' states defined by the abstract distinctions of the philosophers.*"[38]

But while labels like "realism" may be useful for classifying lines of thought, notions such as "composition", "separation", "atom", "particle", "field", "wave" or even "wavefunction" have deeper, archetypical roots. Let me elucidate this by letting C.G. Jung have the last word (my translation):

> "... the quite respectable atomic theory of the old Leukippos and Democritos ... [rests] on a 'mythological' notion of smallest particles which, as atoms of the soul, as smallest living particles, were already known to the still paleolithic people of central Australia. How much the circumstances of the soul are projected into the unknown of the external phenomena is familiar to any expert of the old Natural Science and Natural Philosophy. It is in fact so much that we are not at all capable to ever indicate how the world is really constituted since we are indeed forced to transpose the physical event into a psychical process if we want to talk of knowledge at all. However, who garantees that in this transposition any adequate 'objective' world view will result?"[39]

Geneva University
Switzerland

NOTES

[1] See Anderson 1972, quoted in Silvan Schweber, *Physics, Community and the Crisis in Physical Theory, Physics Today*, November 1993, p. 34.
[2] See Tarozzi 1986, pp. 360-368.
[3] See Rauch *et al.* 1975.
[4] See Schmiedmayer *et al.* 1995.
[5] See, for instance, Popper 1984, p.144.
[6] See Pauli 1994.
[7] See Enz 1985.
[8] See Pauli 1994.
[9] See Agazzi 1969.
[10] See, for instance, Selleri 1989.
[11] See Einstein *et al.* 1935.
[12] See von Meyenn *et al.* 1985: letter [412], p. 404.
[13] See Kant 1993.
[14] See Meier 1992.
[15] See Dehmelt 1990.
[16] See Sauter *et al.* 1986.
[17] See Dehmelt 1990.
[18] See Enz 1991.
[19] See Kastner 1993.
[20] See Kastner 1993.
[21] See Goldman and Su 1995; also Jain 1989, 1990.
[22] See again Goldman and Su 1995.
[23] See Rosen 1985.
[24] See Kiess *et al.* 1985.
[25] See Kiess *et al.* 1993.
[26] See, e.g., d'Espagnat 1989.
[27] See Crommie *et al.* 1993; also Collins 1993.
[28] See Zurek 1991.
[29] See Milburn *et al.* 1994.
[30] See, e.g., Enz 1992, p. 229.
[31] See Agazzi 1969.
[32] See Tarozzi 1986, p. 367.
[33] See d'Espagnat 1987 and 1989.

[34] See Milburn *et al.* 1994.
[35] See von Meyenn 1993: letter [971], p. 559-560.
[36] See Billeter 1989.
[37] See, e.g., Enz 1992.
[38] See, e.g., d'Espagnat 1987, 1995; Niiniluoto 1987.
[39] See Jung 1993.

REFERENCES

E. Agazzi, *Temi e problemi di filosofia della fisica,* Manfredi, Milano 1969.

P.W. Anderson, "More is different", *Science* **177**, 1972, 393.

J.F. Billeter, *L'art chinois de l'écriture,* Albert Skira S.A., Genève 1989.

M.C. Crommie, C.P. Lutz and D.M. Eigler, "Confinement of Electrons to Quantum Corrals on a Metal Surface", *Science* **262**, 1993, 218. See also G.P. Collins "STM rounds up electron waves at the QM corral", *Physics Today,* November 1993.

H. Dehmelt, "Experiments with an Isolated Subatomic Particle at Rest", *Rev. Mod. Phys.* **62**, 1990, 525.

B. d'Espagnat, "Nonseparability and the Tentative Description of Reality", in W. Schommers ed., *Quantum Theory and Pictures of Reality,* Springer, Berlin Heidelberg 1989.

B. D'Espagnat, "Veiled Reality", in P. Lahti and P.Mittelstaedt eds., *Symposium on the Foundations of Modern Physics 1987. The Copenhagen Interpretation 60 years after the Como Lecture,* World Scientific, Singapore 1987; *Veiled Reality. An Analysis of Present-Day Quantum Mechanical Concepts, Frontiers in Physics,* vol. 91, Addison-Wesley, Reading 1995.

A. Einstein, B. Podolsky and N. Rosen, "Can Quantum-Mechanical Description of Physical Reality be considered complete?", *Phys. Rev* **47**, 1935, 777.

C.P. Enz, "Le rôle de l'espace et le problème de localisation en physique moderne, vus en particulier par Wolfgang Pauli", in *Les relations mutuelles entre la philosophie des sciences et l'histoire des sciences,* Entretiens de l'*Académie Internationale de Philosophie des Sciences,* 15-18 mai 1985, Fribourg, *Archives de l'Institut internationale des Sciences théoriques,* vol. 29, Office international de librairie, Bruxelles 1987, Tilger Press, Genova 1987; reprinted as Special Issue of *Epistemologia,* vol. 10, 1987.

C.P. Enz, *Course on Many-Body Theory Applied to Solid State Physics,* World Scientific, Singapore 1992.

C.P. Enz, "Quantum Theory in the Light of Modern Experiments", in G. Schurz and G.J.W. Dorn eds., *Advances in Scientific Philosophy. Essays in Honour of Paul Weingartner on the Occasion of the 60th Anniversary of his Birthday,* Rodopi, Amsterdam, Atlanta, GA, 1991.

V.J. Goldman and B. Su, "Resonant Tunneling in the Quantum Hall Regime: Measurement of Fractional Charge", *Science* **267**, 1995, 1010. See also J.K. Jain. "Composite-Fermion Approach for the Fractional Quantum Hall Effect", *Phys. Rev. Lett.* **63**, 199, 1989; *Phys. Rev.* B **41**, 1990, 7653.

C.G. Jung, "Über den Archetypus mit besonderer Berücksichtigung des Animabegriffes (1936)", in L. Jung-Merker and E. Rüef editors, *C.G. Jung, Gesammelte Werke,* Band 9/I, Walter-Verlag, Olten, 6. Auflage 1985, pp. 69-87; reprinted in C.G. Jung. *Archetypen,* Deutscher Taschenbuch Verlag, München, 4. Auflage 1993, pp. 59-60.

I. Kant, *Kritik der reinen Vernunft,* Reclam, Stuttgart 1993.

M.A. Kastner, "Artificial Atoms", *Physics Today,* January 1993.

T.E. Kiess, Y.H. Shih, A.V. Sergienko and C.O. Alley, "Einstein-Podolsky-Rosen-Bohm Experiment Using Pairs of Light Quanta Produced by Type-II Parametric Down-Conversion", *Phys. Rev. Lett.* **71,** 1993, 3893.

C.A. Meier ed., *Wolfgang Pauli und C.G. Jung. Ein Briefwechsel 1932-1958,* Springer, Berlin Heidelberg 1992, Appendix 3.

G.J. Milburn, K. Jacobs and D.F. Walls, "Quantum-limited measurement with the atomic force microscope", *Phys. Rev.* A **50**, 1994, 5256.

I. Niiniluoto, "Varieties of Realism", in P. Lahti and P. Mittelstaedt eds., *Symposium on the Foundations of Modern Physics 1987. The Copenhagen Interpretation 60 years after the Como Lecture,* World Scientific, Singapore 1987.

W. Pauli, "The Philosophical Significance of the Idea of Complementarity", in C.P. Enz and K. von Meyenn editors, *Wolfgang Pauli. Writings on Physics and Philosphy,* Springer, Berlin /Heidelberg 1994.

W. Pauli, "Matter", in C.P. Enz and K. von Meyenn eds., *Wolfgang Pauli. Writings on Physics and Philosophy,* Springer, Berlin Heidelberg 1994.

K.R. Popper, *Auf der Suche nach einer besseren Welt. Vorträge und Aufsätze aus dreißig Jahren,* Piper, München /

Zürich 1984.
H. Rauch, A. Zeilinger, G. Badurek, A. Wilfing, W. Blauspiess and U. Bonse, "Verification of coherent spinor rotation of fermions", *Phys. Lett.* **54A**, 1975, 425.
N. Rosen, "Quantum Mechanics and Reality", in P. Lahti and P. Mittelstaedt eds., *Symposium on theFoundations of Modern Physics. 50 years of the Einstein-Podolsky-Rosen Gedankenexperiment*, World Scientific, Singapore 1985.
Th. Sauter, W. Neuhauser, R. Blatt and P.E. Toschek, "Observation of Quantum Jumps", *Phys. Rev. Lett.* **57**, 1986, 1696.
J. Schmiedmayer, M.S. Chapman, C.R. Ekstrom, T.D. Hammond, S. Wehinger and D.E. Pritchard, "Index of Refraction of Various Gases for Sodium Matter Waves", *Phys. Rev. Lett.* **74**, 1995, 1043.
F. Selleri, "Wave-Particle Duality: Recent Proposals for the Detection of Empty Waves", in W. Schommers editor, *Quantum Theory and Pictures of Reality*, Springer, Berlin Heidelberg 1989.
G. Tarozzi, "Filosofia della fisica", in E. Agazzi ed., *La filosofia delle scienze in Italia nel '900,* Franco Angeli, Milano 1986.
K. Von Meyenn, with A. Hermann and V.F. Weisskopf eds., *Wolfgang Pauli, Scientific Correspondence with Bohr - Einstein, Heisenberg a.o.*, vol. II, *1930-1939*, Springer, Berlin Heidelberg 1985.
K. Von Meyenn ed., *Wolfgang Pauli. Scientific Correspondence with Bohr, Einstein, Heisenberg, a.o.*, vol. III: *1940-1949*, Springer, Berlin Heidelberg 1993.
W.H. Zurek, "Decoherence and the Transition from Quantum to Classical", *Physics Today*, October 1991.

BERNULF KANITSCHEIDER

QUANTUM MECHANICS, REALISM AND THE ULTIMATE OBSERVER

I want to make a remark on a high speculative hypothesis, a hypothesis that roots in a well known enigma of QM, the cat. Cats seem to play an important role in physics. The theory of general relativity predicts the existence of collapsed systems which have lost all their former properties with the exception of mass, charge and angular momentum. These black holes states of the gravitational field of the former star have been compared with the cheshire-cat of Alice's adventures in wonderland whose grin hang in the tree even long after the cat had disappeared.[1]

In quantum mechanics generations of students and even philosophers of science have been led astray believing that something must be wrong with that theory on account of the untrustsworthy story of Schrödinger's cat.

Everything began when Schrödinger tried to point out the counterintuitive consequences of QM when probability relations between separated systems are regarded.

In distinction to Einstein, who tried to prove the incompleteness of QM, Schrödinger acknowledged the fundamental character of the theory but focussed on certain traits of this theory namely its way of interlocking parts to a whole.

In QM partial systems of a whole are concatenated in a characteristic non-classical way. The interrelationship of the parts of larger quantum mechanical systems described by a total wave function does not allow the factorization to its partial wave functions.

The philosophical gain of this state of affairs was a clarification of the obscure commonsense wisdom, that the whole is more than the total of its parts. This is a instructive case of turning upside down the philosophical tradition, as Prof. Shapere has characterized this direction of influence. Here we encounter an example where quantum theory has done something for philosophy instead the other way round, which normally is taken for granted as the customary procedure, namely that epistemologists impose their preconceived methodological value-system on physics.

On the other hand already Schrödinger realized that there is something strange about the qm situation of entanglement because an observer through measurement of a partial system, can steer the other subsystem to a certain state without physical interaction with the other remote part. This procedure for which Schrödinger[2] coined the concept of disentanglement is according to his view of 'sinister importance' threatening us thereby with at least an infinite regress. Schrödinger had without doubt a presentiment of the runaway situation hidden in the procedure of disentanglement that later on was called the von Neumann catastrophe.

In order to demonstrate the paradoxical situation brought about by a macroscopical amplification of a microscopic measurement, Schrödinger[3] came up with his famous cat. Basically, this illustration is an application of the superposition principle which

E. Agazzi and M. Pauri (eds.), The Reality of the Unobservable, 311-316.

expresses the linearity of QM.

As you surely will know, after one hour the whole sealed steel chamber with the cat in it is described by a wave function, that is the superposition of two states $\Psi = 1/\sqrt{2}\ (\Psi_L + \Psi_D)$. Schrödinger surmised that in our common sense macroscopic world we never would encounter such hazy situations in which two incompatible properties, a live cat Ψ_L and a dead cat Ψ_D are inextricably entangled. The assertion however of the incompatiblity roots obiously in a conviction at that time shared by almost every physicist and expressed rigorously by Niels Bohr that the classical description holds strictly for massive bodies because classical physics is logically prior to quantum mechanics.

The situation changed in 1957 when Bardeen, Cooper, and Schrieffer successfully explained superconductivity in terms of QM. They used a single three-dimensional matter wave to describe the electrons in a superconductor, that matter wave or Ψ-function expresses the state of 10^{18} electrons and thus represents a macroscopic phenomenon.

In the early eighties experimental results on macroscopic tunnel effect came to the fore which disproved the asumption of the prewar physics, that a macroscopic system with two or more macroscopically distinct states at its disposal will at all times be in one or the other of these states.

Thanks to the development of electronic systems of remarkable sensitivity, physics is able to study experimentally the questions that Bohr and Einstein argued philosophically. It is now possible to create individual macroscopic quantum objects in the order of magnitude of centimetres.

The most promising candidates for displaying macroscopic quantum behaviour are various kinds of electronic circuits like e.g. superconducting rings containing 'weak links' as are used in SQUID magnetometers.[4] The SQUID ring experiments show that in principle the ring may lie in a quantum superpositions of states.[5] At the microscopic level charge Q and magnetic flux φ are related by an uncertainly principle $\Delta\varphi \cdot \Delta Q \geq \hbar/2$. The weak-link ring can adapt two quantum modes - a flux mode and a charge mode.

Clearly this depends on the art of the experimenter. Principially only isolated systems are fully described by QM, any coupling to a classical environment tends to reduce it. But the crucial point is, that if the coupling can be made very weak, one might avoid reduction.

Schrödinger in his time however clung to the principle of state distinction. He was convinced that there must be some deficiency in the basic structure of QM, he surmised that the reason of the violations of macrodistinguability could be the concept of time in QM, quantum-mechanical time being a classical parameter and not a unitary operator.

Schrödinger's cat caused quite a stir in the scientific and philosophical community. There was considerable disagreement on the agent who could trigger the collapse of the wavefunction, only the external human observer after having opend the steel box or as argued Hilary Putnam, the cat himself. The latter posibility led him to his epistemological posture of internal realism.

The situation suggest indeed some kind of epistemic relativity. Either we accept only the external observer view as the pertinent framework, or we regard the cat himself as a quantum mechanical observer. We can stick to realism in regard to internal issues e.g. to certain properties of the cat and the diabolic device, but the nature of these properties is dependent of the framework set by the observer.

At any rate with time going on, it became quite clear that the principle that macroscopic variables cannot be in a state of superposition, reflects a dogmatic attitude and limits the validity of QM without necessity to the realm of atomic magnitude. If however the Schrödinger equation applies to macroscopic systems including large mammals (primates) then the dynamics of the combined system of cat and observer develops into the still larger superposition of the two states of a dead cat and an observer who sees this state of affairs and a live cat and an observer recognizing this fact.

This result of extended quantum dynamics seems to be preposterous to most people, because the observer himself has to be aware of contradicting visual impressions.

Only as a sideremark: There are persons who claim to be able to exist quite perfectly in a superposition of macroscopic states - say in love - hate relationship for another person or say half sleep, beeing sleeping and awake at the same time. But apart from those psychological considerations which are out of our present concern, it cannot be objected that even after the observation the case remains undecided. This bizarre situation was the starting point for the mentalistic interpretation of quantum measurement.

Eugene P. Wigner argued that a material measurement apparatus and an intelligent organism react quite differently. A material device is correctly described by a superposition but a conscious observer proceeds immediately to a mixture, that is, be realizes that only *one* of the two possibility has actually occured.

Wigner's solution of the cat paradox has far-reaching consequences for our understanding of nature as a whole. If Wigner were right, linear QM could not be applied to organic and mental systems. The antinaturalistic approach is also incompatible with a neurological theory of mental events. Currently the results of neurology point much more in the direction of a materialistic monism. Therefore the mentalistic solution of the measurement problem of QM is less promising than it was in Wigner's times. This situation caused John Bell[6] to ask the question: "Was the wavefunction of the world waiting to jump for thousands of millions of years until a single-celled living creature appeared or did it have to wait a little longer for some better qualified system ... with a Ph.D?"

Without evoking the intricacies of the mind-body problem John Barrow and Frank Tipler proposed a genial but breathtaking solution. Already Hugh Everett some time ago had pointed to the unpleasant runaway situation which occurs if many observers are connected in series. Under ordinary physical asumptions, without smuggling in mental states of an observer, the measurement process never gets complete. Barrow and Tipler now assume, that every chain of observers that extends into the future meets at a certain point an observer who belongs to an other chain, thereby establishing a network of observers of all intelligent living being in the universe.

> This joining of sequences of observers continues ... until *all* sequences of observations by all observers of all intelligent species that have ever existed and ever will exist, of all events that have ever occurred and will ever occur are finally joined by the *Final Observation* of the *Ultimate Observer*.[7]

But alas! the Ultimate Observer [UO] has a special status. He can afford the final observation and is responsible for bringing the entire Universe into existence because he is located at the final singularity in a closed universe and at the time-like infinity in an open universe. On account of the structure of singularities there can be no further observation, singularities and timelike infinity being boundaries of space-time. Invocating the UO who acts outside the space-time manifold in order to actualize the Universe, seems to be very similar to the Berkelian approach to rescue reality from disappearing into nothingness while it is not observed by the mind of God.

In any case the scenario is extremely idealistic, since consciousness is essential to engender the entire cosmos. Because this can only be done at the final state it is unclear what will happen if by some accidental cataclysm the chain of observers will be interrupted in an intermediate state. What is more, a supernaturalistic interpretation of the UO seems to lurk around the corner, because the very concept of an observer outside of spacetime seems to be a metaphysical notion that transcends the realm of physical ontology.

The hypothesis may however be taken as an example of a slippery slope leading from physics to metaphysics. What generations of philosophers of science have tried to keep apart by rigorous analysis got mixed up in once stroke.

Since the early days of logical empiricism, it had been taken for granted that the domain of factual science and the domain of transcendent metaphysics can be neatly separated; with the advent of physical escatology this distinction seems to become blurred to a increasing degree. This tendency is observed in publications of the type of Frank Tipler's "physics of immortality". His Ω-point theory leads up in a straightforward way to a continues transition of naturalistic physics to a spiritualistic metaphysics. The Ω-point theory as well as the hypothesis of the UO seems to me a profound categorial mistake, a tremendous confusion between the mental, the physical, and the metaphysical, an ontological muddle that casts the lay man in a state of embarassment.

Needless to say that there are ways out to solve the problem of measurement in QM without resort to idealistic metaphysics or theological notions according to which observers, be they customary or ultimate, engender reality.

Already in 1986 Roger Penrose[8] proposed that the collapse of the wave function may be caused by gravitational interaction. What is important about of Penrose approach, is that the state vector is taken seriously in terms of physics and that its collapse is seen as a real dynamic process. His demand for physical objectivity of the state vector and dynamic reality of the reduction cannot be fulfilled within the mathematical framework of conventional QM. As teaches the example of Schrödinger's cat, linearity leads to an unlimited runaway process. Penrose takes the superposition of the dead and live cat as an indication that physical reality is not completely described by the evolution of the state vector according to Schrödinger's equation. At the level of the cat something is missing in the customary account and it becomes the character of approximation. Penrose surmised that quantum-mechanical state reduction is produced by a growth of gravitational entropy. This process cannot however be covered by the unitary dynamics of conventional linear QM, some nonlinear instability being responsible of the transition.

In the meantime a few difficulties arose in defining unambiguously gravitational entropy, but the general tendency to establish a physical dynamical account of quantum-mechanical measurement process is the ruling idea of what is called the modern interpretation. It is characterized by the attempt to incorporate measurement in the larger context of cosmology. QM as a rigorous theory should apply to the universe as a whole, therefore the external observer formulation, the source of the mentalistic turn gets in applicable. The modern interpretation disregards the distinction, fervently defended by Niels Bohr, of a classical and a quantum mechanical realm of reality. QM as the overriding theory giving way to the emergence of nearclassical domains. Murray Gell-Mann and James Hartle put it quite succintly: "Measurements and observers cannot be fundamental notions in a theory that seeks to discuss the early universe when neither existed. There is no reason in general for all classical domain to be fundamental or external in a basic formulation of QM."[9]

The basic notion favoured mostly by James Hartle and Murray Gell-Mann is that of the *alternative histories of the universe*, formerly called many universes, and the distinction of the fine-grained and coarse grained histories.

Within the coarse grained description, the interference terms have disappeared an account of the interaction with the environment of the system. That means applied to the cat example, that their is no interference among the live and the dead cat because the interaction with the surrounding of the steel chamber decoheres the two coarse grained histories.

The approach of decoherence demystifies the enigma of the quantum measurement in a refreshing way. QM has been abused quite often to dematerialize phyical reality. It is relieving to notify however that there exists a kind of conservation law of naturalness that reduces the magical intrusions triggered by the enigmas of the quantum. As Herbert Feigl, one of the founding fathers of analytical philosophy, put it many years ago, quantum matter is still largely material.

Justus-Liebig University at Gießen
Germany

NOTES

1 Cf. Carrol 1960.
2 Cf. Schrödinger 1934.
3 Cf. Schrödinger 1935.
4 Cf. Ball 1990.
5 The weak link in these rings is a point contact; transport of electron 'Cooper' pairs across the contact relies on quantum-mechanical tunnelling.
6 Cf. Bell 1990.
7 Cf. Barrow & Tipler 1986.
8 Cf. Penrose 1986.
9 Cf. Gell-Mann & Hartle 1990.

REFERENCES

Ph. Ball, "Schrödinger's Cat Ensnared", *Nature* **341**, 1990, 330-331.
J.D. Barrow & F.J. Tipler, *The Anthropic Cosmological Principle*, Clarendon Press Oxford 1986, p. 47.
J.S. Bell, "Against measurement", in *Sixty-two Years of Uncertainty: Historical, Philosophical, and Physical Inquiries into the Foundations of Quantum Mechanics*, A.I. Miller ed., Plenum Press, 1990.
L. Carrol, *Alice's Adventures in Wonderland*, New York 1960.
M. Gell-Mann & J.B. Hartle, "Quantum Mechanics in the Light of Quantum Cosmology", in W.H. Zurek ed., *Complexity, Entropy and the Physics of Information*, New York 1990, p. 429.
R. Penrose, "Gravity and State Vector Reduction", in R. Penrose and C.J. Isham eds., *Quantum Concepts in Space and Time*, Clarendon Press, Oxford 1986, pp. 129-146.
E. Schrödinger, "Discussions of Probability Relations between Separated Systems", *Proc. Cambridge Phil. Soc.* **31**,

1934, 555-562.
E. Schrödinger, "Die gegenwertige Situation in der Quantenmechanik", *Die Naturwissenschaften* **23**, 1935, p. 807-849.

PETER MITTELSTAEDT

INDIVIDUALISTIC AND STATISTICAL INTERPRETATION OF QUANTUM MECHANICS

1. INTRODUCTION - HISTORICAL REMARKS

The quantum mechanical formalism which was discovered by Heisenberg[1] and Schrödinger[2] in 1925 was first interpreted by Born[3] in a statistical sense. The formal expressions $p(\varphi, a_i)=|(\varphi, \varphi_{a_i})|^2, i \in \mathbb{N}$ were interpreted as the probabilities that a quantum system S with preparation φ possesses the value a_i which belongs to the state φ_{a_i}. This original Born-interpretation which was formulated for scattering processes was, however, not tenable in the general case. The probabilities must not be related to the system S in state φ since in the preparation φ the value a_i of an observable A is in general not subjectively unknown but objectively undecided. Instead, one has to interpret the formal terms $p(\varphi, a_i)$ as the probabilities for finding the value a_i after the measurement of the observable A of the system S with preparation φ. In this improved version, the statistical or "Born interpretation" is used in present day literature.

Usually the statistical (Born) interpretation of quantum mechanics is taken for granted and the formalism of quantum mechanics is considered as a theory which provides statistical predictions referring to a sufficiently large ensemble of equally prepared systems $S(\varphi)$ after the measurement of the observable in question. On the other hand, the meaning of the same formal terms $p(\varphi, a_i)$ for an individual system is highly problematic. Heisenberg[4] tried to understand the meaning of the terms $p(\varphi, a_i)$ for a single system by means of the Aristotelian concept of "potentia". Popper introduced the new concept of "propensity" for the same reason. However, since both attempts to understand probability on the individual level don't give rise to any observable prediction, in the current literature the individualistic interpretation of quantum mechanics is not considered as an alternative which should be taken seriously.

In contrast to this well known situation, recent results on quantum probability seem to be more in favour of the opposite way of reasoning. Quantum mechanics is then primarily a theory of individual systems, and the statistical predictions of the same theory merely result from the behaviour of a large number of individual systems. If this way of reasoning proves to be correct, the statistical (Born) interpretation of quantum mechanics would turn out to be merely a preliminary way of speaking about a domain of reality which can, in principle, be understood also on the individualistic level. In the present paper this way of reasoning will be elaborated more in detail.

It is obvious that the results mentioned have important consequences for the possibility of a realistic interpretation of quantum mechanics. Indeed, the crucial question

E. Agazzi and M. Pauri (eds.), The Reality of the Unobservable, 317-327.

of whether the theoretical terms $p(\varphi, a_i)$ correspond to the observer's subjective knowledge, to the objective reality, or to something else in between, depend sensitively on the problem of whether quantum mechanical propositions refer to individual objects or to statistical ensembles of equally prepared systems.

The formal results of this article are valid for any sharp quantum mechanical observable and for any unitary measurement. However, for the sake of simplicity, we restrict the formal derivations to observables with discrete and non-degenerate spectra and to unitary repeatable measurements. The main results can then easily be demonstrated without unnecessary mathematical complications. For the proofs of the more general situations concerning continuous and unsharp observables, and arbitrary measurements, we refer to the literature[5].

2. INTERPRETATION OF QUANTUM MECHANICS

Here we distinguish two interpretations of quantum mechanics which connect the formal terms of the theory with experimental results. The two interpretations - the "minimal interpretation" and the "realistic interpretation" are of different explanatory power, and correspond to different kinds of measurements.

a) Minimal Interpretation

The minimal interpretation I_M is the weakest possible interpretation of quantum mechanical formalism. It is in the spirit of David Hume and the positivism of Ernst Mach, and was advocated in particular by Niels Bohr. It avoids any statements about properties of individual objects and instead refers only to observational data, i.e. to the results of a measuring process. Let S be an object quantum system prepared in a pure state φ, and $A=\sum a_i P[\varphi_{a_i}]$ a discrete observable with eigenvalues a_i and eigenstates φ_{a_i}. We can then distinguish three postulates which constitute the "minimal interpretation".

The *calibration postulate* (C_M):

If φ is an eigenstate φ_{a_i} of A, and the system possesses a value a_i of A, then a measurement of A leads to a pointer value Z_i of the measuring apparatus M indicating that the system S had the eigenvalue $a_i = f(Z_i)$ and the state φ_{a_i} before the measurement, where $f(Z)$ is the pointer function. The postulate (C_M) represents the probability-free part I^0_M of the minimal interpretation.

The *pointer objectification postulate (PO)*:

If φ is not an eigenstate of A, then it is not possible to predict with certainty the pointer value after the measurement of A. However, the postulate (PO) requires that after an A-measurement the pointer possesses some objective value Z_i even if it cannot be predicted with certainty.

The *probability reproducibility condition* (RP):

In addition to the pointer objectification, the postulate (RP) requires that the formal probability $p(\varphi, a_i) = tr\{P[\varphi]P[\varphi_{a_i}]\}$ which is induced by φ and A, is reproduced in the statistics of the pointer values $Z_i^{(n)}$ after that A-measurements have been made on N equally prepared systems $S^n(\varphi)$ with $n = 1, 2, ..., N$.

b) Realistic Interpretation

In case of repeatable measurements of the observable A one can apply the stronger realistic interpretation I_R, which refers not only to the pointer values but also to the object system. Again one can distinguish three postulates.

The realistic calibration postulate (C_R):
If φ is an eigenstate φ_{a_i} of A, and the system possesses the value a_i of A, then an A-measurement leads with certainty to a pointer value Z_i which shows that S possesses the value a_i after the measurement. The postulate C_R represents the probability free part I^0_R of the realistic interpretation.

The system objectification postulate (SO):
If φ is not an eigenstate of A, then it is not possible to predict the A-value of S with certainty. However, the postulate (SO) requires that the system possesses some objective value a_i of A after the measurement, even if it cannot be predicted.

The probability reproducibility condition (RS):
In addition to the system objectification (SO), the postulate (RS) requires that the initial probability $p(\varphi, a_i)$ is not only reproduced in the statistics of the pointer values, but also in the statistics of the A-values $a_i^{(n)}$ of the N object systems S_n after the measurement.

The two interpretations I_M and I_R of the quantum mechanical formalism obviously differ with respect to their explanatory power. Whereas in the minimal interpretation the theoretical terms refer only to pointer values Z_i, the realistic interpretation relates the theoretical terms also to A-values a_i of the object system. However, the realistic interpretation can only be applied to formalism in case of repeatable measurements - in contrast to the minimal interpretation, the applicability of which is not restricted to a certain class of measurements.

c) Quantum Theory of Measurement
In order to illustrate the relevance of the measuring process for the possible interpretations, we recall some important features of the measuring process. The object system S, and the measuring apparatus M are considered as proper quantum systems, the initial states of which are given by φ and Φ, respectively. In the first step of the measuring process, the *preparation*, the system S and M are considered as a compound system $S + M$ in the state $\Psi(S + M) = \varphi(S) \otimes \Phi(M)$. In the second step, the *pre-measurement*, the systems S and M are in interaction, which is described here by a unitary operator $U(t)$ acting on state $\Psi(S + M)$ within the time interval $0 \leq t \leq t'$. Hence the compound state after the pre-measurement reads $\Psi'(S + M) = U(t')(\varphi \otimes \Phi)$.

The state $\Psi'(S+M)$, and hence the interaction $U(t)$, is further specified by the calibration postulates (C_M) and (C_R). Indeed, according to C_M (in the interpretation I_M), for a system S with the preparation $\varphi = \varphi_{a_i}$ it should hold $U(\varphi_{a_i} \otimes \Phi) = (\varphi'_i \otimes \Phi_i)$, and, according to C_R (in the interpretation I_R), $(\varphi_{a_i} \otimes \Phi) = (\varphi_{a_i} \otimes \Phi_i)$. Here the states Φ_i are orthonormal eigenstates of the pointer observable $Z = \sum Z_i P[\Phi_i]$ and φ'_i are the correlated states of S which are not further specified. By means of the expansion $\varphi = \sum c_i \varphi_{a_i}$, with $c_i = (\varphi_{a_i}, \varphi)$ in the general case one obtains

$$\Psi' = U(\varphi \otimes \Phi) = \sum c_i U(\varphi_{a_i} \otimes \Phi) = \sum c_i \varphi'_i \otimes \Phi_i$$

with $\varphi'_i = \varphi_{a_i}$ for repeatable measurements.

In the third step of the measuring process, *objectification and reading*, the two systems S and M are again dynamically independent but still correlated. As sub-systems of the compound system $S + M$ in the state Ψ' they can be described by the reduced (mixed) states

$$W'_s = \sum |c_i|^2 P[\varphi_{a_i}], \quad W'_M = \sum |c_i|^2 P[\Phi_i],$$

respectively. According to the *pointer objectification postulate* (PO), after the measurement the pointer should possess some objective value Z_i indicating the measuring result $a_i = f(Z_i)$, where f is again the pointer function. If we are dealing with repeatable measurements, the

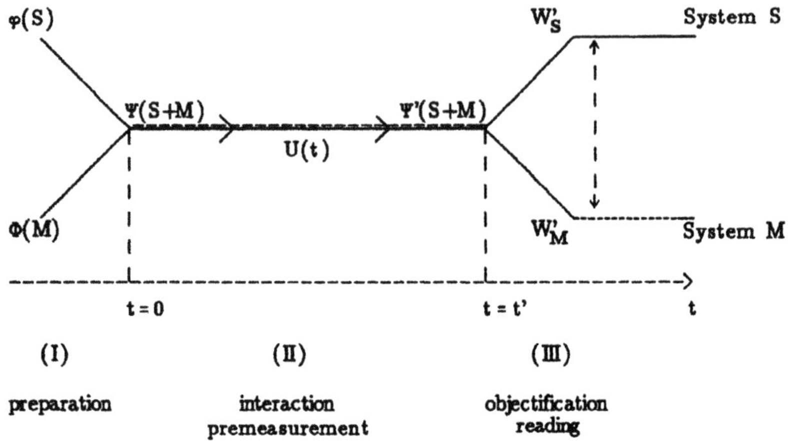

Fig. 1 Schematic representation of the measuring process

The measuring process consists of three subsequent steps:
(I) The preparation of the system *S*(j) and the apparatus *M*(F)
(II) The interaction between system *S* and apparatus *M*,
(III) The objectification and reading of the pointer values.

strong correlations between *S* and *M* after the pre-measurement imply the *system objectification postulate* (*SO*). This means that the object system *S* objectively possesses one of the values a_i also.

3. PROBABILITY THEOREMS

At first glance the postulates (C_M), (*PO*), and (*RP*) of the minimal interpretation I_M, and the corresponding postulates (C_R), (*PS*), and (*RS*) of the realistic interpretation I_R, seem to be independent requirements which characterize the respective interpretations I_M and I_M. However, it turns out that the conditions (*RP*) and (*RS*) , i.e. the probabilistic parts of the interpretations I_M and I_R, follow from the probability-free interpretations I^0_M and I^0_R, respectively. This important and somewhat surprising result is the content of the probability theorems I and II discussed below.

a) *The probability theorem in the interpretation* I_M
Within the interpretation I_M, the calibration postulate C_M corresponds to a probability-free minimal interpretation I^0_M which merely states that for a system which possesses a value a_i of A, a measurement of A leads with certainty to the pointer value $Z_i = f^{-1}(a_i)$ which is correlated with the value a_i. The quantum theory of measurement provides explicit examples of unitary

operators U_A which fulfill C_M postulate requirement.

The calibration postulate C_M of the interpretation I_M implies that the post-measurement state of the apparatus system M reads $W'_M = \sum |\varphi_{a_i}, \varphi|^2 P[\Phi_i]$. However, the meaning of the coefficients $c_i = (\varphi_{a_i}, \varphi)$ is still open. It is easy to show that the real positive numbers $p(\varphi, a_i) = |(\varphi_{a_i}, \varphi)|^2$ are probabilities in the formal sense. The interpretation of the probability distribution $p(\varphi, a_i)$ can then be obtained from the probability reproducibility condition (*RP*). According to this condition (*RP*), the probability distribution $p(\varphi, a_i)$ induced by φ and A is reproduced in the statistics of the pointer values Z_i. In other words, the number $p(\varphi, a_i)$ is the probability to find - after the measuring process - the pointer value Z_i, which indicates that the value $a_i = f(Z_i)$ was measured. This means that if one performed a series of N measurements of the observable A of equally prepared systems S_n ($n = 1,2,...,N$), then the relative frequency $f^N(\varphi Z_i)$ of pointer values Z_i would in some sense approach the probability $p(\varphi, a_i)$ for $N \to \infty$.

The probability reproducibility condition (*RP*), i.e. the probabilistic part of the minimal interpretation I_M, follows from the calibration postulate C_M, i.e. from the probability-free interpretation I^0_M by means of quantum mechanics, provided this theory is assumed to be universally valid. In particular, it must be presupposed that quantum mechanics can equally be applied to a single object system as well as to a compound system $S^{(N)}$ consisting of N equally prepared systems S_i.

In order to derive this interesting result, we consider N independent systems S_i with equal preparations $\varphi^{(i)}$ as a compound system $S^{(N)}$ in the tensor product state $(\varphi)^N$. For each system S_i, a pre-measurement of A transforms the initial state Φ of M into the mixed state

$$W'_M = \sum p(\varphi, a_i)\, P[\Phi_i],$$

with eigenstates Φ_i of the pointer observable Z corresponding to pointer values Z_i... If the observable A is measured for each system S_i, then the full measuring outcome is given by a sequence of pointer values $\{ Z_{l_i} \}$ and pointer states $\{ \Phi_{l_i} \}$, respectively, with an index set $l = \{l_1, l_2,...,l_N\}$. In the N-fold tensor product Hilbert space $H_M^{(N)}$ of the apparatus, the states $\Phi_l{}^N = \Phi_{l_1} \otimes \Phi_{l_2} \otimes ... \otimes \Phi_{l_N}$ form a complete orthonormal basis. For any product state Φ_l^N the relative frequency $f^N(l,k)$ of pointer values Z_k is objectively determined, and one can define a new observable "relative frequency of pointer value Z_k" by $F_k{}^N = \sum_l f^N(k,l)\, P[\Phi_l{}^N]$ where the sum runs over all sequences l.

After a pre-measurement of A the apparatus is in the mixed state W'_M. The sequence of N measurements is then described by the N-fold product $(W'_M)^N$ of these mixed states. In general, this product state is not an eigenstate of the relative frequency operator $F^N{}_k$, which means that the relative frequency of pointer values Z_k is not an objective property in the mixed state $(W'_M)^N$.

However, in the limit of large N, the post-measurement product state $(W'_M)^N$ becomes an eigenstate of the relative frequency observable $F^N{}_k$, and the relative frequency of the pointer value Z_k approaches the probability $p(\varphi, a_k)$. This is the content of the *probability theorem I*, which can formally be expressed by the relation:

$$\lim_{N\to\infty} tr\left\{\left(W'_M\right)^N\left(F_k{}^N - p\left(\varphi, a_k\right)\right)^2\right\} = 0$$

It holds in the minimal interpretation I_M.

b) The probability theorem in the interpretation I_R

In case of repeatable measurements, one can apply the realistic interpretation I_R. The calibration postulate C_R again corresponds to the probability-free part I^0_R of the realistic interpretation. It requires that for a system S which possesses a value a_i of A, a measurement of A leads not only with certainty to a pointer value Z_i which indicates the result a_i, but also to an object value a_i, i.e. the measurement preserves the value a_i of the object system. Clearly, this requirement can only be fulfilled by repeatable measurements.

The calibration postulate C_R implies that the post-measurement state of the object system reads $W'_S = \sum p(\varphi, a_i)\, P[\varphi_{a_i}]$, where $p(\varphi, a_i)$ are again the initial probabilities induced by the preparation φ and the observable A. The interpretation of the formal probabilities is given here by the probability reproducibility condition (RS): $p(\varphi, a_i)$ is the probability that after the measurement of A, the system S possesses the value a_i of A. This means that if one performed A-measurements on N equally prepared systems $S_n(\varphi)$, the relative frequency $f^N(\varphi, a_i)$ of object systems $S_n(\varphi)$ with A-values a_i would in some sense approach the probabiltity $p(\varphi, a_i)$ for $N\rightarrow\infty$.

In order to derive this result one must again assume that quantum mechanics can be applied not only to a single object system S_i, but also to a compound system $S^{(N)}$, which consists of N equally prepared systems S_i. The totality $\{S_i\}^N$ of N systems can then be considered as a compound system $S^{(N)}$ in the tensor product state $(\varphi)^N$. A pre-measurement of A transforms the initial state $\varphi^{(i)}$ of each system S_i into the mixed state $W'_S = \sum p(\varphi, a_i)\, P[\varphi_{a_i}]$, with eigenstates φ_{a_i} of A corresponding to the values $a_i = f(Z_i)$. If A is measured of each system S_i, the measuring result is given by a sequence $\{a_{l_1}, a_{l_2}, ..., a_{l_N}\}$ of object values a_{l_i} and states φ_{a_i} respectively, with the index set $l = \{l_1, l_2 ..., l_N\}$.

In the N-fold tensor product Hilbert space $H_S^{(N)}$ of the compound system $S^{(N)}$, the states $(\varphi)^N_l = \varphi^{(1)}_{l_1} \otimes \varphi^{(2)}_{l_2} \otimes ... \otimes \varphi^{(N)}_{l_N}$ with an index set l form a complete orthonormal basis. In a state $(\varphi)^N_l$ each subsystem S_i has a well defined A-value a_i. The relative frequency $f^N(k,l)$ of values a_k in the state $(\varphi)^N_l$ is then objectively determined and we can construct an operator "relative frequency of systems with values a_k" by defining $f^N_k = \sum_l f^N(k,l)\, P[(\varphi)^N_l]$, where the sum runs over all sequences l. After a pre-measurement, a system S_i is in a mixed state W'_{S_i}. If N pre-measurements of A are performed, the state of the compound system $S^{(N)}$ is given by the N-fold tensor product state $(W'_S)^N = W'_{S_1} \otimes W'_{S_2} \otimes ... \otimes W'_{S_N}$ of these mixed states W'_{S_i}. In general this tensor product state is not an eigenstate of the relative frequency operator f^N_k and thus the relative frequency of values a_k is not an objective property of the system $S^{(N)}$ in the state $(W'_S)^N$. However, in the limit of large values of N, the post-measurement product state of the compound system becomes an eigenstate of f^N_k, and the value of the relative frequency of a_k approaches the probability $p(\varphi, a_k)$. This is the content of the *probability theorem II* which can be expressed by the formula

$$\lim_{N\rightarrow\infty} tr\left\{(W'_S)^N \left(f_k^N - p\,(\varphi, a_k)\right)^2\right\} = 0$$

Its validity is restricted to the realistic interpretation I_R.

c) Discussion of the probability theorems

i) Calibration

For the derivation of the probability theorems the calibration postulates - C_M, C_R - play a two-fold role. On the level of the single object system, the calibration postulate requires that for the objective properties of the system S the result of a measuring process can be predicted with certainty. In this case the property pertains to the system in an objective sense.[6] On the level of the many-body compound system $S^{(N)}$, the calibration postulate is needed again. Here it requires that the property "relative frequency of a value a_k" pertains to the system $S^{(N)}$ whenever the state of the compound system is an eigenstate of a conveniently defined relative frequency observable. The universal validity of quantum mechanics is used here in the sense that the probability-free minimal interpretation, expressed by the calibration postulate, can equally be applied to the single system S as well as to the compound system $S^{(N)}$.

ii) Objectification

It is obvious from the preceeding discussion, that the probability theorems I and II have a physical meaning only if, in addition to the calibration postulates C_M and C_R, the pointer objectification (*PO*) or the system objectification (*SO*) are also presupposed. The probability reproducibility condition (*RP*) requires that the probability $p(\varphi, a_i)$ is reproduced in the statistics of the pointer values. Clearly, this postulate can only be fulfilled if after the measurement, the pointer possesses an objective value Z_i, i.e. if the pointer objectification (*PO*) is given. Otherwise one could not count the pointer values and compute the relative frequency $f^N(\varphi, Z_i)$. Correspondingly, in the realistic interpretation I_R, the probability reproducibility condition (*RS*) requires that the probability distribution $p(\varphi, a_i)$ is reproduced in the statistics of the values a_i which pertain to the object system after the measurement. Again, this requirement can be satisfied only if after the measurement, the object system actually possesses some value a_i of A, i.e. if the system objectification (*SO*) is guaranteed.

4. INDIVIDUALISTIC INTERPRETATION OF QUANTUM MECHANICS

The probability theorems I and II have important consequences for the problem of an individualistic interpretation of quantum mechanics. In order to demonstrate these implications, let us briefly recall the usual probabilistic interpretation: For a system S with preparation φ the observables A with $[A,P[\varphi]]=0$ possess objective values a_i of A. The observables $B=\sum b_k P(b_k)$ with $[P(b_k), P[\varphi]] \neq 0$ have no objective value in the state φ. However, to the system S one can attribute probabilities $p(\varphi, b_k)$ which provide some additional information about the values b_k. Indeed, $p(\varphi, b_k)$ is the probability of obtaining the result b_k after a measurement of the observable B. If the measurement is repeatable, the system S will possess some value b_k after the measurement, otherwise the pointer will have a value Z_k which indicates the result b_k. The experimental verification of these probability propositions is a statistical one: if one performed a large number N of B-measurements on equally prepared systems $S_n(\varphi)$, then the relative frequency f^N_k of outcomes b_k would in some sense approach the value $p(\varphi, b_k)$ in the limit $N\to\infty$.

This means that the probability proposition $P\{p(j,b_k)\}$ is a proposition whose referent is a compound quantum system $S^{(N)}$, composed of N originally equally prepared systems $S_n(j)$, after a measurement of B. The proposition $P\{p(\varphi, b_k)\}$ states that the relative frequency for obtaining a result b_k in the ensemble $S^{(N)}$ approaches the value $p(\varphi, b_k)$ for $N\to\infty$. This is the content of the probability reproducibility postulates (*RP*) and (*RS*) of the interpretations I_M and I_R respectively.

However, this explanation does not answer the question of what the term $p(\varphi, a_i)$ could mean for an individual system S with preparation φ. This question is motivated by the fact that quantum probabilities express some genuine indeterminism and cannot be reduced to the distribution of hidden initial conditions. Clearly, a system $S(\varphi)$ objectively possesses values of all observables A with $[A,P[\varphi]] = 0$. On the other hand, probability statements concerning values of non-objective properties can neither be verified nor falsified by experimental means, since the single system can only be measured once.

In order to attach probabilities $p(\varphi, a_i)$ to a single physical system, two often discussed proposals have been made by Heisenberg and Popper. In the first approach, Heisenberg[7] referred to the philosophy of Aristotle. Being aware that quantum probabilities must not be considered as quantification of the observer's subjective ignorance only, he also tried to understand probabilities as some kind of physical reality which pertains to the object system in question. In this sense he argued that quantum probabilities are a quantitative formulation of the concept of *potentia* in Aristotle's philosophy. Accordingly, in quantum mechanics one should interpret the "possibility" or "tendency" which is quantified by $p(\varphi, a_i)$ as some intermediate layer of reality, halfway between the objective reality of quantum systems and the subjective reality of the observers knowledge.

The second attempt to give some meaning to the probability of a single event was made by Popper[8]. He introduced the concept of *propensity* in order to provide a single case interpretation of probability. However, the propensity concept was not primarily conceived as a means for interpreting quantum mechanics. Instead it was understood by Popper as a new concept which can be used for all kinds of probabilistic problems. For this reason, there is a subjectivistic interpretation which holds no interest for our problems, and an objectivistic interpretation which might be of interest for physics. Within the objectivistic approach, one has to distinguish the deterministic case, e.g. dice tossing experiments, and the indeterministic case, which presumably incorporates quantum mechanical situations. Whereas for deterministic situations the propensity concept is often considered to be void[9], some authors claim that propensities might be particulary useful in quantum mechanical situations which are considered to be indeterministic.[10]

In spite of these intuitively rather convincing arguments, in an actual experimental situation this way of reasoning amounts to saying almost nothing. Firstly, there is no measurable property of a single system which could be interpreted as a probability. Secondly, and this is more important, a pure state φ provides already a maximal information about the system $S(\varphi)$. Indeed, the values of all objective properties can be deduced from the knowledge of φ and these values can be obtained with certainty by a measuring process which fulfills the calibration postulate.

Moreover, by means of probability-theorem I it follows from the general validity of the calibration postulate that the numbers $p(\varphi, a_i)$ are values of a relative frequency observable F^N_k which pertains to an ensemble $\{z_i\}^N$ of N pointer values Z_i obtained by A-measurements on N equally prepared systems $S_n(\varphi)$, in the limit $N \rightarrow \infty$. In other words, on the basis of the probability-free interpretation I^0_M the values $p(\varphi, a_i)$ turn out to be objective properties of the ensemble $\{z_i\}^N$. These arguments show that for an individual system $S(\varphi)$ there is no need for an additional "potentia" or "propensity", since from the state φ one can deduce all objective properties, even the statistical ones which turn out to be *yes-no* properties of a conveniently defined ensemble. For an individual system $S(\varphi)$ the totality of probabilities $p(\varphi, a_i)$, $p(\varphi, b_k)$ expresses nothing but the fact that φ is the state of S.

In case of repeatable measurements one can also apply probability theorem II. Under the assumption that the calibration postulate holds for a single system $S(\varphi)$ as well as for a compound system $S^{(N)}$, one arrives at the following result: the numbers $p(\varphi, a_i)$ are not only the relative frequencies of pointer values $Z_i = f^{-1}(a_i)$ in the ensemble of pointer outcomes, but the $p(\varphi, a_i)$ are also relative frequencies of systems $S_i(\varphi_{a_i})$ in the ensemble $S^{(N)}$ of equally prepared systems after A-measurements. This means that the relative frequency observable f^N_k pertains to the post-measurement compound system $S^{(N)}$ as an objective observable whose value is given by $p(\varphi, a_i)$, at least in the limit $N \rightarrow \infty$.

Summarizing the preceeding arguments we arrive at the following result. A single quantum system $S(\varphi)$ with preparation φ possesses values a_i of all objective observables $A = \sum a_i P(a_i)$. According to the calibration postulate these values can be measured with certainty. For non-objective observables B nothing is known about their value b_k and no prediction can be made for the result of a B-measurement. The probabilities $p(\varphi, b_k)$, induced by φ and B, do not refer to the system $S(\varphi)$ but to an ensemble $\{z_i\}^N$ of pointer values or to an ensemble $\{S_i(\varphi_{a_i})\}$ of N equally prepared systems $S_i(\varphi)$ after measurements of B. According to probability theorems I or II the values $p(\varphi, b_k)$ are eigenvalues of the relative frequency observables of the ensemble $\{z_i\}^N$ or $S^{(N)}$, respectively, in the limit of infinite N.

Obviously, the probability-free interpretations I^0_M and I^0_R are completely sufficient for the interpretation of a quantum system. Any interpretation of an individual system, which goes beyond the calibration postulates making use of "metaphysical" concepts like *potentiality* or *propensity*, is purely hypothetical and no longer justified by quantum mechanics. There is no need and no room for further interpretations of the theoretical terms $p(\varphi, a_i)$. For an individual system S with preparation φ, the values $p(\varphi, a_i)$ give no additional information except that S is in the state φ. Hence, one gets the impression that the problem of a single case interpretation of quantum mechanical probabilities is merely a pseudo-problem.

5. CONCLUDING REMARKS

Since the discovery of quantum mechanics in the twenties, this theory was considered as intimately related to the concepts of uncertainty, indeterminacy, and probability. Within the framework of different interpretations, these concepts were understood as subjective ignorance, as deficiency of information, as objective indeterminacy, as potentiality, or as propensity. Without referring to one of these meanings, the statistical content of the concepts was incorporated into the "minimal interpretation" of quantum mechanics.

In contrast to this well established understanding of quantum theory, probability theorems I and II have led to the surprising result that quantum mechanics is not primarily a probabilistic theory, but a theory which is concerned with yes-no propositions. A result of this kind has already been conjectured by several authors.[11] However, instead of referring to the reduced state of the object system or of the apparatus after the measurement, these authors prove an analogue of probability theorem II for a pure state preparation prior to the measurement. The present investigations have shown that the original conjectures of the authors mentioned were nevertheless correct, and can be justified within the quantum theory of measurement by means of probability theorems I and II.

The considerations we have given show that quantum mechanics allow for a somewhat restricted realistic interpretation. The theoretical terms $p(\varphi, a)$ correspond to properties $P(a)$ of an individual system $S(\varphi)$, if $p(\varphi, a) \in \{0,1\}$, and otherwise to relative frequencies of these properties in an ensemble of equivalent systems. In both cases the theoretical terms refer to elements of the physical reality, either of the individual system S or of the ensemble $S^{(N)}$. The limitation of this interpretation becomes obvious if the formal expressions refer to an ensemble $S^{(N)}$. In this case, i.e. if $p(\varphi, a) \notin \{0,1\}$, it is objectively undecided whether a given individual system $S(\varphi)$ possesses the property $P(a)$ and one cannot predict the result of an experimental test of $P(a)$.

Cologne University
Germany

NOTES

1 See Heisenberg 1925.
2 See Schrödinger 1926.
3 See Born 1926.
4 See Heisenberg 1959.
5 See Busch, Lahti and Mittelstaedt 1996, and Mittelstaedt 1993.
6 In Einstein's terminology, this means that an objective property is an "element of the physical reality"; see Einstein 1935.
7 See Heisenberg 1959.
8 See Popper 1957.
9 See Sklar 1970, Giere 1973, Schneider 1994.
10 See Popper 1982.
11 See, for example, Everett 1957, Finkelstein 1952, Hartle 1958, Graham 1973, De Witt 1971.

REFERENCES

M. Born, "Zur Quantenmechanik der Stoßvorgänge", *Zeitschrift für Physik*, **37**, 1926, 863-867.

P. Busch, P. Lahti and P. Mittelstaedt, *The Quantum Theory of Measurement*, Springer, Heidelberg 1991, 2nd ed. 1996.

A. Einstein, B. Podolsky and N. Rosen, "Can Quantum-Mechanical Description of Physical Reality be considered complete?", *Phys. Rev*, **32**, 1935, 777-780.

H. Everett, *The Theory of the Universal Wave Function. Theses.* Princeton University, 1957. Reprinted in B.S. DeWitt and N. Graham eds., *The Many Worlds Interpretation of Quantum Mechanics*, Princeton University Press, Princeton N.Y. 1973, 1-140.

D. Finkelstein, "The Logic of Quantum Physics", *Transactions of the New York Academy of Science*, **25**, 1962, 621-637.

R. Giere, "Objective Single-Case Probabilities", in P. Suppes ed. et al. *Logic, Methodology and Philosophy of Science*, 1973, 457-483.

N. Graham, "The Measurement of Relative Frequency", in B.S. De Witt and N. Graham eds., *The Many-Worlds Interpretation of Quantum Mechanics*, Princeton University Press, Princeton N.Y. 1973, 229-252.

J.B. Hartle, "Quantum Mechanics of Individual Systems", *Am. J. of Phys.* **36**, 1968, 704-712.

W.Heisenberg, "Über quantentheoretische Umdeutung kinematischer und mechanischer Beziehungen", *Zeitschrift für Physik* **33**, 1925, 879-893.

W. Heisenberg, "Die Plancksche Entdeckung und die physikalischen Probleme der Atomphysik", *Universitas* **14**, 1959, 135-148.

P. Mittelstaedt, "The Measuring Process and the Interpretation of Quantum Mechanics", *Intern. Journ. of Theor. Phys.*

32, 1993, 1763-1775.
R.K. Popper, "The Propensity Interpretation of the Calculus of Probability and the Quantum Theory", in S. Körner ed., *Observation and Interpretation in the Philosophy of Physics*, Butterworth, London 1957.
R.K. Popper, *Quantum Theory and the Schism of Physics*, Postscript III to the *Logic of Scientific Discovery*, 1982. W.W. Bartley IIIrd ed. Hutchinson, London, Melbourne 1988.
C. Schneider, "Two Interpretations of Objective Probabilities", *Philosophia Naturalis* **31**, 1994, 107-131.
E. Schrödinger, "Quantisierung als Eigenwertproblem", *Annalen der Physik* **79**, 1926, 361-376.
L. Sklar, "Is Probability a Dispositional Property?", The Journal of Philosophy **67**, 1970, 355-366.
B.S. De Witt, "The Many-Universes Interpretation of Quantum Mechanics", in *Foundations of Quantum Mechanics*, IL Corso, Academic Press, New York 1971, 167-218.

BRIGITTE FALKENBURG

HOW TO OBSERVE QUARKS

1. INTRODUCTION

Unlike an electron or photon, a quark has never been detected as a "free" particle (a localized quantum of charge or energy) - that is, as an isolated fractional charge. The quark model, however, has been well-established in particle physics for 20 years: recall the recent announcement of experimental evidence for the top quark. In what follows, I shall discuss how the current evidence for quarks fits in with the *generalized concept of observation*, which was proposed by D. Shapere in 1982.

To that end, after some introductory remarks on quarks (section 1), I shall translate Shapere's concept of direct observation (and observability) into the language of physics, or of magnitudes attributed to some physical object or system based on physical theory and measurement (section 2). The translation can easily be applied to the theoretical models which are involved in the analysis of the scattering experiments of particle physics. I shall show that a necessary condition for directly observing an entity in Shapere's sense is the experimental possibility of singling out an individual system which carries the physical magnitudes attributed to the entity.

As I shall show in section 3, this condition was not satisfied by the earliest kinds of experimental evidence for the quark model. In full agreement with Shapere's criterion for direct observation of an entity, particle physicists regarded such evidence as "indirect". Only later did the physicists succeed in performing experiments which gave rise to "more direct" evidence. At present, there is at least one kind of evidence which may count as direct observation in Shapere's sense.

In the context of the debate on scientific realism, however, the proposed analysis of how to observe quarks suggests the following conclusion (section 4): observation (of an entity) and measurement (of properties) are distinct. However, since the construction of entities from properties turns out to be problematic in the relativistic quantum domain (where no sharp criterion for a distinction between particles and fields is available), the interpretation of the experimental evidence in terms of entities is not unambiguous, and the question of whether (and under which conditions) there are "really" quarks needs further investigation.

2. ON QUARKS

According to the standard model of particle physics, matter consists of two types of fundamental constituents: leptons and quarks. The most familiar lepton is the electron

E. Agazzi and M. Pauri (eds.), The Reality of the Unobservable, 329-341.

which can be isolated from an atom in an ionization process. The charge of the electron was first measured in Millikan's famous oil droplet experiment, based on a method which has been useful for precise measurements of charge until today. Until the rise of the quark model, the electron charge was considered to be the smallest quantum of charge which can be attributed to a discrete part of matter, and it was taken as a natural unit, the elementary charge. Quarks, however, are supposed to carry fractional charges. According to both the quark model of 1964 and the current standard model of particle physics, they have ± 1/3 or ± 2/3 of this charge unit, even though the two models do not have the same dynamic structure. According to both models, all hadrons are made up of either two or three quarks[1]; in any hadron, the fractional charges of the quark constituents sum up to an integer multiple of the electric charge. The most familiar hadron is the nucleon, i.e., the constituent of the atomic nucleus (the proton or neutron, unified by the concept of isospin). The other kinds of hadrons are either to be found in cosmic rays, or generated in the scattering experiments of high energy physics which are carried out with big particle accelerators.

Since the atomic nucleus is supposed to be composed of quarks which are discrete with regard to the charges they carry, quarks may well be regarded as *particles* in the traditional sense of *discrete parts of matter*. However, they can also be considered to be particles according to the most general formal definition of the informal expression 'particle'. According to this definition, a *particle* is an *irreducible representation of a spatio-temporal invariance group* such as the Poincaré or Galilei group. Thus a particle 'is' a set-theoretical structure fixed within the framework of group theory[2]. In 1939, Wigner characterized the irreducible representations of the Poincaré group in terms of mass, spin, and parity.[3] Such a definition also works for the Galilei group.

Interacting particles carry charges in the generalized sense of 'charge' according to which the gravitational mass, for example, is the charge of gravitation. The coupling of particles is due to charges which add to the (inertial) mass, spin, and parity of the "free" or unbound particle states. Unbound charged particles may thus form multiplets of particles belonging to a certain representation of the Poincaré group, or to certain values of mass, spin, and parity. The charges of elementary particles belong to an additional 'internal' ('dynamical') symmetry, that is, to the gauge group of a gauge invariant quantum dynamics. Hadrons known around 1960 were classified in 1961 according to multiplets corresponding to the representations of such a gauge group; they displayed a SU(3) symmetry of conserved quantum numbers such as isospin and strangeness. The classification gave rise to the quark model of 1964 once the SU(3) was associated with an attempt to explain the hadron multipletts as resulting from the composition of quarks which carry the same spin and parity, approximately the same mass, and different charges.

According to the formal definition of 'particle' given above, quarks are particles understood to have the following physical properties (which correspond to the representations of the spatio-temporal or dynamic invariance groups of some fundamental dynamics):

masses:	m_i (non-degenerate)
spin:	$s = 1/2$
charges:	$q_e \in \{\pm 1/3, \pm 2/3\}$
	$q_f \in \{u, d, s, c, b, t\}$
	$q_c \in \{\text{"colors"}\}$

Here, mass and spin are associated with the irreducible representations of the Poincaré group, that is, with spatio-temporal invariance; the electric charges q_e and the "flavors" q_f are associated with the gauge group U(1) x SU(2) of the unified quantum field theory of the electroweak interactions, the Salam-Weinberg-Theory (SWT). The "colors" q_c are the charges of the quantum field theory of strong interactions, quantum chromodynamics (QCD), which is an SU(3) invariant gauge theory. (Expressed in terms of the actual standard model of particle physics, however, the hadronic SU(3) of the original quark model of 1964 was based on the first three "flavors" of the weak interactions, and not on the SU(3) "color" symmetry of the strong interactions.)

The main problem of the quark model, in 1964 as well as today, is that there is *no measurement of such particles*: all attempts to detect isolated fractional charges have been fruitless - except one measurement based on Millikan's method. But the measurement could not be reproduced experimentally, and so it is considered to be unreliable. In the case of quarks, the gap between the (axiomatic) theoretical definition of a particle, and the experimental notion of a particle is substantial. An (operational) experimental definition of 'particle' may be given as follows: A *particle* is a *collection of local properties which are always detected together*. According to such an operational definition, a particle is an empirical substance in Locke's sense, that is, a collection of properties, or magnitudes, which constantly go together. All presently known stable (i.e.non-decaying) particles, and all unstable particles with a lifetime sufficiently long for detection, are particles in this operational sense - except the quarks.

The solution of the problem, now generally accepted, is the *confinement hypothesis*, a phenomenological hypothesis which was first suggested on theoretical grounds: bound states of quarks come only in singlets, according to the quark model and to the phenomenological structure of the hadron multiplets. It was proposed that there are no free or unbound quarks since no bound quark states except the "colorless" (singlet) states are stable. The confinement hypothesis suggests that quarks have the following *paradoxal dynamic behaviour*: they are *asymptotically free in bound states*, whereas the potential *V*(*r*), which ties them together, rises with increasing distance *r*. Today, the potential *V*(*r*) governing quark interactions is phenomenologically described by the following expression:[4]

$$V(r) = -\frac{4}{3}\frac{\alpha_S}{r} + kr ,$$

where the first term on the right side is a Coulomb-like term dominating for short distances, and the second term expresses a constant force dominating at greater distances.

The confinement hypothesis is an assumption which models quantum chromodynamics according to the phenomena, but which was not yet derived from the theory. So the suspicion may arise that it is only *ad hoc*. Unlike any ad hoc hypothesis, however, it has enormous explanatory power.

Today, the quark model is based on a very rich pattern of more or less indirect experimental evidence, although, according to the confinement hypothesis, *direct* evidence for quarks must not be expected if 'direct evidence' means 'evidence for isolated quarks'. Milestones in providing the quark model with an empirical basis were: the discovery of "scaling behaviour" in deep-inelastic electron-nucleon scattering, that is, a certain kind of energy dependence of a measured scattering cross section which was interpreted as evidence for point-like scattering centers inside the proton or neutron (1968); the detection of unexpected narrow hadronic resonances such as the J/Y(1974) or the ° (1977), indicating bound "quarkonium" states consisting of a quark-antiquark pair; and the "fragmentation" of quark-antiquark states of sufficiently high energies into "jets" (1975). Retrospectively, two leading particle physicists conclude from such evidence as follows:

> If the quarks are never free, how can they be observed? Of course they were *observed indirectly* in deep inelastic scattering. However, the parton model and QCD indicated that *more direct evidence* should be obtained by studying certain reactions, the simplest being e^+e^- annihilation. While the produced quarks could not be seen, the initial quarks should materialize into jets of hadrons moving nearly along the directions of the quarks.[5]

To see whether this way of distinguishing between "indirect" and "more direct" evidence matches perfectly with Shapere's account of what counts as direct observation or not, let us look now at that account.

3. A GENERALIZED CONCEPT OF OBSERVATION

In 1982, Dudley Shapere proposed to disentangle the use of the term 'observation' in empirical science from the traditional epistemic concept of observation, which is tied to perception. One might expect a concept of observation based on scientific practice, and cut off from perception, to disappoint philosophers in some very important respects. However, such a concept is perfectly suited to take into account the theory dependence of scientific experience, and to determine when such "theory-ladenness" of scientific experience is crucial, and when it is not. This is precisely Shapere's main goal when he gives the following criteria for the direct observation (observability) entity x:[6]

x is directly observed (observable) if:

(1) information is received (can be received) by an appropriate receptor; and

(2) this information is (can be) transmitted directly, i.e. without interference, to the receptor from the entity x (which is the source of the information).

The concept of observation underlying this criterion is a generalization of the corresponding empiricist concept. It is obtained in two steps: first, by disregarding the specific mind-dependent features of sensory perception (i.e., the distinction between primary and secondary qualities, the nature of ideas, of conciousness, etc.); and secondly, by drawing an analogy between a physicalist account of what is going on in an act of perception, and the transmission of information. According to this analogy, the eye is only one kind of information receptor among many. Obviously, the concept is based on applying the principles of a naturalized epistemology to an analysis of the empiricists' concept of observation.

Shapere's criterion is a *criterion* in that it is expressed as a *sufficient condition* for direct observation, or observability, of an entity x. According to this criterion, three theories may be involved in the analysis of how to directly observe an entity x: theories of (*i*) the source x, (*ii*) the receptor, (*iii*) the act of transmission. However, the criterion implies a *definiens* for what makes an observation *direct*: the act of transmission must be direct, that is, *without loss of information.*

(Shapere says: "without interference", taking into account the possibility of neutrino oscillations or other ways in which quantum states may interfere, since his concept of 'direct observation' stems from a detailed case study concerning the question of whether the center of the sun is observable by means of neutrinos. However, we need a general definition of 'direct transmission'. Therefore I say: "without loss of information.").

That the transmission be direct is a *necessary condition* for a direct observation in a generalized sense, and this necessary condition relies on a theory of transmission of information. In what follows, I will call it the *reliability condition* for direct observation in Shapere's sense:

Reliability condition:

The observation of an entity x is *direct* only if the transmission from x to a receptor of information *preserves* the original information.

The reliability of some kind of transmission of information depends on the physical nature of a concrete act of transmission. To take Shapere's example: the propagation of the neutrinos, which are emitted from the center of the sun, is a physical process in which the neutrinos either oscillate, or they do not (given that there *are* neutrinos, as far as physics is concerned according to two rival theories). If they oscillate significantly, the reliability condition is violated. If they do so, but we do not know that they do, our theory of transmission is wrong and gives rise to an epistemically wrong use of 'direct observation'.

The epistemic reliability of the available background knowledge which is discussed in Shapere's paper, however, should be separated from the non-epistemic content of such a reliabilty condition. (Do the neutrinos oscillate? Or are there no neutrino oscillations? Particle physics does not yet have an answer. Experiments can only give upper limits on the relative frequencies of neutrinos which change their "flavor", and neither of the current rival theories is supported by conclusive experimental evidence.) If it was based only on epistemic grounds, Shapere's criterion for direct observation would obviously yield no sharp distinction between *direct observation*, and its opposite, that is, *inference* based on observation - Shapere's expression for what an experimenter would call "indirect evidence". Since the distinction depends on background knowledge, the line between what counts as an observation or as an inference is shifting as physics develops.[7]

However, due to the non-epistemic content of the *reliability condition* given above, the line between direct observation and theoretical inference can not be shifted arbitrarily in such a way that, in the last analysis, *any* experimental result may count as a direct observation of some entity x. To treat the distinction of observation and non-observation as completely arbitrary and vague (as did, for example, van Fraassen for purely epistemic reasons, when he criticized Maxwell's famous defence of theoretical entities[8]) is therefore not convincing.[9]

The above consideration suggests that we should ask whether there are *further* necessary conditions for the direct observation of an entity x, on the basis of Shapere's analysis - and whether these conditions might help us to determine how it is possible to observe quarks. Accordingly, I propose to translate Shapere's concept of observation into the language of physics, on the basis of the following translation rules:

1. *information* means *some measured magnitudes* $\{A, B, ...\}$ that is, measured values of physical quantities such as mass,charge, etc.;
2. *receptor* means *measuring device*, that is, any apparatus or system of detectors, computers, etc., which may be used to obtain numerical results from an experiment performed in a physics laboratory and to interpret them unambiguously in terms of physical quantities;
3. *transmission* means *a sequence of interactions* in, or between, physical systems that ultimately yields a *measurement* of the magnitudes $\{A, B,...\}$;
4. the *source* x means a *physical system* understood to have *two kinds of properties*: *permanent* properties $\{p_k\}$, and *non-permanent* properties $\{r_j\}$.

According to these translation rules, a direct observation in Shapere's sense is a measurement of some permanent and non-permanent properties, which can be attributed unambiguously to a physical system. A direct transmission underlying any direct observation is a sequence of interactions that yields a measurement which preserves the original information - that is, a measurement which permits us to relate unambiguously the measured magnitudes $\{A, B, ...\}$ to the properties $\{p_k, r_j\}$ of the system.

In the language of physics, we must express the *reliability condition* for direct observation in two parts, namely: (*i*) the *general description* of physical systems and measurements in terms of physical magnitudes, and (*ii*) the possibility of ascribing such magnitudes to an actual *individual system* understood as having been measured in an actual experiment with a definite numerical result. The first part simply requires that we give a mapping from the measured quantities $\{A, B, ...\}$ to a system's permanent and non-permanent properties $\{p_k, r_j\}$ (with much more emphasis on the permanent properties which characterize a physical system). The second part deals with the possibility of ascribing such properties to an individual physical system. It cannot be translated into the language of physics, since any talk of physical systems, magnitudes, etc., refers to *universals*: properties (magnitudes); collections of properties that always go together (e.g., the set of quantum numbers which characterizes a given quantum state); classes of properties (physical quantities); or classes of collections of properties that always go together (kinds of physical systems). But how do we express the fact that physical properties constantly go together in an actual physical system *here and now*? In the language of physics, there are no indexicals, and there is no talk of *individuals* - except in terms of spatio-temporal coordinates, which are themselves universals (numbers). Stories about individual physical systems are always expressed in an informal language which extends the usual use of 'here and now' in *causal* terms to a domain of unobservables. So I suggest the following reliability condition for the direct observation of a physical system:

Reliability condition:

The observation of a physical system x is *direct* only if

1. a mapping from the measured magnitudes $\{A, B, ...\}$ to at least the *permanent properties* $\{p_k\}$ of the system x can be given, and
2. an *individual causal story* can be told which shows that the measured properties can indeed be attributed to an individual system located in a certain space-time region, accessible for an actual measuring device.

The second part of the reliability condition requires that in a direct observation *à la* Shapere, an individual system can be singled out from the experimental results of a given measurement. Let us see now how this condition fits in with the different kinds of evidence for quarks.

3. HOW TO OBSERVE QUARKS

From Shapere's generalized concept of observation and the reliability condition given in the last section, we obtain the following necessary and sufficient conditions for the direct observation of a particle $\{m, s, q_i\}$ with the permanent properties mass m, spin s and charges q_i:[10]

A particle $\{m, s, q_i\}$ is *directly observed* if and only if:

1. some magnitudes $\{A, B, ...\}$ are measured by a particle detector (or system of particle detectors);

2. it can be shown on the basis of reliable background knowledge that the measured magnitudes $\{A, B, ...\}$ result from a sequence of interactions, and that $\{A, B, ...\}$ are unambiguously related to the collection of properties $\{m, s, q_i\}$

According to (1) and (2), the individuation of an observed particle is due to an actual measuring device or apparatus in which particle tracks, scattering events, showers, etc. are detected. (It has to be noted that 'individuation' means only 'position measurement' here, or 'localization within a certain space-time region', but not full spatio-temporal individuation. So we do not run into troubles with quantum theory.) The rest of the causal story, according to which the measurement results from the detected tracks or events belongs to the permanent properties $\{m, s, q_i\}$ of a particle, is based on well-established physical theory, that is, on epistemically reliable background knowledge in Shapere's sense.

Now can quarks be observed in the above sense, even though they are confined inside hadrons? Since they do not exist in unbound states according to the confinement hypothesis, the crucial question here is whether a measurement method is specific enough to give rise to such a causal story. In the history of the quark model, it turned out that more and more specific measurement methods were invented, once there was *any* specific evidence of quarks. In what follows, however, I can briefly sketch only the most important measurement methods.

3.1. *Probing the nucleon structure*

According to the measurement method which was the first step towards an empirical validation of the quark model, quarks are *pointlike structures* inside the nucleon. In 1968, a scattering experiment with an electron beam and a fixed target, giving rise to electron-nucleon interactions, produced a scattering cross section with an unexpected "scaling behaviour" in the so-called "deep inelastic region".[11] The *deep inelastic kinematic region* of a relativistic scattering process[12] is characterized by an energy transfer between the scattered particles which is large enough to create a shower of secondary hadrons. The *scaling behaviour* of the measured cross section was a phenomenological feature indicating that a *pointlike structure* had been found inside the nucleon, contrary to the belief that the nucleon consists of a continuous extended charge distribution described by electromagnetic form factors.

A pointlike structure inside the target of a scattering experiment is supposed to give rise to a cross section that does not depend on any magnitude of the dimension 'length' which can be attributed to the scattering center. On the basis of a well-known formal analogy between the quantum mechanics of scattering and the classical diffraction of a wave, and on the grounds of a very general dimensional argument, it can be shown that such a cross section, if normalized appropriately to a dimensionless quantity, cannot depend on the energy of the scattered beam. Such *scaling behaviour* of a cross section (that is, an independence of the energy scale at which a scattering experiment with particles of a given type is performed) can easily be derived, for example, from the famous Rutherford formula. On the basis of well-understood relativistic generalisations of the Rutherford formula, for any process of "elastic" scattering (in which *no* shower of secondary hadrons is observed) one may calculate an empirical form factor $F(q)$ from a measured cross section, which depends on the momentum transfer q of a scattering process, and which is a direct measure of the charge distribution $\rho(r)$ of a scattering center, e.g. of an atomic nucleus.[13] On the basis of some additional general principles of relativistic quantum theory,[14] the charge distribution of the nucleon can be described by two electromagnetic form factors $F_{1,2}(q^2)$ which were measured in the 1950s by Hofstädter.[15] A form factor which describes a non-pointlike structure depends on the momentum transfer of a scattering process and destroys the scaling behaviour of the cross section. So the scaling behaviour of a measured cross section indicates a pointlike scattering center, and a momentum dependent form factor indicates a non-pointlike structure to be involved in the scattering.[16]

For the deep inelastic scattering of electrons and protons or neutrons, two electromagnetic "structure functions" $W_{1,2}(q^2, \nu)$ can be defined on the basis of similar background knowledge. They are related to electromagnetic form factors which depend on two magnitudes: the 4-momentum transfer q (respectively its square) and the relative energy transfer ν. They express the relevant information about the dynamic structure of the nucleon (here, the physicist's use of 'information' comes close to the terms in which Shapere's concept of observation is expressed). The crucial experiment in 1968 showed that the structure functions depend not on both magnitudes independently, but rather on their ratio, indicating scaling behaviour - in straightforward analogy to the famous experiments in 1909/11 which laid the ground for the derivation of the Rutherford formula.[17] I can not go into any detail about the theoretical analysis which was given by Bjorken and Feynman, when they developed the "parton model" and elaborated it as the "quark-parton model" in the years that followed. I just want to emphasize that neither the discovery and its subsequent explanation in terms of the quark-parton model, nor any later experiment which was performed according to that measurement method, may count as a direct observation of quarks in Shapere's sense.

According to the quark-parton model in its modern form,[18] quarks are dynamic parts of the nucleon. The structure functions $W_n(q^2, \nu)$ an be expressed in terms of weighted sums of the momentum distributions of quarks which are characterized by different "flavors". The weights in the sums derive from the charges involved in the scattering, and the nucleons carry an integer charge and an isospin which result from the sums of the electric and "flavor" charges in the quark-parton model. There are additional, though more complicated sum rules for momentum and spin, which indicate, however, that the discrete partons, which were predicted by the original quark-parton model, can not be the *only* dynamic constituents of the nucleon, since they do not carry the total amount of the nucleon's momentum and spin. The measurement results and the predictions of the quark-parton model are in full agreement. But it is significant that the reliability condition which was given at the end of section 2 is violated in two ways:

First, although there is perfect correspondence between the measured structure functions and the momentum distributions of quarks, the measurement method does not establish an unambiguous relation between a single measured quark momentum distribution and one particular collection of properties $\{m, s, q_e, q_f\}$ of a certain mass m, spin s, fractional electric charge q_e, and "flavor" q_f (not to speak of the "colors" q_c, which always come in quantum superposition, a singlet state of bound quarks).

Secondly, the measurement method does not permit us to tell a causal story relating a definite part of a measured structure function to an individual quark involved in the scattering of an actual beam in an actual apparatus. The causal story which can be told is only about the *total* quantity of quarks within the nucleon, or about the *total* effect of *several* pointlike structures which do not give rise to *separate* effects.

3.2. *Inference to individual quarks*

Nevertheless, the quark-parton model suggested several ways of performing new experiments and providing experimental evidence for discrete, or individual, quarks in agreement with the required conditions for a direct observation. Recall the distinction between "indirect" and "more direct" evidence of quarks already quoted above:

> ... the parton model and QCD indicated that *more direct evidence* should be obtained by studying certain reactions, the simplest being e^+e^- annihilation.[19]

e^+e^- annihilation is a process which occurs when the particles interact during the course of a *collider experiment* in the crossing area of an electron and a positron beam, the so-called interaction zone around which a big detector system is built. The reverse process, e^+e^- production, can be investigated in a variety of fixed target or collider experiments. One of the most spectacular results of e^+e^- annihilation or production was the discovery of the J/Ψ resonance in 1974. The J was observed (in Shapere's sense) in e^+e^- *production*, and the Ψ in e^+e^- *annihilation*.[20] Both resonances indicated the same state of an unstable particle, characterized by a mass related to the energy at which the resonance peak in the energy-dependent total cross section occurs, and a lifetime which is related to the width of the resonance. The unstable particle was interpreted as a bound quark-antiquark system, a "quarkonium" state of a certain mass. Further resonances were found later which could be associated with the resonance spectrum of the excited states of such a positronium-like[21] compound system. This discovery gave rise to the so-called *November revolution* which convinced most members of the particle physics scientific community[22] that the quark model, and some associated quantum dynamics, must be the correct account of what is going on inside a hadron.

The discovery of the J/Ψ was an observation in Shapere's sense of an unstable particle $\{m,\tau\}$ with mass m and lifetime τ. But it was not at all an observation of a quark or of a compound system of quarks. To see why, let us quote again:

> What was lacking was proof that its constituents were indeed the charmed quarks first proposed by Bjorken and Glashow.[23]

The measurement was not specific enough to identify the most important permanent property of the quark content of the discovered particle, that is, the new (and unexpected) quark "flavor" *charm*, which was predicted in a theoretical model explaining the phenomenological absence of certain kinds of particle reaction. The "*charm*" content of the "charmonium" (as this kind of a quark-antiquark system has been called) can be identified only in another type of experiment: the "flavors" involved in the secondary decays of the unstable particle can be measured from an analysis of the individual tracks, on the basis of the selection rules and conservation laws which are associated with quantized properties such as "strangeness" or "charm".

The so-called fragmentation of hadrons into two "jets" or sharply bundled showers of secondary hadrons, however, is a much better candidate for a direct observation in Shapere's sense (recall the end of the first quotation, at the end of section 1)[24]. The crucial model from which the interpretation of such experimental evidence derives is based on the confinement hypothesis. Due to the exceptional kind of force which gives rise to confinement, the quarks in a bound quark-antiquark state will be more distant the more energy the state has. Therefore, by increasing the energy of the colliding beams in an e^+e^- annihilation experiment, the probability increases that a quark-antiquark state would be created with such a large internal distance, that another quark-antiquark pair would be created which would recombine with the single quarks of the former bound state into two new bound states of lower energy. These would appear as two unstable mesons at the 'poles' of the former quark-antiquark state, decaying into two spatially separated hadron showers which would be sharply centered and directed in two opposite directions.[25] From an analysis of the individual tracks making up each of both showers or "jets", the quark "flavors" of each 'pole' of the primary quark-antiquark state can be unambiguously identified, at least for *some* of the several possible decay modes, one of which must be realised in an individual "jet" event which is due to the decay of an individual quark-antiquark state. For the analysis of such an event, both parts of the reliability condition for a direct observation are fulfilled (given the theoretical explanation is sufficiently convincing with regard to statistical significance):

First, from an analysis of the single tracks of the secondary hadrons, the quark content of the initial quark-antiquark state can be unambiguously determined, at least in principle. And this means that the measured magnitudes of each of the two jets can be unambiguously related to a particular kind of quark $\{m, s, q_f\}$.

Secondly, the impressive pictures which were obtained from the computer analysis of what was going on at any given moment in the concrete system of particle detectors give rise to a perfect causal story about what was going on inside the interaction zone of a collider experiment. The story relates each of both jets unambiguously to an individual quark, which, as well-established theory tells us, is one of the two constituents of a compound system which is produced in a certain region of the interaction zone of two concrete beams, in a certain moment, inside a given measurement device.

Therefore, the inference from hadron jets to quarks may be regarded as based on the reliable background knowledge which is admittedly involved in a direct observation of quarks.

4. OBSERVATION AND MEASUREMENT

Several conclusions may be drawn from my analysis of Shapere's concept of observation, in the light of the available evidence for quarks. The first (and most obvious one) is that Shapere's criterion for a direct observation agrees perfectly with the way in which particle physicists ask for "direct evidence". The second conclusion is that there are good reasons to think that quarks have indeed been observed as the cause of hadron "jets"-a result which has some import for the debate on scientific realism. There are other measurements, however, in which individual quarks *can not be singled out* from experimental evidence, for example due to the production of more and more non-discrete nucleon content which is supposed to give rise to scaling violations in deep inelastic lepton-nucleon scattering. Such measurements show that the distinction between pointlike structures within the nucleon and an additional field content of the nucleon depends on the energy at which a scattering experiment is performed. In the final analysis, the distinction is to a certain degree arbitrary. So my result concerning the observability of quarks is not general, and instead of concluding that quarks do indeed exist as the constituents of all empirical bodies around us, I want to defend the more modest claim that they exist under the specific conditions which are realised in certain kinds of experiment in high energy physics. (It should be noted, however, that the question of whether quarks are still there when they are *not* observed in a scattering experiment is in striking analogy to the question of whether a macroscopic red body is still red when it is not exposed to light.)

My third conlusion is related to the distinction between *particles* and *fields*, and to the *locality condition* which is implicitly involved in Shapere's criterion for a direct observation. The distinction between particles and fields is meaningless according to the group theoretical axiomatic definition of 'particle' which applies primarily to *fields*.[26] A general distinction between particles and fields, which is not restricted to the classical domain, may be drawn only from an operational point of view: a *particle* is a *collection of local properties which are always detected together*. (A field, on the other hand, may be conceived as a non-local distribution of properties which do not necessarily come in constant collections[27]). Shapere's concept of direct observation also relies on locality. The locality of an entity which can be directly observed was made explicit in part (ii) of the reliability condition given at the end of section 2: a direct observation necessarily implies that an individual causal story can be told about an entity with certain measured properties, located in a certain space-time region. Putting the locality of an observation and that of a particle together, we must conclude as follows: particles are easily observed (since the observation of a collection of local properties which have always been detected together is the observation of a particle), but it is impossible to observe fields (given that fields are non-local). For example, no direct observation in Shapere's sense would be able to decide whether the interaction between two bodies is due to a field, or to action-at-a-distance.[28]

However, a sharp distinction between observation and measurement, based on Shapere's concept of observation, might help us to clarify some important issues in the debate on scientific realism. Such a distinction *is* possible, contrary to van Fraassen's claim that observation is a notoriously vague concept. The meaning of 'observation' is no longer vague, once one succeeds in picking out the structural features of what may count as an observation. The most important distinguishing marks of observation and measurement seem to be that observation requires *an individual entity* to be observed, whereas measurement requires only *physical magnitudes* to be measured. Observation confirms the existence of an entity of a certain kind within a certain space-time region. Measurement, on the contrary, confirms only the validity (or adequacy) of a model which is expressed in terms of magnitudes. The validity of a model does not depend on the question of whether it is possible to construct an individual physical system from properties which were measured in an experiment. My analysis of Shapere's concept of observation has shown that even a quantum system can be directly observed in a generalized sense, given that it is localizable as a collection of dynamic properties in a certain space-time region. We can adduce good grounds in support of such a claim - even for quarks.

Heidelberg University
Germany

NOTES

[1] Thus mesons (hadrons with a mass which is small compared to the nucleon mass) are made up of a quark and an antiquark, and baryons (hadrons with a mass which is at least of the order of the nucleon mass) are made up of three quarks. According to the actual composite model of hadrons, however, two or three quarks are no longer considered to be the only constituents of a hadron. In addition, it has to be assumed that there are constituents carrying a measurable amount of momentum and spin, which are identified with gluons (the vector bosons mediating the strong interaction) and quark-antiquark pairs.

[2] See e.g. Streater 1988, p. 144.

[3] Wigner 1939. Wigner, however, does not use the term 'particle'; he classifies the solutions of ***any*** relativistic (i.e. Poincaré invariant) field equation with group theoretical methods, speaking only of "wave functions".

[4] See Perkins 1987, p. 20. a_s is the coupling constant of the strong interactions, k is a constant.

[5] Cahn & Goldhaber 1989, p. 307 (my emphasis). The book is a collection of the crucial publications in experimental particle physics, starting with the discoveries of the neutron and the positron, ending with the evidence for the fifth quark, and for the vector bosons of the weak interactions. The quotations are taken from the editors' introduction.

[6] Shapere 1982, p. 492.

[7] See Shapere 1982, p. 492, where Shapere emphasizes that "*specification of what counts as directly observed (observable), and therefore of what counts as an observation, is a function of the current state of physical knowledge*". See also p. 517: "The *epistemically* important line between the non-inferential and the inferential is drawn in terms of the distinction between that which we have specific reasons to doubt ... and that upon which we can build confidently."

[8] See van Fraassen 1980, p. 14 ff., and Maxwell 1962.

[9] Shapere himself, however, does not make explicit the necessary condition for direct observation which is contained in his criterion for direct observation, showing that the latter is indeed much more than a *criterion*.

[10] I give a definition instead of a sufficient criterion for the direct observation of particles, since there is no way of observing particles in *another* sense than the one explained here.

[11] For the following, see Cahn & Goldhaber 1989, p. 217 ff., and the articles reprinted there.

[12] According to the use of the term 'kinematics' in high energy physics, the kinematic magnitudes are those which characterize the 4-momentum state of a free particle, i.e., momentum and energy.

[13] A non-relativistic form factor is simply the Fourier transform of a spatial charge distribution. For relativistic form factors, this relation still holds for a special frame of reference (the Breit frame) in which there is no energy transfer, and in which the relativistic 4-momentum transfer thus reduces to its 3-momentum components, establishing correspondence to the non-relativistic case.

[14] Only uncontroversial principles such as Lorentz invariance or gauge invariance are involved. See Drell 1961.

[15] McAllister & Hofstadter 1956; reprinted in Cahn & Goldhaber 1989.

[16] See Falkenburg 1993, where the argument is analysed in more detail. The different kinds of form factor belong to the non-relativistic or relativistic domains.

[17] Bloom et al. 1969, Breidenbach et al. 1969; both reprinted in Cahn & Goldhaber 1989. - One of the phenomenological features of the measured cross section was large angle scattering, in perfect analogy to the backward scattering discovered by Marsden in 1909 in Rutherford's laboratory. But backward scattering may be explained in various ways, as Rutherford knew when he spent two years calculating multiple scattering effects (before he announced his model of the atom). In 1968, however, the rival theoretical models which might account for backward scattering from a nucleon were much less reliable, and the discovered scaling behaviour of the scattering cross section was more spectacular.

[18] The original model turned out to be a crude idealisation, in the light of subsequent experimental evidence.

[19] Cahn & Goldhaber 1989, p. 307 (my emphasis).

[20] See Aubert et al. 1974 and Augustin et al. 1974, both reprinted in Cahn & Goldhaber 1989.

[21] A bound state of an electron and a positron is called positronium.

[22] An exception was, for example, the prominent physicist Heisenberg who, since the experimental results of 1968, has attempted to incorporate *scaling invariance* into his *Ansatz* of a unified field theory; see Heisenberg 1976.

[23] Cahn & Goldhaber 1989, p. 263 f.

[24] See Cahn & Goldhaber 1989, p. 306 ff., and the articles reprinted there.

[25] It should be noted that there are also events with three and more jets, which are interpreted as evidence for gluons being involved in the scattering. Do they give rise to a direct observation of gluons?

[26] See Wigner 1939, and note 3.

[27] It should be noted, however, that this concept is *not* operational.

[28] I am indebted to Dennis Dieks for drawing my attention to this point.

REFERENCES

J. J. Aubert, *et al.*, "Experimental Observation of a Heavy Particle J", *Phys. Rev. Letters* **33**, 1974, 1404-06.

J. E. Augustin, *et al.*, "Discovery of a Narrow Resonance in e^+e^- Annihilation", *Phys. Rev. Letters* **33**, 1974, 1406-08.

R.N. Cahn & G.Goldhaber, *The Experimental Foundations of Particle Physics*, Cambridge 1989.

E.D. Bloom, *et al.*, "High-Energy Inelastic e-p Scattering at 6° and 10°", *Phys. Rev. Letters* **23**, 1969, 930-4.

M. Breidenbach, *et al.*, "Observed Behavior of highly Inelastic Electron-Positron Scattering", *Phys. Rev. Letters* **23**, 1969, 935-9.

R.W. McAllister, & R. Hofstadter, "Elastic Scattering of 188-MeV Electrons from the Proton and the Alpha Particle", *Phys. Rev.* **102,** 1956, 851-6.

S.D. Drell, & F. Zachariasen, *Electromagnetic Structure of Nucleons*, Oxford 1961.

B. Falkenburg, "The Concept of Spatial Structure in Microphysics", *Philosophia naturalis* **30**, 1993, 208-228.

W. Heisenberg, "Was ist ein Elementarteilchen?", *Die Naturwissenschaften* **1**, 1976, 1-7. Reprinted in W. Blum et al., eds., *Gesammelte Werke C III.*, München 1985, 507-13.

G. Maxwell, "The Ontological Status of Theoretical Entities", in H. Feigl & M. Scriven eds., *Minnesota Studies in the Philosophy of Science*, Minneapolis 1962.

D.H. Perkins, *Introduction to High Energy Physics,* third edition, Addison-Wesley, 1987.

D. Shapere, "The Concept of Observation in Science and Philosophy", *Philosophy of Science* **49**, 1982, 485-525.

R.F. Streater, "Why Should Anyone Want to Axiomatize Quantum Field Theory?", in H.R. Brown & R. Harré eds., *Philosophical Foundations of Quantum Field Theory*, Oxford 1988, pp.135-48.

B.C. van Fraassen, *The Scientific Image*, Oxford 1980.

E.P. Wigner, "On Unitary Represantations of the Inhomogeneous Lorentz Group", *Annals of Mathematics* **40**, 1939, 149-204.

GIOVANNI MARIA PROSPERI

COMMON EXPERIENCE AND QUANTUM THEORY

Observables and Beables

By Classical Physics we usually mean the physics that was developed before the formulation of Quantum Theory or that works in that same conceptual framework.

Classical Physics proceeds in terms of idealized models directly related to our common experience and assumes that the properties of an object are specified by a set of variables having a well defined value at any time and changing continuously in time. Such variables are also in general assumed to satisfy systems of differential equations which control their time evolution and in principle enable us to predict their values at a subsequent time, once they are known at a previous time. In this sense Classical Physics is a *deterministic theory*, in which any statement is supposed to have an immediate meaning and no explicit reference to actual observation procedures is necessary.

On the contrary Quantum Theory consists in a set of abstract mathematical rules, which have no intuitive meaning in themselves, but permit us, given certain initial information about a system, to evaluate the *probability* that a successive experiment will give a certain result. An explicit reference, therefore, to the observation procedures and to the times in which they are applied is essential and no meaning can be given to any statement on the system independently of such a reference. In fact any attempt of adopting an intuitive model for the object with properties specified as in Classical Physics is bound to meet with irremediable contradictions, if applied without restriction and out of specific limit conditions.

A typical illustration of the above situation is given by the diffraction and interference phenomena occurring with particles. If we think of a Young two slit experiment performed by a beam of electrons, it is clear that the idea of a single electron passing through a single slit is inconsistent with the possibility of obtaining an interference figure. Actually we must admit that any attempt to detect the slit that the electron went through would destroy the interference pattern. It is consequently the very idea of a particle as a small body having a well defined position at any time that does not hold any more.

In Quantum Theory the term of *observable* is used to stress the impossibility of ascribing to a *quantity* a value independently of any actual observation. On the contrary, J. Bell has proposed to use the word *beable* for a quantity which is supposed to have always intrinsically a well defined value. So Quantum Theory works in terms of *observables*, Classical Physics in terms of *beables*.

The most significant aspects of Quantum Theory from the logical point of view can be summarized in three postulates.

i) The *state* of an object is specified by a unitary vector $\psi(t)$ (*state vector*) in a

E. Agazzi and M. Pauri (eds.), The Reality of the Unobservable, 343-351.

an appropriate infinite dimensional linear space $\mathfrak{R}$ (an Hilbert space). The time evolution of $\psi(t)$ is controlled by a linear differential equation (*Schroedinger equation*), $i\hbar\, \partial_t \psi(t) = H\psi(t)$, H being an appropriate Hermitian linear operator in $\mathfrak{R}$.

ii) Any observable A referring to the object is associated with a Hermitian linear operator $\hat{A}$. The eigenvalues $\alpha_1, \alpha_2, \alpha_3, \ldots$ and the eigenvectors $\varphi_1, \varphi_2, \varphi_3, \ldots$ of $\hat{A}$ (defined by the equation $\hat{A}\varphi_s = \alpha_s \varphi_s$) give the possible values assumed by the quantity A and the states (eigenstates) for which a measurement of A gives the corresponding eigenvalue with certainty. If, at a certain time t, the object is in the *superposition* state $\psi = \sum_s c_s \varphi_s$, then the probability that a measurement of A gives the value α_s is $P(A = \alpha_s; t) = |c_s|^2$.

iii) If a measurement of A has been made at the time t_0 and the value $A = \alpha_s$ has been found, the state vector has to be redefined immediately as $\psi(t_0) \rightarrow \psi(t_0 + \varepsilon) = \varphi_s$ (*state vector reduction* or *wave function collapse*).

Obviously, from the mathematical point of view the impossibility to ascribe intrinsically to A a value independently of any observation derives from the contrast between the linearity of the evolution equation (i) and the quadratic dependence of the probability on the state vector (ii). However the most intriguing rule is considered (iii), according to which an abrupt change takes place in the state vector as a consequence of the observation.

Naturally, if we want to talk of experiments performed on an object and of the results of such experiments, it is necessary for us to talk of the experimental apparatus and of the modifications occurring in them in a well defined way.

According to Bohr, in fact, Quantum Physics provides necessarily only an open description of its objects[1]. The experimental arrangement and the result of its action must be described in a way independent of any observation, i.e. in classical language. The classical language has to be considered as the proper language for the world of our common experience, while Quantum Theory talks only of the modifications produced in such world by otherwise abstract objects we call particles or systems of particles. The duration itself of an experiment, the times and the types of possible intermediate measurements should be specified in advance and any modification of such circumstances would amount to a modification of the experimental conditions; it would therefore require an independent treatment from the beginning and give an independent answer. In short we can say that observables presuppose beables.

The apparatus, however, as all objects of our ordinary life, are thought to be built by the same atoms, nucleus and electrons to which Quantum Theory is supposed to apply and actually many properties of macroscopic matter have been understood by applying such theory to systems formed by a great number of the above constituents. So the problem of the consistency of Borh's interpretation can not be avoided. It is this problem that von Neumann has tried to solve with his quantum theory of measurement.

In applying Quantum Theory also to the measurement apparatus, von Neumann tried to idealize the process of measurement of a quantity A in the following way[2]:

a) the interaction between the object and the apparatus is treated as a collision and is supposed to last so short a time that there is practically no free evolution of the two systems during it;

b) the apparatus is assumed to be in a certain initial state Φ_0;

c) if the object is in an eigenstate φ_s of A, the result of the collision of the object and of the apparatus is assumed to be $\varphi_s \Phi_0 \rightarrow \varphi_s \Phi_s$ where Φ_s are the eigenstates of some quantity M referring to the apparatus (e.g. the position of a pointer).

Obviously, if we denote the object by I and the apparatus by II, it is assumption c) that qualifies system II as an appàratus for the measurement of the quantity A referring to I. As a consequence of c) a correspondence is established between the eigenstates of A and M in such way that, if immediately before the interaction I is in the state φ_s (corresponding to the value α_s), the response of II is necessarily $M=\mu_s$, without any modification in A. More generally, if I is initially in the superposition state $\psi = \sum_s c_s\varphi_s$, we have $\psi\Phi_0 = \sum_s c_s\varphi_s\Phi_0 \rightarrow \sum_s c_s\varphi_s\Phi_s$, due to the linearity of the Schroedinger equation. Therefore, the probability of observing $M=\mu_s$ on the apparatus at the end of the interaction is the same as that of observing $A=\alpha_s$ on the object when the apparatus is not explicitly taken into consideration.

Notice, however, that, having applied Quantum Theory to II too, in principle any statement concerning the quantity M should again refer to a specific measurement procedure carried out on II by a second apparatus, which we shall denote by III. For instance, if we think of I as a set of particles in a collision experiment and of II as the system of detectors we employ in the experiment, III could be identified with the electronic equipment we use to register and elaborate the response of the detectors. Obviously III has to be treated as a classical system.

The argument can be repeated. Thus, according to von Neumann, we have to think of an entire chain of apparatuses, each "observing" the preceding one. Such a chain necessarily includes the sense organs and the nervous system of the experimenter, ending up with the "abstract ego" of the human observer. Any statement of Quantum Theory concerning I is actually a statement on the effects produced by I on II and any statement on II is a statement on the effects that II produces on III. The demarcation line between what we consider the object and what we consider the "observer" is largely arbitrary. What is important is that there is consistency between the case in which I is treated as the object and II+III as the observer (and so Quantum Theory is applied only to I) and the case in which I+II is treated as the object and III is considered as the observer (Quantum Mechanics is applied to I+II). Ultimately, however, what decides about what has happened is the conscious act of the abstract ego.

Apart from the somewhat paradoxical aspects related to the need for an explicit reference to the consciousness of the human person, two criticism can be made to von Neumann's theory. The first criticism concerns assumption c), which is definitely too strong and not sufficiently realistic, the second concerns the fact that no specific attention is paid either to the size of the object and or to the macroscopic character of the apparatus.

Concerning assumption c) it must be stressed that the effect on the apparatus of the interaction with the object is not left to our choice. Such effect is the result of the microscopic structure of the two systems and of the nature of the forces existing within their elementary components. Consequently, it is not sufficient to assume that different eigenstates of an observable referring to the object couple with different eigenstates referring to the apparatus. It is necessary to show that this coupling can be really attained in the specific case. Wigner has actually shown that the existence of certain additive constants of motion for the compound system I+II (like angular momentum for spin measurements) prevents the possibility of satisfying exactly c) for any of such quantities[3]. Furthermore it is essential, in von Neumann's treatment, that the states of the object that couple with the states of the apparatus should be true physical states represented by normalizable vectors. On the contrary observables with purely continuos spectrum, like position and momentum, or with mixed spectrum, like energy, would not be true observables. In these cases one should resort, quite artificially, to purely discrete observables which approximate the preceding ones.

The above difficulty is in fact related to an imperfect formulation of ordinary Quantum Theory and can be fully overcome within the context of a more satisfactory theory worked out manly by G. Ludwig and collaborators, E. B. Davies, A. S. Holevo (L.D.H.)[4]. In order to understand the ideas at the basis of the L.D.H. theory, it is convenient to remember that in the ordinary formulation of Quantum Theory the probability of observing a certain value $A=\alpha_s$ of A can also be expressed in terms of the orthogonal projection of the state vector ψ over the eigenvector φ_s or (in case of degeneration) the eigenspace corresponding to that eigenvalue. Precisely we can write $P(A=\alpha_s;\, t) = |c_s|^2 = \|P_s\psi\|^2 = <\psi\,|\,P_s\psi>$, where P_s denotes the above mentioned projection operation: $P_s\psi = \varphi_s < \varphi_s\,|\,\psi>$. Obviously, the set of all the projectors P_s referring to A satisfies the property $\sum_s P_s = 1$, which corresponds to the completeness of the eigenvectors and guarantes that the total probability is 1. In the L.D.H. formulation the projectors are replaced by a more general type of operators called *effects*. An effect is an Hermitian operator F satisfying the relation $0 \leq F \leq 1$ (obviously a projector is a particular type of effect). Then an observable A' is generally associated with a set of effects F_s which are again related to certain values α'_s and have the property $\sum_s F_s = 1$. The probability of observing a specific value α'_s is supposed to be given by $P\,(A' = \alpha'_s;\, t) = <\psi\,|\,F_s\psi>$.

Generalized observables have many interesting properties. In particular they allow us to give a precise mathematical formulation of ideas like that of imprecise measurement of a quantity or even of simultaneous imprecise observation of incompatible quantities (e.g. the position and the momentum of a particle). As a consequence, they also enable us to put a sequence of observations separated in time on the same formal footing as a single observation made at a definite time. What interests us here is that they allow a completely satisfactory reformulation of von Neumann consistency theorem. In the new context, indeed, it can be shown that in all generality (without any *ad hoc* assumption) the observation of any quantity on the apparatus, after the interaction with the object, is mathematically equivalent to the observation of a well defined corresponding quantity on the object. Obviously the quantity we observe on the object depends on the apparatus, on the nature of the interaction between the object and the apparatus, on the state in which the apparatus has been prepared and on the particular quantity we have chosen to observe on the apparatus. What has to be stressed, however, is that in general the observable on the object resulting in this way is a L.D.H. *generalized observable*, even if the observable chosen for the apparatus was an ordinary one.

Naturally, in L.D.H. formulation, even the reduction postulate has to be modified. I do not want to enter into details here, however it is worthwhile to mention that it is no longer possible to remain within the framework of the *pure states* alone (specified by single vectors in the Hilbert space of the system), but it is necessary to resort to the formalism of the *mixed states* (specified by the so called *density operators* and corresponding to less complete information on the object). A new type of mathematical operation can then be introduced which is called *instrument* and has the effect of replacing the (pure or mixed) state of the object before the interaction by a new *mixed* state depending on the outcome of the experiment. The specific *instrument* to be associated with a certain experiment is again a consequence of the nature of the apparatus and of its interaction with the object. In this perspective it is important to stress that, if the apparatus II is included in the treatment, the observation of M on II at a time t_0 after the interaction becomes compatible with the observation of a second quantity B on I at a later time t_1. Then the reduction of the state of I as a consequence of the observation of A at t_0 can be simply described as the replacement of the joint probability of observing certain values of M at t_0 and of B at t_1 by the corresponding conditional probability. In this way even the reduction process loses much of the paradoxical aspect it had in the old formulation.

Let us now consider the second criticism to von Neumann's theory of measurement. Obviously, the macroscopic description of a large body (i.e. of a body constituted by a large number of atoms or molecules) should be the proper domain of Classic Physics. Such a body is here typically idealized as a continuum and its *macroscopic state* is specified in terms of certain *macroscopic quantities* (like density, chemical composition, local velocity, density of energy, electric and magnetic polarization etc.) that are treated as *beables*, in our language, and are supposed to obey the laws of classical Mechanics, Thermodynamics, Electromagnetism. One feels that this circumstance should have specific consideration in the theory and that the problem of a link between the classical and the quantum theory at this level of description cannot be avoided.

Obviously, the most natural framework within which to look for that link is Quantum Statistical Mechanics. Indeed, Statistical Mechanics or Statistical Physics is the chapter of Physics which sets out to explain the macroscopic behaviour of a large body on the basis of the properties of its elementary constituents. Historically, we had first Classical and then Quantum Statistical Mechanics, according to whether one applied classical or quantum physics to the elementary constituents. Many problems are still open, both in Classical and in Quantum Statistical Mechanics. Many interesting results, however, have already been obtained. In particular in the context of Quantum Statistical Mechanics, as I have already mentioned, it has been possible to understand many aspects of the behavior of matter, even in conditions very far from those of our normal experience, as they occur in stars or at the very low temperatures we can obtain in our laboratories today.

In Quantum Statistical Mechanics, the macroscopic variables are expressed as appropriate averages over small intervals of time and sufficiently large groups of molecules. The time evolution of such quantities, the relations among them by means of which equilibrium is attained, and the very existence of equilibrium states should be a consequence of the dynamic behavior of the elementary constituents. Obviously, a *macroscopic state* of our large body corresponds to a very poor specification of its *microscopic state*, in the sense that there are many quantum (or classical) *pure* states that correspond to the same macroscopic state. However, an appropriate assumption of *a priory* equi-probability of the various compatible pure states makes it possible to associate a practically unique *mixed* state with any macroscopic state. As a result we can typically expect the *expectation values* of the macroscopic quantities to obey deterministic classical types of equations, while covariances remain very small. If this is the case and as long as we operate on a scale at which the corresponding fluctuations cannot be detected, macroscopic variables can be treated as *beables*.

To a practical minded physicist the above circumstance could even appear to provide a basis for a satisfactory solution of the problem of measurement in Quantum Theory. Once the macroscopic level has been reached, an explicit reference to Quantum Theory would seem to become inessential, making it possible for von Neumann's chain to be broken. To my knowledge, the first author who suggested a similar point of view was P. Jordan[5] and, in the recent past, also our group in Milan did some work to make this idea explicit[6]. The above results, however, while preventing obvious inconsistencies, cannot provide a real solution of the problem. As stressed in particular by Wigner, interference effects among different assignments of the macroscopic variables would remain present in principle, even if small and difficult to detect, and an explicit reference to an observer does not seem to be avoidable in this way[7].

Notice that an essential consequence of the Bohr-von Neumann interpretation is the need to divide the world in two parts, the first one considered as the object and treated by Quantum Theory, the second one regarded as the observer, described in classical terms and including necessarily all conscious beings. In perspective this prevents applying Quantum Theory to the entire universe. In view of a quantization of Einstein's theory of gravitation, difficulties would arise even for a consistent definition of space and time in a physical framework.

These circumstances, together with the philosophical prejudice that a satisfactory scientific theory should be necessarily deterministic, explain the efforts that have been made with respect to the so-called *hidden variables theories*. In such theories the statistical character of Quantum Theory should be the result of an average over additional variables that do not appear in our usual formulation. As is well known, however, J. Bell was able to derive certain inequalities (*Bell inequalities*), concerning correlation at-a-distance of appropriate quantities, that any such theory must satisfy under very general hypotheses[8]. In particular the assumption is made that all interactions are local and propagate with a finite velocity, according to Relativity. In fact, such inequalities are violated by ordinary Quantum Theory in specific situations and enable us to discriminate this theory from the hidden variable ones. Along these lines various accurate experiments have been made, all essentially concerning the polarization's of pairs of photons or other particles emitted in opposite directions in a zero angular momentum state. On the whole, such experiments have largely disproved hidden variables theories and confirmed Quantum Theory.

Obviously, the above discussion has been made upon the assumption that Quantum Theory has an unrestricted validity, in the sense that it applies to a system formed by a small and by a large number of particles indifferently. However, the possibility exists that the present theory has to be seen as the limit case for a small number of particles of a more general theory in which the *macroscopic variables* have *intrinsically* the role of *beables*. Among the various attempts in this line, I find particularly interesting a proposal by Ludwig, in which the time evolution of the macroscopic variables is described in terms of well defined world lines[9]. In contrast with ordinary Classical Physics, however, such world lines are not supposed to be determined by a set of differential equations, but rather specified in terms of probability distributions in an appropriate functional space. The proposal is made in the context of an axiomatic theory in which such distributions of probability are expressed in a formalism similar, under many respects, to that used for observables at a definite time in ordinary Quantum Theory. In fact, in his treatment Ludwig remained on a very general level and was not able to give specific prescriptions for an actual realization of his abstract ideas and for establishing an explicit connection with present quantum theory. New perspectives seem to arise, however, if one tries to combine Ludwig's ideas with the idea of observations continuously extended in time.

To my knowledge, the idea of a continuous monitoring in Quantum Theory was considered for the first time by E. B. Davies for a counting type experiment in 1970.[10] A general formulation of the problem has been independently developed by our group in Milan since 1982[11], while recently a different (but in my opinion less satisfactory) approach was proposed also by M. Gellmann and J. B. Hartle[12] (also the work of G. C. Ghirardi, A. Rimini and T. Weber[13] could be in some way related to this problematic). I shall try to explain the basic concepts of the Milan formulation and its relevance for the problem of interpretation of Quantum Theory.

In Quantum Theory the main difficulty in considering observations continuously extended in time comes from the so-called Zeno paradox, discovered by B. Misra and E. C. G. Sudarshan. Let us consider for a certain system an observable A and assume that we make repeated measurements of A at certain times t_1, t_2, t_3,... Then, using the rules of ordinary Quantum Theory, it is possible to evaluate the probability for a certain sequence of results α_1, α_2, α_3,... to occur. However, if we work in the context of ordinary Quantum Theory, it can be shown that in the limit $t_n - t_{n-1} \rightarrow 0$ the quantity A is *frozen,* i.e. the probability of always observing an identical value becomes 1 and, obviously, the probability of observing any other sequence becomes 0. In other words, in ordinary Quantum Theory the continuous observation of a quantity disturbs the system so much that it prevents it from evolving. The situation changes in the L.D.H. formulation we discussed above. First of all a repeated observation does not correspond to a projector, but it can be shown that it corresponds to an *effect*; therefore in the generalized formulation a repeated observation can be put on the same footing as a single observation. Secondly, by replacing the projectors corresponding to the original *exact* observation of A by appropriate *effects*, we can consider the case in which we make repeated *approximate* measurements of it. Then it is also possible to consider a limit situation in which $t_n - t_{n-1} \rightarrow 0$ and, simultaneously, the measure becomes more and more imprecise (the precision being controlled by an appropriate parameter introduced into the effects) in such a way that the result remains finite. In this way the Zeno paradox is circumvented and we can obtain a probability distribution for a kind of *fuzzy A-world lines*. Obviously, the formalism so obtained differs from Feynman's path integral formalism in that Feynman's formalism works in terms of probability amplitudes and not of probability distributions.

The formalism of the *continuous observations* has been applied usefully to the discussion of various types of experiments and phenomena like photon counting, laser technique for electromagnetic-magnetic field measurements *(heterodyne detection)*, detection of atomic transitions induced by laser light *(shelfing effect)*, gravitational wave detection. However, what is important for me here is the possibility of applying it to the a realization of Ludwig's ideas by the introduction of the world lines of appropriate macroscopic *beables M* in the context of Quantum Theory. Indeed, from the formal point of view one can try to treat a *beable* as a quantity continuously kept under observation. Naturally, in order to do this consistently, additional requirements are necessary and this makes the mathematics very stringent. Furthermore, it is necessary to decide *a priori*, once for all, which quantities are *beables* and this assumption has to be considered an intrinsic element of the theory. Obviously, all this amounts to a true modification of the theory, due to the permanent perturbation produced by the *continuous observation*. In particular it is necessary to introduce at least a new fundamental constant γ, that typically can be chosen with the dimensions $1/t$. Thus, the structure of the theory becomes such that only average values $\overline{M}$ over short time intervals Δt of the beables M have a meaning. Also a kind of intrinsic complementarity principle also exists between the magnitude of the fluctuations of $\overline{M}$ and the value of Δt, in the sense that a term proportional to

$1/\Delta t$ appears in the expression of the covariance of $\overline{M}$.

If we describe matter in the non-relativistic second quantized formalism, a possible choice of the *beables* is given by the *macroscopic* densities $n'(x,t)$, $n''(x,t)$, of the various types of particles, which then become the fundamental quantities of the theory[14]. Notice that with such a choice, in principle, macroscopic physics can be reconstructed entirely in terms of expectation values and correlation functions. In fact, if the state of the system is close to local thermal equilibrium, we expect n', n'', ... to have small fluctuations around their expectation values and such values to evolve practically according to deterministic equations (as in the case of Statistical Mechanics, but now without the possibility, even in principle, of detecting interference effects among different world lines). On the contrary, if the state contains one particle or a small number of particles S far from the local equilibrium, the statistical aspect of the theory would emerge and the value of n', n'', ... would be modified as a consequence of the interaction with S, essentially according to the probability laws of ordinary Quantum Theory as applied to S. Notice that in this picture the so-called *elementary constituents* and their related observables would appear simply as a *description* of the *deviations* of the basic macroscopic quantities from a deterministic behavior. Only n', n'', ... would have a direct meaning and a connection with our common experience. Any other *quantity* would ultimately refer to modifications produced in them in a given situation.

Unfortunately, however, in the above model, the *continuity equations* satisfied by $\langle n'\rangle$, $\langle n''\rangle$,..., contain certain additional dissipation terms so that the total numbers N', N'', ... of the various types of particles are no longer conserved, but decay in a time of the order $1/\gamma$. A similar circumstance occurs for the expectation value of the microscopic energy. On the other hand, in order that the above mentioned intrinsic fluctuations remain small, one has to assume $\bar{n}\,\Delta V \Delta t >> 1/\gamma$, where $\bar{n}$ is the average value of the density n over the volume ΔV and the time interval Δt. Then, e.g., for $\Delta V \approx 10^{-3}\,\mathrm{cm}^3$ and $\Delta t \approx 10^{-2}$ s, taking into account the fact that, for a gas in normal conditions, it would be $\bar{n} \approx 3\cdot 10^{19}\,\mathrm{cm}^{-3}$, we should have $1/\gamma < 10^{15}\,s$ (to be compared with the life of the universe $\approx 10^{17}\,s$). Even for a person open to the idea of a cosmological particle and energy decays, this value for $1/\gamma$ is definitely too small and in contradiction with the bounds set at present by the so called passive physics experiments. Furthermore, the model can have a relativistic extension only at the price of introducing an indefinite metric in the Hilbert space of the system and this does not seem easily interpretable. In conclusion, in spite of its many appealing aspects, the above choice could provide an interesting illustration of the underlying concepts, but certainly not a realistic model.

An alternative possibility would be to treat as beables appropriate *macroscopic* Bose fields like the *force fields*. Notice that these include the electromagnetic and the gravitational fields in terms of which our brain state and space-time geometry can be respectively expressed. Thus, in this case a very interesting situation would emerge. In particular there would no longer be any difficulty in applying Quantum Theory to the entire universe or in the definition of space and time.

In this perspective, so far, only the model of a self-interacting scalar field has been treated in detail. This model can be made relativistically invariant, but there are still problems with the definition of *macroscopic* energy (expressed in terms of the macroscopic fields) and with a consistent renormalization of the theory[15]. Even the extension of the theory to the case of vectorial and tensorial fields, possibly in interaction with Fermi fields, is not obvious at all.

In conclusion, I think that, if one does not want to accept von Neumann's (explicitly *subjectivistic*) point of view, a really satisfactory interpretation of Quantum Theory does not exists at present, in spite of the efforts made. The idea that the present theory should refer to the limit case of a small number of particles is, however, very suggestive to me and in this perspective the theory of continuous measurement can perhaps offer a useful illustration of what a future more general theory could be.

Milan University
Italy

NOTES

[1] See Wheeler and Zurek 1983.
[2] See Wheeler and Zurek 1983.
[3] See d'Espagnat 1971.
[4] See d'Espagnat, *op. cit.*, Hartkämper and Neumann 1974, and Ludwig 1982.
[5] See Jordan 1947.
[6] See Wheeler and Zurek, *op. cit.*, d'Espagnat, *op. cit.*, Hartkämper and Neumann, *op. cit.*.
[7] See d'Espagnat, *op. cit.*.
[8] See d'Espagnat, *op. cit.* and Kamefuchi *et al.*, 1984.
[9] See Hartkämper and Neumann, Ludwig, *op. cit.*.
[10] See Davies 1976.
[11] See Kamefuchi *et al.*, *op. cit.*, Prosperi 1994.
[12] See Omnès 1995.
[13] See Ghirardi *et al.*, 1986.
[14] See Prosperi 1994.
[15] See Kamefuchi *et al.*, *op. cit.*.

REFERENCES

E. B. Davies, *Quantum Theory of Open Systems*, Academic Press, London 1976.
B. d'Espagnat ed., *Foundations of Quantum Mechanics*, Academic Press, New York and London 1971.
G.C. Ghirardi, A. Rimini and T. Weber, *Phys. Rev.*, **D34**, 1986, 470.
A. Hartkämper and H. Neumann eds., *Foundations of Quantum Mechanics and Ordered Linear Spaces*, Springer-Verlag, Berlin, Heidelberg, New York 1974.
P. Jordan, *Das Bild der modernen Physik*, Stromverlag, Hamburg 1947.
S. Kamefuchi et al. eds., *Foundation of Quantum Mechanics in the Light of new Technology*, Soc. Phys. of Japan, Tokyo 1984.
G. Ludwig, *Foundation of Quantum Mechanics*, Springer-Verlag, Berlin, Heidelberg, New York 1982.
G. M. Prosperi, *Journal of Theoretical Physics*, **33**, 1994, 115.
R. Omnès, *Rev. of Mod. Physics*, **64**, 1995, 339.
J. A. Wheeler and W. H. Zurek eds., *Quantum Theory and Measurement*, Princeton Univ. Press, New Jersey 1983.

ENRICO BELTRAMETTI AND SLAWOMIR BUGAJSKI

ON THE RELATIONSHIPS BETWEEN CLASSICAL AND QUANTUM MECHANICS

1. INTRODUCTION

Almost a century since the birth of quantum mechanics, in presence of an impressive and still developing accumulation of empirical evidence of quantum behaviour from the atomic scale down to the subnuclear scale and up to the astronomical one, much attention is still focused on the foundations of quantum theory. In particular there are still open problems about the relationship between classical and quantum features. For instance, why putting together quantum objects (like atoms) to get a compound system one progressively looses the quantum behaviour? Why we never encounter macroscopic objects in those nonlocalised states (think, e.g., of the states staying on both arms of an interferometry device) in which its atoms, when isolated, would be able to live? The difficulty to find convincing answers is behind the puzzling aspects of the quantum theory of measurement, and behind the so-called paradoxes of quantum theory. Interesting conjectures can be found in the literature: for instance the GRW model based on the assumption of a dynamical process of spontaneous localization, or the idea that for open systems the interaction with the environment might be responsible for the transition from quantum to classical.

The goal of this paper is however in a different direction. We shall focus on the possible relationships between the mathematical edifices that host classical and, respectively, quantum mechanics. In this perspective let us remind some basic elements which are common to both cases. A first element is the notion of physical system, or physical object: it is understood as a portion of the physical universe, whose interaction with the rest of the universe can be neglected or, in any case, is such as to cause no trouble about the identity of the separated portion. In other words, the notion of physical system embodies the permanent attributes of the physical situation one is dealing with. Another basic concept is that of state of the physical system: it collects all those attributes that are accidental, in the sense that they may be different in different situations and may change with time. By state it is meant the result of the experimental procedures used to isolate and prepare the physical system. It should be clear that the point where the definition of system ceases and the definition of state begins is, to some extent, a matter of convention, and may depend on the historical development of scientific knowledge. A third basic ingredient is the notion of observable, or physical quantity: it means any quantity that can be measured. Energy, position, momentum are familiar examples.

E. Agazzi and M. Pauri (eds.), The Reality of the Unobservable, 353-359.

In the classical context, to a given physical system one associates a measurable space Ω, the so-called phase space, which carries the information about the degrees of freedom of the physical system. In the quantum context to the physical system is associated a more structured mathematical object, namely a separable, generally complex, Hilbert space $\mathcal{H}$. The states are classically represented as probability measures on Ω: among them the measures concentrated at a point of Ω, which correspond to a maximal information about the preparation of the system and are called pure states. In the quantum case the states are represented by density operators in $\mathcal{H}$ (that is linear, bounded, self-adjoint, positive, trace class operators of trace one): among them the one dimensional projectors that are identified with the pure states. Coming to the observables, classical mechanics describes them as real valued functions on Ω, so that a pure state (a point of Ω) determines a well defined value of each observable, as determinism requires. Quantum mechanics describes observables by self-adjoint operators in $\mathcal{H}$, and the probability that the value of an observable B lies in the real interval X when the system is in the state D is $Tr(DP_{B,X})$ where $P_{B,X}$ is the projector associated to the pair (B,X) by the spectral theorem for self-adjoint operators: this outlines the quantum indeterminism since $Tr(DP_{B,X})$ is in general not concentrated in $\{0,1\}$ even when D is a pure state.

As well known, efforts to reconcile quantum mechanics with determinism are almost as old as quantum mechanics itself. The traditional way is the one of conjecturing some incompleteness in the quantum description of pure states: additional variables, hidden to our present knowledge, are supposed to exist which would complete the description offered by quantum mechanics. According to this view, the indeterminism of quantum theory would reflect our unabilty to "see" these hidden variables, while determinism would be restored when they are taken into account. Deep mathematical results were obtained along the years putting constraints on the admissible forms of hidden variable theories that are able to reproduce the predictions of quantum mechanics when the average is taken over the unobserved hidden parameters. In 1964 Bell[1] put a milestone on this issue by proving that the so-called "local" hidden variable theories are unable to reproduce all the quantum predictions. The particular physical context of Bell's analysis was the one of the Einstein-Podolsky-Rosen correlation and he proved that the empirical probabilities emerging in that correlation have to fulfill certain inequalities, since then called Bell inequalities, in order to admit an interpretation in terms of local hidden variable theories. A well known experiment then came in the early eighties violating Bell inequalities while agreeing with the quantum prediction[2].

In the sequel we shall follow the approach that gives to the states of the physical system a privileged role of primitive undefined entities. The basic mathematical structure of the set of states is the convexity, that translates the physical fact that states can be mixed. With reference to the convex structure, the pure states will be viewed as extremal elements, as elements that cannot be thought of as convex combinations of distinct elements. The departure between the classical and the quantum case can be already recognized at this level: in both cases every mixed state admits a convex decomposition in terms of pure states but this decomposition is unique in the classical case while it is never unique in the quantum case. In other words, the convex set of states of a classical system is a simplex, but it is not so for a quantum system.

In Section 2 we shall introduce a physically natural definition of observable that encompasses both the traditional definitions used in standard classical and quantum mechanics. By use of this definition we shall develop the idea of extension of a descriptive model showing that the quantum model admits an extension based on a set of states which is a simplex, hence a classical extension that proves to be able to host the typical features of quantum observables. In Section 3 we focus on Bell inequalities viewing their violation as a particular case of a more general issue that we call the Bell phenomenon: it will be shown that the occurrence of this phenomenon is not restricted to the quantum model but can be hosted also in a classical context.

2. CLASSICAL EXTENSION OF QUANTUM MECHANICS

Let S be the convex set formed by the states of the physical system under discussion. According to the physical intuition the notion of observable has to carry the information about the possible measurement outcomes, and about the probability of their occurrence for each state of the physical system. Thus we are led to define an observable as an affine map of S into the convex set $M_1^+(\Xi)$ of all the probability measures on some measurable space Ξ in which the observable takes values (typically the real line). This definition has been already used in the literature and encompasses the standard definitions used in the quantum as well as in the classical case[3].

In the quantum case S is the set S_Q of all density operators on a separable Hilbert space $\mathcal{H}$. Typical of S_Q is the nonunique decomposition of mixed states into pure ones, which mirrors the fact that S_Q is not a simplex. Our notion of observable, when referred to S_Q, gives back[4] the so-called positive operator valued (POV) measures, which are typical ingredients of the "operational" approach to quantum mechanics: we quote the recent volume[5] also for reference to the relevant literature. Notice that when positive operators are restricted to projectors and the measurable space Ξ is specified as the real line $\boldsymbol{R}$ then the POV measures become the familiar PV measures on $\boldsymbol{R}$, hence they correspond to the self adjoint operators of $\mathcal{H}$ (via the spectral theorem). Therefore, in the case of Hilbert-space quantum mechanics, the definition of observable we are here adopting is fully equivalent to the operational one, and it includes as a special case the traditional observables represented by self-adjoint operators.

In the classical case the set S of states takes the typical structure of the set $M_1^+(\Omega)$ of all the probability measures on the "phase space" Ω of the physical system. Of course Ω is understood as a measurable space, and we additionally assume that the one-point subsets of Ω are measurable. The measures concentrated at one point of Ω (the Dirac measures) correspond to the pure states and the unique decomposability of mixed states into pure states becomes the distinguishing feature that mirrors the simplex nature of the convex set $M_1^+(\Omega)$. According to our definition an observable taking values in the measurable space Ξ is now an affine map of $M_1^+(\Omega)$ into $M_1^+(\Xi)$. The observable $B : M_1^+(\Omega) \to M_1^+(\Xi)$ will be called regular[6] if

$$B\mu = \int_\Omega (B\delta_\omega)\, d\mu(\omega)$$

for any $\mu \in M_1^+(\Omega)$, where δ_ω denotes the Dirac measure concentrated at $\omega \in \Omega$.

We shall say that a regular observable B is sharp if it maps Dirac measures on Ω into Dirac measures on Ξ; if it is not sharp we shall say that it is fuzzy.

In general a physical system admits different levels of description. For instance one might include or ignore some degrees of freedom; as another example, the quantum description of a system which is part of a compound system might be done on the basis of the Hilbert space of the subsystem or on the basis of the Hilbert space of the compound system. In other words, different levels of coarse graining can be adopted. We say that the descriptive model based on the convex set S of states admits an extension based on a set $\bar{S}$ of states when there exists an affine surjective map $R: \bar{S} \to S$. Such a map will be called the reduction map, for it reduces the $\bar{S}$-based model to the S-based one. So, the extended model makes use of a richer set of states, and the reduction map R determines in general a many-to-one correspondence between $\bar{S}$ and S: the elements of $\bar{S}$ mapped into the same element of S form a "coarse grain". Since $\bar{S}$ is richer than S, the set of observables on $\bar{S}$ will be, in turn, richer than the set of observables on S. Indeed, if $B{:}S \to M_1^+(\Xi)$ is an observable of the S-based model, then the map composition $\bar{B} := B \circ R$ is obviously an affine mapping of $\bar{S}$ into $M_1^+(\Xi)$, hence an observable of the $\bar{S}$-based model. Thus every observable on S finds a representative among the observables on $\bar{S}$ (not the converse, of course).

It is remarkable the fact that we can have S quantum, that is of the form S_Q, and $\bar{S}$ classical, that is of the form $M_1^+(\Omega)$ for some measurable space Ω. In this sense we speak of classical extension of quantum mechanics. Explicitly, if one considers the measurable space ∂S_Q formed by the extremal elements of S_Q (that is the pure states, or the one-dimensional projectors on the Hilbert space $\mathcal{H}$) then there exists[7] an affine surjective map $R_M : M_1^+(\partial S_Q) \to S_Q$ (1) which carries the canonical classical extension of quantum mechanics. The reduction map has been denoted R_M to remind that it was already studied by Misra[8] some twenty years ago.

It turns out[9] that the statistical distributions of results of a quantum observable and of its classical representative are the same, and the typical quantum features are preserved in the classical extension: so, e.g., whenever two quantum observables obey an uncertainty relation the same do their classical representatives (with the same uncertainty limit). This is not paradoxical because the classical representatives of the quantum observables are regular but not sharp: they are fuzzy. As it will be seen in the next Section, this is also why the violation of Bell-type inequalities, typical of the quantum context, can survive in a classical framework.

3. THE BELL PHENOMENON

Consider a finite collection $\{\Xi_1, \Xi_2, ..., \Xi_n\}$ of measurable spaces. If $\{i_1, i_2, ..., i_r\} \subseteq \{1,2,...,n\}$, let us write $\Xi^{(i_1, i_2, ..., i_r)}$ for the Cartesian product $\Xi_{i_1} \otimes \Xi_{i_2} \otimes ... \otimes \Xi_{i_r}$.

Definition: A consistent family of observables on S is a collection $\mathcal{C}$ of affine mappings of S into some of the convex sets $M_1^+(\Xi^{(i_1, i_2, ..., i_r)})$ such that

(i) to any sequence $\{i_1, i_2, ..., i_r\}$ there corresponds at most one (i.e., one or none) affine mapping of S into $M_1^+(\Xi^{(i_1, i_2, ..., i_r)})$ that we denote $B^{(i_1, i_2, ..., i_r)}$.

(ii) if $B^{(i_1, i_2, ..., i_r)}, B^{(i_1, i_2, ..., i_s)} \in \mathcal{C}$ and $\{i_1, i_2, ..., i_r\} \cap \{j_1, j_2, ..., j_s\} = \{k_1, k_2, ..., k_t\} \neq \emptyset$, then the marginal projetcions of $B^{(i_1, i_2, ..., i_r)}$ and $B^{(j_1, j_2, ..., j_s)}$ over $M_1^+(\Xi^{(k_1, k_2, ..., k_t)})$ coincide.

(iii) for every $i \in \{1,2,...,n\}$ there is in $\mathcal{C}$ an element $B^{(i)}$ such that $B^{(i)}: S \to M_1^+(\Xi_i)$: the observables $B^{(1)}, B^{(2)}, ..., B^{(n)}$ will be called the basic elements of $\mathcal{C}$.

In case the consistent family of observables $\mathcal{C}$ contains an element $B^{(1,2,...,n)}: S \to M_1^+(\Xi^{(1,2,...,n)})$, then this observable generates all the members of the family by the mechanism of marginal projections. In such a case the family $\mathcal{O}$ is said to be complete and $B^{(1,2,\dots,n)}$ is called the generating element of $\mathcal{O}$. If $\mathcal{O}$ is not complete, then the question arises whether there exists an element $B^{(1,2,\dots,n)}: S \rightarrow M_1^+(\Xi^{(1,2,\dots,n)})$ such that $\mathcal{O} \cup \{B^{(1,2,\dots,n)}\}$ is a complete consistent family of observables. When such a $B^{(1,2,\dots,n)}$ does exist then it generates all the members of $\mathcal{O}$ by the mechanism of marginal projections, and we say that $\mathcal{O}$ admits a generating observable[10].

The elements of a consistent family of observables $\mathcal{O}$ are defined as affine mappings of S into the set of probability measures (in the sequel we shall simply say measures always understanding probability measures) on some measurable space: so, if one evaluates all the members of $\mathcal{O}$ at some fixed $\alpha \in S$ one gets a family of measures, to be denoted $\mathcal{M}(\mathcal{O}, \alpha)$ and called a consistent family of measures. It is now clear that if $\mathcal{O}$ admits a generating observable $B^{(1,2,\dots,n)}$ then, for every $\alpha \in S$, $B^{(1,2,\dots,n)}\alpha$ is a generating measure for $\mathcal{M}(\mathcal{O}, \alpha)$. Of course, by this we mean that $B^{(1,2,\dots,n)}\alpha$ generates all the members of the family $\mathcal{M}(\mathcal{O}, \alpha)$ by the mechanism of marginal projections.

When a consistent family of measures contains a generating measure we say that the family is complete. The relevant and not so obvious fact is that there are consistent families of measures which do not admit a generating measure: in other words they cannot be thought of as subfamilies of complete families. When a consistent family of measures $\mathcal{M}(\mathcal{O}, \alpha)$ does not admit a generating measure we say that it exhibits the *Bell phenomenon*. Similarly, we say that the consistent family of observables $\mathcal{O}$ shows the *Bell phenomenon for observables* if at some $\alpha \in S$ the family $\mathcal{M}(\mathcal{O}, \alpha)$ exhibits the Bell phenomenon for measures, i.e., $\mathcal{M}(\mathcal{O}, \alpha)$ has no generating measure.

The reason for the above terminology lies in the fact that the Bell phenomenon encompasses as a particular case the violation of the familiar Bell inequalities. This can be seen[11] by considering the EPR situation in the quantum context and by checking that the usual family of quantum observables of the EPR correlation, when evaluated at the singlet state of the two spin-$^1/_2$ particles, determines a family of measures that does not admit a generating measure, hence exhibiting the Bell phenomenon. In more general terms, the connection between what we have called the Bell phenomenon and what are referred to as Bell-type inequalities[12] has been discussed in Beltrametti and Bugajski 1996.

Let us now review some properties related to the notion of Bell phenomenon.

(1) When referred to the basic elements $B^{(1)}, B^{(2)},\dots, B^{(n)}$ of $\mathcal{O}$ (or to some of them) the definition of generating observable reduces to the familiar definition of joint observable. When a joint observable exists it is also customary[13] to say that the $B^{(i)}$'s are mutually co-measurable (or coexistent[14]). The notion of generating observable generalizes the notion of joint observable: indeed, if $B^{(1,2,\dots,n)}$ is a joint observable of $B^{(1)}, B^{(2)},\dots, B^{(n)}$ then it is a generating observable of the consistent family of observables $\{B^{(1)}, B^{(2)},\dots, B^{(n)}\}$. And conversely, whenever a consistent family of observables $\mathcal{O}$ admits a generating observable, the latter is a joint observable of the basic elements of $\mathcal{O}$.

(2) Within the level of generality here considered (in particular about S) the existence of a joint observable is by no means guaranteed. Think, e.g., of the quantum context where $S = S_Q$, and consider two quantum observables associated to noncommuting self-adjoint operators: they do not have, in this context, any joint observable.

(3) Even in the case when the basic observables $B^{(1)}, B^{(2)},\dots, B^{(n)}$ of a consistent family $\mathcal{O}$

admit a joint observable, this need not be a generating observable for the whole family $\mathcal{C}$ when the latter contains other observables beyond the basic ones.

(4) A generating observable of a consistent family $\mathcal{C}$ need not be unique. Actually, when there are more than one, then there are infinitely many because any convex combination of two generating observables is still a generating observable.

(5) If for any $\{i_1,i_2,...,i_r\} \subseteq \{1,2,...,n\}$ the set $\{B^{(i_1)}, B^{(i_2)},...,B^{(i_r)}\}$ of basic observables of the consistent family of observables $\mathcal{C}$ possesses only one joint observable, then $\mathcal{C}$ admits a unique generating observable. The proof of this result can be found in Beltrametti and Bugajski 1996.

Up to now the notion of Bell phenomenon has been introduced without any special assumption about the underlying set S of states. Come now to the relevant case in which S is classical, that is S takes the form of the simplex $M_1^+(\Omega)$ for some measurable space Ω. An observable with value space Ξ now becomes an affine mapping of $M_1^+(\Omega)$ into $M_1^+(\Xi)$. As already said in Section 2, the classical frame contains the class of regular observables: these are the observables we shall be concerned with. They can be divided into the sharp and the fuzzy ones, according to the definition given in Section 2. Now we have the following further facts.

(6) A joint observable for the collection of regular observables $\{B^{(i)}: M_1^+(\Omega), M_1^+(\Xi_i) \mid i=1,2,...,n\}$ is now provided by the product $B^{(1)} \times B^{(2)} \times ... \times B^{(n)}$ [15].

(7) If the regular observables $B^{(1)}, B^{(2)}, ..., B^{(n)}$ are sharp then there is no other joint observable besides their product.

(8) If a consistent family $\mathcal{C}$ of regular observables on $M_1^+(\Omega)$, having basic elements $B^{(1)}$, $B^{(2)}$, ..., $B^{(n)}$, contains - besides its basic elements - only product observables then the product $B^{(1)} \times B^{(2)} \times ... \times B^{(n)}$ is a generating observable for the whole family.

(9) A consistent family of regular observables on $M_1^+(\Omega)$ need not admit a generating observable, but a consistent family $\mathcal{C}$ of regular sharp observables on $M_1^+(\Omega)$ admits a unique generating observable which coincides with the product $B^{(1)} \times B^{(2)} \times ... \times B^{(n)}$ of all the basic elements of $\mathcal{C}$.

As discussed in Beltrametti and Bugajski 1995(a), the standard classical statistical models make use only of sharp observables: in view of the above item (9) these standard models cannot exhibit the Bell phenomenon. But classical models can host also fuzzy observables: for these models the Bell phenomenon is not excluded. A typical situation occurs with the classical extension of quantum mechanics that has been shortly recalled in Section 2. If B is an observable on S_Q then $\overline{B} := B \circ R_M$ is an observable on $M_1^+(\partial S_Q)$. If $\mathcal{C}$ is a consistent family of observables on S_Q then the family $\overline{\mathcal{C}}$ obtained from $\mathcal{C}$ by composing all its members with R_M is a consistent family of observables on $M_1^+(\partial S_Q)$. Moreover, all probability distributions are preserved by the classical extension, in the sense that given any $\alpha \in S_Q$ we have $B\,\alpha = \overline{B}\,\beta$ for every β in the counterimage of α under R_M. From these facts it follows that whenever a consistent family $\mathcal{C}$ of observables on S_Q exhibits the Bell phenomenon the same does the family $\overline{\mathcal{C}}$ of the classical representatives of the elements of $\mathcal{C}$. In other words, all occurrences of the Bell phenomenon in quantum mechanics are carefully reproduced in the classical extension.

Comparing the above conclusion with item (9), the consequence comes that the classical representatives of the quantum observables cannot be regular and sharp: indeed it is known that they are *regular* and *fuzzy*[16].

For explicit examples of occurrence of the Bell phenomenon for classical observables, we refer to Beltrametti and Bugajski 1996, and Aerts 1991.

University of Genoa
Italy

University of Silesia
Poland

NOTES

[1] See Bell 1964.
[2] See Aspect *et al.* 1982.
[3] See, for instance, Ali *et al.* 1977, Holevo 1982, Singer and Stulpe 1992, Bugajski 1993, Beltrametti and Bugajski 1993, Cassinelli and Lahti 1993, Beltrametti and Bugajski 1995(a), Beltrametti and Bugajski 1996.
[4] See Beltrametti and Bugajski 1995(a), Beltrametti and Bugajski 1996.
[5] Busch *et al.* 1995.
[6] See Beltrametti and Bugajski 1995(a).
[7] See Beltrametti and Bugajski 1995(a).
[8] See Misra 1974.
[9] See Beltrametti and Bugajski 1995(a).
[10] For a more exhaustive discussion we refer to Beltrametti and Bugajski 1996.
[11] See Beltrametti and Bugajski 1996.
[12] See Pitowsky 1989, Beltrametti and Maczynzki 1991, 1993(a), 1993(b).
[13] See Beltrametti and Bugajski 1995(a), 1995(b).
[14] See Busch *et al.* 1995.
[15] See Beltrametti and Bugajski 1996.
[16] See Beltrametti and Bugajski 1995(a).

REFERENCES

D. Aerts, *Helv. Phys. Acta* **64**, 1991, 1-23.
S.T. Ali and E. Prugovecki, *J. Math. Phys* **18**, 1977, 219-228.
A. Aspect, P. Grangier and G. Roger, *Phys. Rev. Lett.* **49**, 1982, 91-94.
J.S. Bell, *Physics* **1**, 1964, p. 195.
E.G. Beltrametti and S. Bugajski, *Int. J. Theor. Phys.* **32**, 1993, 2235-2244.
E.G. Beltrametti and S. Bugajski, *Int. J. Theor. Phys.* **34**, 1995(a), 1221-1229.
E.G. Beltrametti and S. Bugajski, *J. Phys. A: Math. Gen.* **28**, 1995(b), 3329-3343.
E.G. Beltrametti and S. Bugajski, *J. Phys. A: Math. Gen.* **29**, 1996, 247-261.
E.G. Beltrametti and M.J. Maczynski, *J. Math. Phys.* **32**, 1991, 1280-1286; *J. Math. Phys.* **34**, 1993(a), 4919-4929.
E.G. Beltrametti and M.J. Maczynski, "On the Intrinsic Characterization of Classical and Quantum Probabilities", in P. Busch, P. Lahti, and P. Mittelstaedt eds., *Symposium on the Foundations of Quantum Physics,* World Scientific, Singapore 1993(b).
S. Bugajski, *Int. J. Theor. Phys.* **32**, 1993, 389-398.
P. Busch, M. Grabowski, and P. Lahti, *Operational Quantum Physics,* Springer-Verlag, Berlin 1995.
G. Cassinelli and P. Lahti, *J. Math. Phys.* **34**, 1993, 5468-5475.
A.S. Holevo, *Probabilistic and Statistical Aspects of Quantum Theory,* North-Holland, Amsterdam 1982.
B. Misra, in C.P. Enz and J. Mehra eds, *Physical Reality and Mathematical Description*, Reidel, Dordrecht 1974, pp.455-476.
I. Pitowsky, *Quantum Probability-Quantum Logic,* Lecture Notes in Physics **321**, Springer-Verlag, Berlin 1989.
M. Singer and W. Stulpe, *J. Math. Phys.* **33**, 1992, 131-142.

INDEX OF NAMES

Boston Studies in the Philosophy of Science

Editor: Robert S. Cohen, *Boston University*

1. M.W. Wartofsky (ed.): *Proceedings of the Boston Colloquium for the Philosophy of Science, 1961/1962.* [Synthese Library 6] 1963 ISBN 90-277-0021-4
2. R.S. Cohen and M.W. Wartofsky (eds.): *Proceedings of the Boston Colloquium for the Philosophy of Science, 1962/1964.* In Honor of P. Frank. [Synthese Library 10] 1965 ISBN 90-277-9004-0
3. R.S. Cohen and M.W. Wartofsky (eds.): *Proceedings of the Boston Colloquium for the Philosophy of Science, 1964/1966.* In Memory of Norwood Russell Hanson. [Synthese Library 14] 1967 ISBN 90-277-0013-3
4. R.S. Cohen and M.W. Wartofsky (eds.): *Proceedings of the Boston Colloquium for the Philosophy of Science, 1966/1968.* [Synthese Library 18] 1969 ISBN 90-277-0014-1
5. R.S. Cohen and M.W. Wartofsky (eds.): *Proceedings of the Boston Colloquium for the Philosophy of Science, 1966/1968.* [Synthese Library 19] 1969 ISBN 90-277-0015-X
6. R.S. Cohen and R.J. Seeger (eds.): *Ernst Mach, Physicist and Philosopher.* [Synthese Library 27] 1970 ISBN 90-277-0016-8
7. M. Čapek: *Bergson and Modern Physics.* A Reinterpretation and Re-evaluation. [Synthese Library 37] 1971 ISBN 90-277-0186-5
8. R.C. Buck and R.S. Cohen (eds.): *PSA 1970.* Proceedings of the 2nd Biennial Meeting of the Philosophy and Science Association (Boston, Fall 1970). In Memory of Rudolf Carnap. [Synthese Library 39] 1971 ISBN 90-277-0187-3; Pb 90-277-0309-4
9. A.A. Zinov'ev: *Foundations of the Logical Theory of Scientific Knowledge (Complex Logic).* Translated from Russian. Revised and enlarged English Edition, with an Appendix by G.A. Smirnov, E.A. Sidorenko, A.M. Fedina and L.A. Bobrova. [Synthese Library 46] 1973 ISBN 90-277-0193-8; Pb 90-277-0324-8
10. L. Tondl: *Scientific Procedures.* A Contribution Concerning the Methodological Problems of Scientific Concepts and Scientific Explanation.Translated from Czech. [Synthese Library 47] 1973 ISBN 90-277-0147-4; Pb 90-277-0323-X
11. R.J. Seeger and R.S. Cohen (eds.): *Philosophical Foundations of Science.* Proceedings of Section L, 1969, American Association for the Advancement of Science. [Synthese Library 58] 1974 ISBN 90-277-0390-6; Pb 90-277-0376-0
12. A. Grünbaum: *Philosophical Problems of Space and Times.* 2nd enlarged ed. [Synthese Library 55] 1973 ISBN 90-277-0357-4; Pb 90-277-0358-2
13. R.S. Cohen and M.W. Wartofsky (eds.): *Logical and Epistemological Studies in Contemporary Physics.* Proceedings of the Boston Colloquium for the Philosophy of Science, 1969/72, Part I. [Synthese Library 59] 1974 ISBN 90-277-0391-4; Pb 90-277-0377-9
14. R.S. Cohen and M.W. Wartofsky (eds.): *Methodological and Historical Essays in the Natural and Social Sciences.* Proceedings of the Boston Colloquium for the Philosophy of Science, 1969/72, Part II. [Synthese Library 60] 1974 ISBN 90-277-0392-2; Pb 90-277-0378-7
15. R.S. Cohen, J.J. Stachel and M.W. Wartofsky (eds.): *For Dirk Struik.* Scientific, Historical and Political Essays in Honor of Dirk J. Struik. [Synthese Library 61] 1974 ISBN 90-277-0393-0; Pb 90-277-0379-5
16. N. Geschwind: *Selected Papers on Language and the Brains.* [Synthese Library 68] 1974 ISBN 90-277-0262-4; Pb 90-277-0263-2
17. B.G. Kuznetsov: *Reason and Being.* Translated from Russian. Edited by C.R. Fawcett and R.S. Cohen. 1987 ISBN 90-277-2181-5

Boston Studies in the Philosophy of Science

18. P. Mittelstaedt: *Philosophical Problems of Modern Physics*. Translated from the revised 4th German edition by W. Riemer and edited by R.S. Cohen. [Synthese Library 95] 1976
ISBN 90-277-0285-3; Pb 90-277-0506-2
19. H. Mehlberg: *Time, Causality, and the Quantum Theory*. Studies in the Philosophy of Science. Vol. I: *Essay on the Causal Theory of Time*. Vol. II: *Time in a Quantized Universe*. Translated from French. Edited by R.S. Cohen. 1980 Vol. I: ISBN 90-277-0721-9; Pb 90-277-1074-0
Vol. II: ISBN 90-277-1075-9; Pb 90-277-1076-7
20. K.F. Schaffner and R.S. Cohen (eds.): *PSA 1972*. Proceedings of the 3rd Biennial Meeting of the Philosophy of Science Association (Lansing, Michigan, Fall 1972). [Synthese Library 64] 1974 ISBN 90-277-0408-2; Pb 90-277-0409-0
21. R.S. Cohen and J.J. Stachel (eds.): *Selected Papers of Léon Rosenfeld*. [Synthese Library 100] 1979 ISBN 90-277-0651-4; Pb 90-277-0652-2
22. M. Čapek (ed.): *The Concepts of Space and Time*. Their Structure and Their Development. [Synthese Library 74] 1976 ISBN 90-277-0355-8; Pb 90-277-0375-2
23. M. Grene: *The Understanding of Nature*. Essays in the Philosophy of Biology. [Synthese Library 66] 1974 ISBN 90-277-0462-7; Pb 90-277-0463-5
24. D. Ihde: *Technics and Praxis*. A Philosophy of Technology. [Synthese Library 130] 1979
ISBN 90-277-0953-X; Pb 90-277-0954-8
25. J. Hintikka and U. Remes: *The Method of Analysis*. Its Geometrical Origin and Its General Significance. [Synthese Library 75] 1974 ISBN 90-277-0532-1; Pb 90-277-0543-7
26. J.E. Murdoch and E.D. Sylla (eds.): *The Cultural Context of Medieval Learning*. Proceedings of the First International Colloquium on Philosophy, Science, and Theology in the Middle Ages, 1973. [Synthese Library 76] 1975 ISBN 90-277-0560-7; Pb 90-277-0587-9
27. M. Grene and E. Mendelsohn (eds.): *Topics in the Philosophy of Biology*. [Synthese Library 84] 1976 ISBN 90-277-0595-X; Pb 90-277-0596-8
28. J. Agassi: *Science in Flux*. [Synthese Library 80] 1975
ISBN 90-277-0584-4; Pb 90-277-0612-3
29. J.J. Wiatr (ed.): *Polish Essays in the Methodology of the Social Sciences*. [Synthese Library 131] 1979 ISBN 90-277-0723-5; Pb 90-277-0956-4
30. P. Janich: *Protophysics of Time*. Constructive Foundation and History of Time Measurement. Translated from German. 1985 ISBN 90-277-0724-3
31. R.S. Cohen and M.W. Wartofsky (eds.): *Language, Logic, and Method*. 1983
ISBN 90-277-0725-1
32. R.S. Cohen, C.A. Hooker, A.C. Michalos and J.W. van Evra (eds.): *PSA 1974*. Proceedings of the 4th Biennial Meeting of the Philosophy of Science Association. [Synthese Library 101] 1976 ISBN 90-277-0647-6; Pb 90-277-0648-4
33. G. Holton and W.A. Blanpied (eds.): *Science and Its Public*. The Changing Relationship. [Synthese Library 96] 1976 ISBN 90-277-0657-3; Pb 90-277-0658-1
34. M.D. Grmek, R.S. Cohen and G. Cimino (eds.): *On Scientific Discovery*. The 1977 Erice Lectures. 1981 ISBN 90-277-1122-4; Pb 90-277-1123-2
35. S. Amsterdamski: *Between Experience and Metaphysics*. Philosophical Problems of the Evolution of Science. Translated from Polish. [Synthese Library 77] 1975
ISBN 90-277-0568-2; Pb 90-277-0580-1
36. M. Marković and G. Petrović (eds.): *Praxis*. Yugoslav Essays in the Philosophy and Methodology of the Social Sciences. [Synthese Library 134] 1979
ISBN 90-277-0727-8; Pb 90-277-0968-8

37. H. von Helmholtz: *Epistemological Writings.* The Paul Hertz / Moritz Schlick Centenary Edition of 1921. Translated from German by M.F. Lowe. Edited with an Introduction and Bibliography by R.S. Cohen and Y. Elkana. [Synthese Library 79] 1977
ISBN 90-277-0290-X; Pb 90-277-0582-8
38. R.M. Martin: *Pragmatics, Truth and Language.* 1979
ISBN 90-277-0992-0; Pb 90-277-0993-9
39. R.S. Cohen, P.K. Feyerabend and M.W. Wartofsky (eds.): *Essays in Memory of Imre Lakatos.* [Synthese Library 99] 1976 ISBN 90-277-0654-9; Pb 90-277-0655-7
40. Not published.
41. Not published.
42. H.R. Maturana and F.J. Varela: *Autopoiesis and Cognition.* The Realization of the Living. With a Preface to "Autopoiesis' by S. Beer. 1980 ISBN 90-277-1015-5; Pb 90-277-1016-3
43. A. Kasher (ed.): *Language in Focus: Foundations, Methods and Systems.* Essays in Memory of Yehoshua Bar-Hillel. [Synthese Library 89] 1976
ISBN 90-277-0644-1; Pb 90-277-0645-X
44. T.D. Thao: *Investigations into the Origin of Language and Consciousness.* 1984
ISBN 90-277-0827-4
45. F.G.-I. Nagasaka (ed.): *Japanese Studies in the Philosophy of Science.* 1997
ISBN 0-7923-4781-1
46. P.L. Kapitza: *Experiment, Theory, Practice.* Articles and Addresses. Edited by R.S. Cohen. 1980 ISBN 90-277-1061-9; Pb 90-277-1062-7
47. M.L. Dalla Chiara (ed.): *Italian Studies in the Philosophy of Science.* 1981
ISBN 90-277-0735-9; Pb 90-277-1073-2
48. M.W. Wartofsky: *Models.* Representation and the Scientific Understanding. [Synthese Library 129] 1979 ISBN 90-277-0736-7; Pb 90-277-0947-5
49. T.D. Thao: *Phenomenology and Dialectical Materialism.* Edited by R.S. Cohen. 1986
ISBN 90-277-0737-5
50. Y. Fried and J. Agassi: *Paranoia.* A Study in Diagnosis. [Synthese Library 102] 1976
ISBN 90-277-0704-9; Pb 90-277-0705-7
51. K.H. Wolff: *Surrender and Cath.* Experience and Inquiry Today. [Synthese Library 105] 1976
ISBN 90-277-0758-8; Pb 90-277-0765-0
52. K. Kosík: *Dialectics of the Concrete.* A Study on Problems of Man and World. 1976
ISBN 90-277-0761-8; Pb 90-277-0764-2
53. N. Goodman: *The Structure of Appearance.* [Synthese Library 107] 1977
ISBN 90-277-0773-1; Pb 90-277-0774-X
54. H.A. Simon: *Models of Discovery* and Other Topics in the Methods of Science. [Synthese Library 114] 1977 ISBN 90-277-0812-6; Pb 90-277-0858-4
55. M. Lazerowitz: *The Language of Philosophy.* Freud and Wittgenstein. [Synthese Library 117] 1977 ISBN 90-277-0826-6; Pb 90-277-0862-2
56. T. Nickles (ed.): *Scientific Discovery, Logic, and Rationality.* 1980
ISBN 90-277-1069-4; Pb 90-277-1070-8
57. J. Margolis: *Persons and Mind.* The Prospects of Nonreductive Materialism. [Synthese Library 121] 1978 ISBN 90-277-0854-1; Pb 90-277-0863-0
58. G. Radnitzky and G. Andersson (eds.): *Progress and Rationality in Science.* [Synthese Library 125] 1978 ISBN 90-277-0921-1; Pb 90-277-0922-X
59. G. Radnitzky and G. Andersson (eds.): *The Structure and Development of Science.* [Synthese Library 136] 1979 ISBN 90-277-0994-7; Pb 90-277-0995-5

Boston Studies in the Philosophy of Science

60. T. Nickles (ed.): *Scientific Discovery.* Case Studies. 1980 ISBN 90-277-1092-9; Pb 90-277-1093-7
61. M.A. Finocchiaro: *Galileo and the Art of Reasoning.* Rhetorical Foundation of Logic and Scientific Method. 1980 ISBN 90-277-1094-5; Pb 90-277-1095-3
62. W.A. Wallace: *Prelude to Galileo.* Essays on Medieval and 16th-Century Sources of Galileo's Thought. 1981 ISBN 90-277-1215-8; Pb 90-277-1216-6
63. F. Rapp: *Analytical Philosophy of Technology.* Translated from German. 1981 ISBN 90-277-1221-2; Pb 90-277-1222-0
64. R.S. Cohen and M.W. Wartofsky (eds.): *Hegel and the Sciences.* 1984 ISBN 90-277-0726-X
65. J. Agassi: *Science and Society.* Studies in the Sociology of Science. 1981 ISBN 90-277-1244-1; Pb 90-277-1245-X
66. L. Tondl: *Problems of Semantics.* A Contribution to the Analysis of the Language of Science. Translated from Czech. 1981 ISBN 90-277-0148-2; Pb 90-277-0316-7
67. J. Agassi and R.S. Cohen (eds.): *Scientific Philosophy Today.* Essays in Honor of Mario Bunge. 1982 ISBN 90-277-1262-X; Pb 90-277-1263-8
68. W. Krajewski (ed.): *Polish Essays in the Philosophy of the Natural Sciences.* Translated from Polish and edited by R.S. Cohen and C.R. Fawcett. 1982 ISBN 90-277-1286-7; Pb 90-277-1287-5
69. J.H. Fetzer: *Scientific Knowledge.* Causation, Explanation and Corroboration. 1981 ISBN 90-277-1335-9; Pb 90-277-1336-7
70. S. Grossberg: *Studies of Mind and Brain.* Neural Principles of Learning, Perception, Development, Cognition, and Motor Control. 1982 ISBN 90-277-1359-6; Pb 90-277-1360-X
71. R.S. Cohen and M.W. Wartofsky (eds.): *Epistemology, Methodology, and the Social Sciences.* 1983. ISBN 90-277-1454-1
72. K. Berka: *Measurement.* Its Concepts, Theories and Problems. Translated from Czech. 1983 ISBN 90-277-1416-9
73. G.L. Pandit: *The Structure and Growth of Scientific Knowledge.* A Study in the Methodology of Epistemic Appraisal. 1983 ISBN 90-277-1434-7
74. A.A. Zinov'ev: *Logical Physics.* Translated from Russian. Edited by R.S. Cohen. 1983 [*see also* Volume 9] ISBN 90-277-0734-0
75. G-G. Granger: *Formal Thought and the Sciences of Man.* Translated from French. With and Introduction by A. Rosenberg. 1983 ISBN 90-277-1524-6
76. R.S. Cohen and L. Laudan (eds.): *Physics, Philosophy and Psychoanalysis.* Essays in Honor of Adolf Grünbaum. 1983 ISBN 90-277-1533-5
77. G. Böhme, W. van den Daele, R. Hohlfeld, W. Krohn and W. Schäfer: *Finalization in Science.* The Social Orientation of Scientific Progress. Translated from German. Edited by W. Schäfer. 1983 ISBN 90-277-1549-1
78. D. Shapere: *Reason and the Search for Knowledge.* Investigations in the Philosophy of Science. 1984 ISBN 90-277-1551-3; Pb 90-277-1641-2
79. G. Andersson (ed.): *Rationality in Science and Politics.* Translated from German. 1984 ISBN 90-277-1575-0; Pb 90-277-1953-5
80. P.T. Durbin and F. Rapp (eds.): *Philosophy and Technology.* [*Also* Philosophy and Technology Series, Vol. 1] 1983 ISBN 90-277-1576-9
81. M. Marković: *Dialectical Theory of Meaning.* Translated from Serbo-Croat. 1984 ISBN 90-277-1596-3
82. R.S. Cohen and M.W. Wartofsky (eds.): *Physical Sciences and History of Physics.* 1984. ISBN 90-277-1615-3

83. É. Meyerson: *The Relativistic Deduction.* Epistemological Implications of the Theory of Relativity. Translated from French. With a Review by Albert Einstein and an Introduction by Milić Čapek. 1985 ISBN 90-277-1699-4
84. R.S. Cohen and M.W. Wartofsky (eds.): *Methodology, Metaphysics and the History of Science.* In Memory of Benjamin Nelson. 1984 ISBN 90-277-1711-7
85. G. Tamás: *The Logic of Categories.* Translated from Hungarian. Edited by R.S. Cohen. 1986 ISBN 90-277-1742-7
86. S.L. de C. Fernandes: *Foundations of Objective Knowledge.* The Relations of Popper's Theory of Knowledge to That of Kant. 1985 ISBN 90-277-1809-1
87. R.S. Cohen and T. Schnelle (eds.): *Cognition and Fact.* Materials on Ludwik Fleck. 1986 ISBN 90-277-1902-0
88. G. Freudenthal: *Atom and Individual in the Age of Newton.* On the Genesis of the Mechanistic World View. Translated from German. 1986 ISBN 90-277-1905-5
89. A. Donagan, A.N. Perovich Jr and M.V. Wedin (eds.): *Human Nature and Natural Knowledge.* Essays presented to Marjorie Grene on the Occasion of Her 75th Birthday. 1986 ISBN 90-277-1974-8
90. C. Mitcham and A. Hunning (eds.): *Philosophy and Technology II.* Information Technology and Computers in Theory and Practice. [*Also* Philosophy and Technology Series, Vol. 2] 1986 ISBN 90-277-1975-6
91. M. Grene and D. Nails (eds.): *Spinoza and the Sciences.* 1986 ISBN 90-277-1976-4
92. S.P. Turner: *The Search for a Methodology of Social Science.* Durkheim, Weber, and the 19th-Century Problem of Cause, Probability, and Action. 1986. ISBN 90-277-2067-3
93. I.C. Jarvie: *Thinking about Society.* Theory and Practice. 1986 ISBN 90-277-2068-1
94. E. Ullmann-Margalit (ed.): *The Kaleidoscope of Science.* The Israel Colloquium: Studies in History, Philosophy, and Sociology of Science, Vol. 1. 1986 ISBN 90-277-2158-0; Pb 90-277-2159-9
95. E. Ullmann-Margalit (ed.): *The Prism of Science.* The Israel Colloquium: Studies in History, Philosophy, and Sociology of Science, Vol. 2. 1986 ISBN 90-277-2160-2; Pb 90-277-2161-0
96. G. Márkus: *Language and Production.* A Critique of the Paradigms. Translated from French. 1986 ISBN 90-277-2169-6
97. F. Amrine, F.J. Zucker and H. Wheeler (eds.): *Goethe and the Sciences: A Reappraisal.* 1987 ISBN 90-277-2265-X; Pb 90-277-2400-8
98. J.C. Pitt and M. Pera (eds.): *Rational Changes in Science.* Essays on Scientific Reasoning. Translated from Italian. 1987 ISBN 90-277-2417-2
99. O. Costa de Beauregard: *Time, the Physical Magnitude.* 1987 ISBN 90-277-2444-X
100. A. Shimony and D. Nails (eds.): *Naturalistic Epistemology.* A Symposium of Two Decades. 1987 ISBN 90-277-2337-0
101. N. Rotenstreich: *Time and Meaning in History.* 1987 ISBN 90-277-2467-9
102. D.B. Zilberman: *The Birth of Meaning in Hindu Thought.* Edited by R.S. Cohen. 1988 ISBN 90-277-2497-0
103. T.F. Glick (ed.): *The Comparative Reception of Relativity.* 1987 ISBN 90-277-2498-9
104. Z. Harris, M. Gottfried, T. Ryckman, P. Mattick Jr, A. Daladier, T.N. Harris and S. Harris: *The Form of Information in Science.* Analysis of an Immunology Sublanguage. With a Preface by Hilary Putnam. 1989 ISBN 90-277-2516-0
105. F. Burwick (ed.): *Approaches to Organic Form.* Permutations in Science and Culture. 1987 ISBN 90-277-2541-1

Boston Studies in the Philosophy of Science

106. M. Almási: *The Philosophy of Appearances.* Translated from Hungarian. 1989 ISBN 90-277-2150-5
107. S. Hook, W.L. O'Neill and R. O'Toole (eds.): *Philosophy, History and Social Action.* Essays in Honor of Lewis Feuer. With an Autobiographical Essay by L. Feuer. 1988 ISBN 90-277-2644-2
108. I. Hronszky, M. Fehér and B. Dajka: *Scientific Knowledge Socialized.* Selected Proceedings of the 5th Joint International Conference on the History and Philosophy of Science organized by the IUHPS (Veszprém, Hungary, 1984). 1988 ISBN 90-277-2284-6
109. P. Tillers and E.D. Green (eds.): *Probability and Inference in the Law of Evidence.* The Uses and Limits of Bayesianism. 1988 ISBN 90-277-2689-2
110. E. Ullmann-Margalit (ed.): *Science in Reflection.* The Israel Colloquium: Studies in History, Philosophy, and Sociology of Science, Vol. 3. 1988 ISBN 90-277-2712-0; Pb 90-277-2713-9
111. K. Gavroglu, Y. Goudaroulis and P. Nicolacopoulos (eds.): *Imre Lakatos and Theories of Scientific Change.* 1989 ISBN 90-277-2766-X
112. B. Glassner and J.D. Moreno (eds.): *The Qualitative-Quantitative Distinction in the Social Sciences.* 1989 ISBN 90-277-2829-1
113. K. Arens: *Structures of Knowing.* Psychologies of the 19th Century. 1989 ISBN 0-7923-0009-2
114. A. Janik: *Style, Politics and the Future of Philosophy.* 1989 ISBN 0-7923-0056-4
115. F. Amrine (ed.): *Literature and Science as Modes of Expression.* With an Introduction by S. Weininger. 1989 ISBN 0-7923-0133-1
116. J.R. Brown and J. Mittelstrass (eds.): *An Intimate Relation.* Studies in the History and Philosophy of Science. Presented to Robert E. Butts on His 60th Birthday. 1989 ISBN 0-7923-0169-2
117. F. D'Agostino and I.C. Jarvie (eds.): *Freedom and Rationality.* Essays in Honor of John Watkins. 1989 ISBN 0-7923-0264-8
118. D. Zolo: *Reflexive Epistemology.* The Philosophical Legacy of Otto Neurath. 1989 ISBN 0-7923-0320-2
119. M. Kearn, B.S. Philips and R.S. Cohen (eds.): *Georg Simmel and Contemporary Sociology.* 1989 ISBN 0-7923-0407-1
120. T.H. Levere and W.R. Shea (eds.): *Nature, Experiment and the Science.* Essays on Galileo and the Nature of Science. In Honour of Stillman Drake. 1989 ISBN 0-7923-0420-9
121. P. Nicolacopoulos (ed.): *Greek Studies in the Philosophy and History of Science.* 1990 ISBN 0-7923-0717-8
122. R. Cooke and D. Costantini (eds.): *Statistics in Science.* The Foundations of Statistical Methods in Biology, Physics and Economics. 1990 ISBN 0-7923-0797-6
123. P. Duhem: *The Origins of Statics.* Translated from French by G.F. Leneaux, V.N. Vagliente and G.H. Wagner. With an Introduction by S.L. Jaki. 1991 ISBN 0-7923-0898-0
124. H. Kamerlingh Onnes: *Through Measurement to Knowledge.* The Selected Papers, 1853-1926. Edited and with an Introduction by K. Gavroglu and Y. Goudaroulis. 1991 ISBN 0-7923-0825-5
125. M. Čapek: *The New Aspects of Time: Its Continuity and Novelties.* Selected Papers in the Philosophy of Science. 1991 ISBN 0-7923-0911-1
126. S. Unguru (ed.): *Physics, Cosmology and Astronomy, 1300–1700.* Tension and Accommodation. 1991 ISBN 0-7923-1022-5

Boston Studies in the Philosophy of Science

127. Z. Bechler: *Newton's Physics on the Conceptual Structure of the Scientific Revolution.* 1991 ISBN 0-7923-1054-3
128. É. Meyerson: *Explanation in the Sciences.* Translated from French by M-A. Siple and D.A. Siple. 1991 ISBN 0-7923-1129-9
129. A.I. Tauber (ed.): *Organism and the Origins of Self.* 1991 ISBN 0-7923-1185-X
130. F.J. Varela and J-P. Dupuy (eds.): *Understanding Origins.* Contemporary Views on the Origin of Life, Mind and Society. 1992 ISBN 0-7923-1251-1
131. G.L. Pandit: *Methodological Variance.* Essays in Epistemological Ontology and the Methodology of Science. 1991 ISBN 0-7923-1263-5
132. G. Munévar (ed.): *Beyond Reason.* Essays on the Philosophy of Paul Feyerabend. 1991 ISBN 0-7923-1272-4
133. T.E. Uebel (ed.): *Rediscovering the Forgotten Vienna Circle.* Austrian Studies on Otto Neurath and the Vienna Circle. Partly translated from German. 1991 ISBN 0-7923-1276-7
134. W.R. Woodward and R.S. Cohen (eds.): *World Views and Scientific Discipline Formation.* Science Studies in the [former] German Democratic Republic. Partly translated from German by W.R. Woodward. 1991 ISBN 0-7923-1286-4
135. P. Zambelli: *The Speculum Astronomiae and Its Enigma.* Astrology, Theology and Science in Albertus Magnus and His Contemporaries. 1992 ISBN 0-7923-1380-1
136. P. Petitjean, C. Jami and A.M. Moulin (eds.): *Science and Empires.* Historical Studies about Scientific Development and European Expansion. ISBN 0-7923-1518-9
137. W.A. Wallace: *Galileo's Logic of Discovery and Proof.* The Background, Content, and Use of His Appropriated Treatises on Aristotle's *Posterior Analytics.* 1992 ISBN 0-7923-1577-4
138. W.A. Wallace: *Galileo's Logical Treatises.* A Translation, with Notes and Commentary, of His Appropriated Latin Questions on Aristotle's *Posterior Analytics.* 1992 ISBN 0-7923-1578-2
Set (137 + 138) ISBN 0-7923-1579-0
139. M.J. Nye, J.L. Richards and R.H. Stuewer (eds.): *The Invention of Physical Science.* Intersections of Mathematics, Theology and Natural Philosophy since the Seventeenth Century. Essays in Honor of Erwin N. Hiebert. 1992 ISBN 0-7923-1753-X
140. G. Corsi, M.L. dalla Chiara and G.C. Ghirardi (eds.): *Bridging the Gap: Philosophy, Mathematics and Physics.* Lectures on the Foundations of Science. 1992 ISBN 0-7923-1761-0
141. C.-H. Lin and D. Fu (eds.): *Philosophy and Conceptual History of Science in Taiwan.* 1992 ISBN 0-7923-1766-1
142. S. Sarkar (ed.): *The Founders of Evolutionary Genetics.* A Centenary Reappraisal. 1992 ISBN 0-7923-1777-7
143. J. Blackmore (ed.): *Ernst Mach – A Deeper Look.* Documents and New Perspectives. 1992 ISBN 0-7923-1853-6
144. P. Kroes and M. Bakker (eds.): *Technological Development and Science in the Industrial Age.* New Perspectives on the Science–Technology Relationship. 1992 ISBN 0-7923-1898-6
145. S. Amsterdamski: *Between History and Method.* Disputes about the Rationality of Science. 1992 ISBN 0-7923-1941-9
146. E. Ullmann-Margalit (ed.): *The Scientific Enterprise.* The Bar-Hillel Colloquium: Studies in History, Philosophy, and Sociology of Science, Volume 4. 1992 ISBN 0-7923-1992-3
147. L. Embree (ed.): *Metaarchaeology.* Reflections by Archaeologists and Philosophers. 1992 ISBN 0-7923-2023-9
148. S. French and H. Kamminga (eds.): *Correspondence, Invariance and Heuristics.* Essays in Honour of Heinz Post. 1993 ISBN 0-7923-2085-9
149. M. Bunzl: *The Context of Explanation.* 1993 ISBN 0-7923-2153-7

Boston Studies in the Philosophy of Science

150. I.B. Cohen (ed.): *The Natural Sciences and the Social Sciences.* Some Critical and Historical Perspectives. 1994 ISBN 0-7923-2223-1
151. K. Gavroglu, Y. Christianidis and E. Nicolaidis (eds.): *Trends in the Historiography of Science.* 1994 ISBN 0-7923-2255-X
152. S. Poggi and M. Bossi (eds.): *Romanticism in Science.* Science in Europe, 1790–1840. 1994 ISBN 0-7923-2336-X
153. J. Faye and H.J. Folse (eds.): *Niels Bohr and Contemporary Philosophy.* 1994 ISBN 0-7923-2378-5
154. C.C. Gould and R.S. Cohen (eds.): *Artifacts, Representations, and Social Practice.* Essays for Marx W. Wartofsky. 1994 ISBN 0-7923-2481-1
155. R.E. Butts: *Historical Pragmatics.* Philosophical Essays. 1993 ISBN 0-7923-2498-6
156. R. Rashed: *The Development of Arabic Mathematics: Between Arithmetic and Algebra.* Translated from French by A.F.W. Armstrong. 1994 ISBN 0-7923-2565-6
157. I. Szumilewicz-Lachman (ed.): *Zygmunt Zawirski: His Life and Work.* With Selected Writings on Time, Logic and the Methodology of Science. Translations by Feliks Lachman. Ed. by R.S. Cohen, with the assistance of B. Bergo. 1994 ISBN 0-7923-2566-4
158. S.N. Haq: *Names, Natures and Things.* The Alchemist Jābir ibn Ḥayyān and His *Kitāb al-Aḥjār* (Book of Stones). 1994 ISBN 0-7923-2587-7
159. P. Plaass: *Kant's Theory of Natural Science.* Translation, Analytic Introduction and Commentary by Alfred E. and Maria G. Miller. 1994 ISBN 0-7923-2750-0
160. J. Misiek (ed.): *The Problem of Rationality in Science and its Philosophy.* On Popper vs. Polanyi. The Polish Conferences 1988–89. 1995 ISBN 0-7923-2925-2
161. I.C. Jarvie and N. Laor (eds.): *Critical Rationalism, Metaphysics and Science.* Essays for Joseph Agassi, Volume I. 1995 ISBN 0-7923-2960-0
162. I.C. Jarvie and N. Laor (eds.): *Critical Rationalism, the Social Sciences and the Humanities.* Essays for Joseph Agassi, Volume II. 1995 ISBN 0-7923-2961-9
Set (161–162) ISBN 0-7923-2962-7
163. K. Gavroglu, J. Stachel and M.W. Wartofsky (eds.): *Physics, Philosophy, and the Scientific Community.* Essays in the Philosophy and History of the Natural Sciences and Mathematics. In Honor of Robert S. Cohen. 1995 ISBN 0-7923-2988-0
164. K. Gavroglu, J. Stachel and M.W. Wartofsky (eds.): *Science, Politics and Social Practice.* Essays on Marxism and Science, Philosophy of Culture and the Social Sciences. In Honor of Robert S. Cohen. 1995 ISBN 0-7923-2989-9
165. K. Gavroglu, J. Stachel and M.W. Wartofsky (eds.): *Science, Mind and Art.* Essays on Science and the Humanistic Understanding in Art, Epistemology, Religion and Ethics. Essays in Honor of Robert S. Cohen. 1995 ISBN 0-7923-2990-2
Set (163–165) ISBN 0-7923-2991-0
166. K.H. Wolff: *Transformation in the Writing.* A Case of Surrender-and-Catch. 1995 ISBN 0-7923-3178-8
167. A.J. Kox and D.M. Siegel (eds.): *No Truth Except in the Details.* Essays in Honor of Martin J. Klein. 1995 ISBN 0-7923-3195-8
168. J. Blackmore: *Ludwig Boltzmann, His Later Life and Philosophy, 1900–1906.* Book One: A Documentary History. 1995 ISBN 0-7923-3231-8
169. R.S. Cohen, R. Hilpinen and R. Qiu (eds.): *Realism and Anti-Realism in the Philosophy of Science.* Beijing International Conference, 1992. 1996 ISBN 0-7923-3233-4
170. I. Kuçuradi and R.S. Cohen (eds.): *The Concept of Knowledge.* The Ankara Seminar. 1995 ISBN 0-7923-3241-5

171. M.A. Grodin (ed.): *Meta Medical Ethics*: The Philosophical Foundations of Bioethics. 1995 ISBN 0-7923-3344-6
172. S. Ramirez and R.S. Cohen (eds.): *Mexican Studies in the History and Philosophy of Science.* 1995 ISBN 0-7923-3462-0
173. C. Dilworth: *The Metaphysics of Science.* An Account of Modern Science in Terms of Principles, Laws and Theories. 1995 ISBN 0-7923-3693-3
174. J. Blackmore: *Ludwig Boltzmann, His Later Life and Philosophy, 1900–1906* Book Two: The Philosopher. 1995 ISBN 0-7923-3464-7
175. P. Damerow: *Abstraction and Representation.* Essays on the Cultural Evolution of Thinking. 1996 ISBN 0-7923-3816-2
176. M.S. Macrakis: *Scarcity's Ways: The Origins of Capital.* A Critical Essay on Thermodynamics, Statistical Mechanics and Economics. 1997 ISBN 0-7923-4760-9
177. M. Marion and R.S. Cohen (eds.): *Québec Studies in the Philosophy of Science.* Part I: Logic, Mathematics, Physics and History of Science. Essays in Honor of Hugues Leblanc. 1995 ISBN 0-7923-3559-7
178. M. Marion and R.S. Cohen (eds.): *Québec Studies in the Philosophy of Science.* Part II: Biology, Psychology, Cognitive Science and Economics. Essays in Honor of Hugues Leblanc. 1996 ISBN 0-7923-3560-0
Set (177–178) ISBN 0-7923-3561-9
179. Fan Dainian and R.S. Cohen (eds.): *Chinese Studies in the History and Philosophy of Science and Technology.* 1996 ISBN 0-7923-3463-9
180. P. Forman and J.M. Sánchez-Ron (eds.): *National Military Establishments and the Advancement of Science and Technology.* Studies in 20th Century History. 1996 ISBN 0-7923-3541-4
181. E.J. Post: *Quantum Reprogramming.* Ensembles and Single Systems: A Two-Tier Approach to Quantum Mechanics. 1995 ISBN 0-7923-3565-1
182. A.I. Tauber (ed.): *The Elusive Synthesis: Aesthetics and Science.* 1996 ISBN 0-7923-3904-5
183. S. Sarkar (ed.): *The Philosophy and History of Molecular Biology: New Perspectives.* 1996 ISBN 0-7923-3947-9
184. J.T. Cushing, A. Fine and S. Goldstein (eds.): *Bohmian Mechanics and Quantum Theory: An Appraisal.* 1996 ISBN 0-7923-4028-0
185. K. Michalski: *Logic and Time.* An Essay on Husserl's Theory of Meaning. 1996 ISBN 0-7923-4082-5
186. G. Munévar (ed.): *Spanish Studies in the Philosophy of Science.* 1996 ISBN 0-7923-4147-3
187. G. Schubring (ed.): *Hermann Günther Graßmann (1809–1877): Visionary Mathematician, Scientist and Neohumanist Scholar.* Papers from a Sesquicentennial Conference. 1996 ISBN 0-7923-4261-5
188. M. Bitbol: *Schrödinger's Philosophy of Quantum Mechanics.* 1996 ISBN 0-7923-4266-6
189. J. Faye, U. Scheffler and M. Urchs (eds.): *Perspectives on Time.* 1997 ISBN 0-7923-4330-1
190. K. Lehrer and J.C. Marek (eds.): *Austrian Philosophy Past and Present.* Essays in Honor of Rudolf Haller. 1996 ISBN 0-7923-4347-6
191. J.L. Lagrange: *Analytical Mechanics.* Translated and edited by Auguste Boissonade and Victor N. Vagliente. Translated from the *Mécanique Analytique, novelle édition* of 1811. 1997 ISBN 0-7923-4349-2
192. D. Ginev and R.S. Cohen (eds.): *Issues and Images in the Philosophy of Science.* Scientific and Philosophical Essays in Honour of Azarya Polikarov. 1997 ISBN 0-7923-4444-8

Boston Studies in the Philosophy of Science

193. R.S. Cohen, M. Horne and J. Stachel (eds.): *Experimental Metaphysics.* Quantum Mechanical Studies for Abner Shimony, Volume One. 1997 ISBN 0-7923-4452-9
194. R.S. Cohen, M. Horne and J. Stachel (eds.): *Potentiality, Entanglement and Passion-at-a-Distance.* Quantum Mechanical Studies for Abner Shimony, Volume Two. 1997 ISBN 0-7923-4453-7; Set 0-7923-4454-5
195. R.S. Cohen and A.I. Tauber (eds.): *Philosophies of Nature: The Human Dimension.* 1997 ISBN 0-7923-4579-7
196. M. Otte and M. Panza (eds.): *Analysis and Synthesis in Mathematics.* History and Philosophy. 1997 ISBN 0-7923-4570-3
197. A. Denkel: *The Natural Background of Meaning.* 1999 ISBN 0-7923-5331-5
198. D. Baird, R.I.G. Hughes and A. Nordmann (eds.): *Heinrich Hertz: Classical Physicist, Modern Philosopher.* 1999 ISBN 0-7923-4653-X
199. A. Franklin: *Can That be Right?* Essays on Experiment, Evidence, and Science. 1999 ISBN 0-7923-5464-8
200. D. Raven, W. Krohn and R.S. Cohen (eds.): *The Social Origins of Modern Science.* 2000 ISBN 0-7923-6457-0
201. Reserved
202. Reserved
203. B. Babich and R.S. Cohen (eds.): *Nietzsche, Theories of Knowledge, and Critical Theory.* Nietzsche and the Sciences I. 1999 ISBN 0-7923-5742-6
204. B. Babich and R.S. Cohen (eds.): *Nietzsche, Epistemology, and Philosophy of Science.* Nietzsche and the Science II. 1999 ISBN 0-7923-5743-4
205. R. Hooykaas: *Fact, Faith and Fiction in the Development of Science.* The Gifford Lectures given in the University of St Andrews 1976. 1999 ISBN 0-7923-5774-4
206. M. Fehér, O. Kiss and L. Ropolyi (eds.): *Hermeneutics and Science.* 1999 ISBN 0-7923-5798-1
207. R.M. MacLeod (ed.): *Science and the Pacific War.* Science and Survival in the Pacific, 1939-1945. 1999 ISBN 0-7923-5851-1
208. I. Hanzel: *The Concept of Scientific Law in the Philosophy of Science and Epistemology.* A Study of Theoretical Reason. 1999 ISBN 0-7923-5852-X
209. G. Helm; R.J. Deltete (ed./transl.): *The Historical Development of Energetics.* 1999 ISBN 0-7923-5874-0
210. A. Orenstein and P. Kotatko (eds.): *Knowledge, Language and Logic.* Questions for Quine. 1999 ISBN 0-7923-5986-0
211. R.S. Cohen and H. Levine (eds.): *Maimonides and the Sciences.* 2000 ISBN 0-7923-6053-2
212. H. Gourko, D.I. Williamson and A.I. Tauber (eds.): *The Evolutionary Biology Papers of Elie Metchnikoff.* 2000 ISBN 0-7923-6067-2
213. S. D'Agostino: *A History of the Ideas of Theoretical Physics.* Essays on the Nineteenth and Twentieth Century Physics. 2000 ISBN 0-7923-6094-X
214. S. Lelas: *Science and Modernity.* Toward An Integral Theory of Science. 2000 ISBN 0-7923-6303-5

Boston Studies in the Philosophy of Science

215. E. Agazzi and M. Pauri (eds.): *The Reality of the Unobservable.* Observability, Unobservability and Their Impact on the Issue of Scientific Realism. 2000 ISBN 0-7923-6311-6

Also of interest:

R.S. Cohen and M.W. Wartofsky (eds.): *A Portrait of Twenty-Five Years Boston Colloquia for the Philosophy of Science, 1960-1985.* 1985 ISBN Pb 90-277-1971-3

Previous volumes are still available.

KLUWER ACADEMIC PUBLISHERS – DORDRECHT / BOSTON / LONDON

www.ingramcontent.com/pod-product-compliance
Ingram Content Group UK Ltd.
Pitfield, Milton Keynes, MK11 3LW, UK
UKHW020134160926
475607UK00003B/40

* 9 7 8 0 7 9 2 3 6 3 1 1 8 *